Semiconductor Interfaces, Microstructures and Devices:
Properties and Applications

Semiconductor Interfaces, Microstructures and Devices: Properties and Applications

Edited by

Zhe Chuan Feng

National University of Singapore

Institute of Physics Publishing
Bristol and Philadelphia

05474061

PHYSICS

British Library Cataloguing-in-Publication Data

A catalogue record for this book is available from the British Library.

ISBN 0 7503 0180 5

Library of Congress Cataloging-in-Publication Data are available

Published by Institute of Physics Publishing, wholly owned by The Institute of Physics, London

Institute of Physics Publishing, Techno House, Redcliffe Way, Bristol BS1 6NX, UK

US Editorial Office: Institute of Physics Publishing, The Public Ledger Building, Suite 1035, Independence Square, Philadelphia, PA 19106, USA

Printed in the UK by Galliard (Printers) Ltd, Great Yarmouth, Norfolk

Contents

Part II: III–V Compound Semiconductors

Part III: II–VI Compound Semiconductors

Preface

Semiconductor Interfaces, Microstructures and Devices: Properties and Applications contains eleven review chapters and is grouped into four parts: *General Topics, III–V Compound Semiconductors, II–VI Compound Semiconductors, and IV and IV–IV Semiconductors*. It includes: new developments in growth methods; electric, optical, magnetic and structural characterization and properties; related theories; devices (such as infrared photodetectors, blue diode lasers, etc) and applications. Each chapter focuses on a special subject and is prepared by one or more experts in the field. Each one is a comprehensive review on the recent achievements and creative contributions in the corresponding field. This book concentrates on the interface problems for various semiconductor materials, devices and structures, explored from different concepts or perspectives, and studied by various techniques and methods. Some introductory material is also included to make the book suitable for graduates and new scientists in the field.

Part I includes five chapters. Chapter 1 is written by R Tsu, one of the pioneers in semiconductor superlattice microstructures, and deals with the *Electric field induced localization in superlattices*. The author shows the derivation of the transport solution in superlattices, which is the basis of superlattices, and discusses Stark-ladders, Bloch oscillation, the effects of scattering on localization and the difference between superlattices and real solids.

In Chapter 2, *Intersubband transitions and quantum well infrared photodetectors*, K K Choi reviews the basic physics and the operating principles of quantum well infrared photodetectors, and discusses various design issues, the electrical and infrared properties of multiple quantum well structures, several useful techniques, such as infrared absorption spectroscopy, thermally stimulated hot-electron spectroscopy and infrared photoelectron tunneling spectroscopy. Infrared technology has played an important role in modern science and technology, which has been impressively demonstrated in the past Middle-East war. Infrared photodetectors are the heart of infrared technology. Quantum well structures may become an important device for infrared detection and we expect a rapid development in the near future.

Because of the complexity of modern semiconductor device structures and the stringent demands on their specifications, techniques for real time monitoring of semiconductor characteristics during preparation and processing are very important. Chapter 3 by R W Collins *et al* from Pennsylvania State University is concerned with the development of a novel multichannel ellipsometer, a powerful spectroscopic tool for real time monitoring of semiconductor growth and etching. The new instrument provides ellipsometric spectra over the photon energy range between 1.5 and 4.5 eV with a minimum time resolution of 40 ms. The authors present an overview of recent applications of real time spectroscopic ellipsometry for monolayer-level characterization, including plasma-enhanced chemical vapor deposition of amorphous semiconductors, and ion beam etching of III–V semiconductors.

Chapter 4, written by P F Miceli from Bellcore, focuses on the study of *X-ray reflectivity from heteroepitaxial layers*. Recent work has shown that the extended range reflectivity, which is measured from grazing angles out through the Bragg reflections, can achieve monolayer sensitivity for buried interfaces. Theoretical models are used to interpret the experimental data. From these, information such as layer thicknesses, thickness fluctuations, strains, degree of crystallinity and detection of the presence of reacted layers, can be obtained. The author shows that x-ray scattering plays a vital role in the understanding of surface and interface physics.

In Chapter 5, *Spontaneous and stimulated emissions from optical microcavity structures*, H Yokoyama *et al* from the NEC Opto-Electronics Research Laboratory in Japan describe the alteration of spontaneous emission of materials in optical microcavities with dimensions of the order of emitted wavelength. Theoretical and experimental studies are presented with particular emphasis on planar microcavity structures containing GaAs MQWs. The influence of the cavity on light emission and laser oscillation is discussed. A rate equation analysis is used to study the coupling of the spontaneous emission into the laser mode. A threshold-less laser can be achieved if all the spontaneous emission is confined in a single cavity mode. Nearly threshold-less laser operation was demonstrated using planar optical microcavities confining an organic dye solution. The application of cavity quantum electro-dynamics for optical devices will hopefully produce novel kinds of light sources beyond conventional lasers.

In Part II there are two chapters. Zh I Alferov, Academician of Science, Director of the A. F. Ioffe Physico-Technical Institute, Russia, and his colleague D Z Garbuzov, have contributed Chapter 6, *Radiative and nonradiative recombination in AlGaAs and InGaAsP heterostructures and some features of the corresponding quantum well laser diodes*. The authors review the results of photoluminescence (PL) and electroluminescence (EL) studies of MOCVD- and MBE-grown AlGaAs- and InGaAsP-based laser structures carried out in the last few years at the Ioffe Institute. Three types of quantum well structures, InGaAsP/InP, InGaAsP/GaAs and AlGaAs/GaAs, were studied. The quantum well photoluminescence efficiency, spontaneous emission of laser diodes, threshold current densities, differential efficiency and light-current characteristics are discussed. For more work from Russian and previous Soviet scientists, interested readers may refer to the references at the end of the chapter or a book, *Semiconductor heterostructures: physical processes and applications*, edited by Zh I Alferov (1989, MIR Publishers, Moscow).

M O Manasreh from Wright Laboratory has written Chapter 7, *Far-infrared cyclotron resonance of a 2-dimensional electron gas in III–V semiconductor heterostructures*. Cyclotron resonance (CR) measurements for InAs/AlSb and InAs/AlGaSb single quantum wells as well as GaAs/AlGaAs heterostructures were obtained at liquid helium temperature. The author presents the electron effective mass calculated from the CR transmission peak position energy, effective g factors of the 2-dimensional electron gas, relaxation (scattering) time, and full width at half maximum of the CR spectra as a function of the magnetic field strength B. The oscillatory behavior of these parameters is observed as a function of B which can be explained by Landau level filling factors.

In Part III, Chapter 8, *Optics in lower-dimensional quantum confined II–VI heterostructures*, A V Nurmikko and R L Gunshor review the recent progress in wide-gap II–VI heterostructures with emphasis on optical properties ensuing from quantum confinement in multilayer structures; primarily ZnSe-based QWs. The blue and green diode lasers and light emitting diodes in the (Zn,Cd)Se-type I heterostructures have been realized. An important element in these structures are quasi-2-dimensional exciton effects which are likely to be of relevance as practical optoelectronic devices emerge at short visible wavelengths.

The last section, *IV and IV–IV Semiconductors*, consists of three chapters. Chapter 9 by H-J Gossmann and D J Eaglesham from AT & T Bell Laboratories is concerned with the *Growth and doping of silicon by low-temperature molecular beam epitaxy*. Traditional concepts of epitaxial growth assume that there is a temperature, T_{epi}, separating the regime of epitaxial, single-crystalline growth of the film from the regime of amorphous growth. This concept is not applicable to Si MBE. Instead growth at a given temperature always proceeds epitaxially for a certain limiting thickness, h_{epi}, before the film becomes amorphous. Segregation and low incorporation can be avoided at growth temperatures below ~400°C. The growth of arbitrarily complex doping profiles by thermal, co-evaporative doping thus becomes possible. The characterization and control of defects as well as applications of low-temperature Si MBE are discussed.

E H Poindexter in Chapter 10, *Point defects and charge traps in the Si/SiO$_2$ system and related structures*, reviews the recent and significant advances of our understanding for atomic-scale defects in the metal-oxide-silicon (MOS) system. This topic is quite important for Si-based devices and integrated circuits dominating modern electronics. The author reviews 4 areas of MOS defect study: E' centers with a positive oxide charge in radiation-damaged structures, the P_{b1} center at the Si/SiO$_2$ interface, a new electrochemical model for the negative-bias-temperature instability, and some new ideas in radiation damage mechanisms.

Three world-known authorities on SiC, J A Powell, P Pirouz and W J Choyke co-contribute Chapter 11, *Growth and characterization of silicon carbide polytypes for electronic applications*. Recent advances in bulk and thin film crystal growth have accelerated the development of SiC as a useful semiconductor for sensors and devices capable of operation in high temperature, high radiation environments, high frequency and high power applications. The authors review processes for producing Boule and thin film single-crystal SiC, some theories and models for these processes, properties of SiC polytypes and various types of defects, and some recent optical and electrical measurements relating to donors and acceptors in SiC polytypes.

The study of semiconductor interfaces, microstructures and devices involves many different research groups working on the theoretical and experimental aspects of materials and structures. There have been many developments in recent years. The American Physical Society (APS) Materials Physics Division invited me and Drs L C Feldman (AT & T Bell Laboratories), F K Le Goues (IBM Thomas J Watson Research Center), and H H Wieder (University of California, San Diego) to organize a focused session on *Semiconductor Interfaces and Microstructures* for the 1991 APS March meeting. The quality of information and breadth of topics presented gave rise to this book which aims to provide a succinct overview of recent research.

Finally, I would like to acknowledge the authors' contributions, which have made this book successful. I would also like to complement all the scientists and engineers who are working in the frontiers of this field. Due to their contributions, it is possible for this book to appear. We look forward to more achievements in the future.

Zhe Chuan Feng
Department of Physics,
National University of Singapore

Part I

General Topics

Chapter 1

Electric field induced localization in superlattices

Raphael Tsu

University of North Carolina at Charlotte
Charlotte, NC 28223, USA

ABSTRACT: Since the introduction of the man-made superlattice and quantum well structures, the electric field induced localization in solids known as the Stark-ladder and the closely related subject, Bloch oscillation, may be experimentally studied. Some important questions: the existence of Stark-ladders, the effects of scattering on localization, oscillations at Bloch frequency, the similarity and difference between superlattice and real solids, etc., are discussed.

1. INTRODUCTION

In 1969, Esaki and Tsu (1969, 1970) introduced the man-made superlattice consisting of alternate layers of semiconductors A and B to mimic real solids with a period much greater than the lattice constants of real solids. They showed that a negative differential conductance occurs at an applied constant electric field F such that the energy gain by the carriers in a period d, efd, exceeds the energy broadening given by h/τ with h and τ being the Planck constant $\div 2\pi$ and the scattering time, current will decrease with increase in F. In other words, with the definition of Bloch frequency $\omega_B = eFd/h$, the condition of NDC is written in the familiar form $\omega_B \tau > 1$. Furthermore, in the limit for large τ, the current disappears, leaving the carriers oscillating at the Bloch frequency. Although the purpose of their study was motivated by a new tera-Hertz electronic device, the physics is intimately related to the subject of Wannier Stark-ladder, introduced by Wannier(1960, 1962). Actually the concept, if $\Psi(x)$ is a solution of the wave equation for energy E of a periodic system with a period a, then $\Psi(x-na)$ is a solution for energy E-neFa, with n being any integer, was first explicitly discussed by James(1949). The explicit localization induced by the application of a constant electric field was given by Kane(1959). Since these earlier treatments, there appeared many arguments against the existence of (Stark-ladder) SL principly raised by Rabinovitch and Zak (1971), and arguments in favor of SL and localization by Shockley(1971). Basically, these objections arise because of finite dimensions of the solid, failure of representation of the wave function with an operator x in the Hamiltonian by a superposition of Bloch functions, and tunneling into other bands. Using a vector potential to preserve symmetry, Krieger and Iafrate (1986) were able to remove these objections. Unlike the case in which a scaler potential is used to represent the effect of the field, the field dependent terms are continuous functions of time, permitting the eigenfunctions of the unperturbed Hamiltonian to be the eigenfunction of the Hamiltonian at the instant the field is turned on. The time development of the system for times small compared to the inverse tunneling rate is represented by a linear combination of the eigenfunctions of the instantaneous Hamiltonian. Actually the issue is similar to the fact that the states of a hydrogen atom in a constant field are not discrete: strictly speaking stationary states do not exist, yet there is no confusion in treating the problem in terms of transitions between stationary states. In fact, Krieger and Iafrate side-stepped the issue of stationary states, instead, they showed that optical transitions involving a selection rule which is consistent with the notion of the Stark-ladder.

The present treatment tries to remove some of the confusions, for example: the relationship between localization and Bloch oscillation; what is the degree of localization, the effects of tunneling into other bands, finite length and finite mean free path, relationship to quantum well structure, etc. Furthermore, the difference between superlattice and real solids will be discussed in terms of experimental observations.

2.DERIVATION OF TRANSPORT SOLUTION IN SUPERLATTICES

The derivation given by Esaki and Tsu (1969) in terms of Pippard's (1965) impulse method is essentially correct. However, Lebwohl and Tsu (1970) have succeeded in obtaining the exact Green's function of the Boltzmann equation with constant electric field for scattering representable by a collision time. They showed that their result for Fermi-Dirac distribution is identical to Budd's (1963) proof of Chamber's (1952) path integral method for Maxwellian distribution. It will be clear why this work is reviewed here.

Most of us are quite familiar with the solutions of Boltzmann transport equation with an applied electric field constituting a small departure from the equilibrium function. There is in fact an exact solution first obtained by Lebwohl and Tsu.

Taking the electric field **F** in the x-direction, the Boltzmann transport equation without the time and spatial variation becomes

$$k_0 \frac{\partial f}{\partial k_x} + f = f_0 \qquad (1)$$

in which $k_0 = eF\tau/\hbar$, and $f_0(\mathbf{k})$ is the equilibrium distribution function. The periodicity of the crystal along **F** is described by the reciprocal lattice vector **K**. The energy bands are assumed to be

$$E = (\hbar^2 k_\perp^2/2m) + E_x(k_x) \qquad (2)$$

where k   is the component of the wavevector perpendicular to the direction of the superlattice, and $E_x(k_x)$ is periodic in k_x with period K. Equation (1) has the general solution

$$f(k_\perp, k_x) = \int_{-K/2}^{K/2} G(k_x, k'_x) f_0(k_\perp, k'_x) dk'_x \qquad (3)$$

in which the Green's function G satisfies periodic boundary conditions with period K. The adjoined equation that G satisfies is

$$-k_0 \frac{\partial G}{\partial k_x} + G = \sum_{-\infty}^{\infty} \delta(k_x - k'_x - mK) \quad . \qquad (4)$$

Figure 1 shows the deformation of contour from c to c' in the complex z-plane.

$$G = \frac{1}{K} \sum_{-\infty}^{\infty} \frac{e^{i2\pi m(k_x - k'_x)/K}}{1 - i2\pi m k_0/K}$$

$$= \pm \frac{1}{2\pi i K} \int_c \frac{e^{(k_x - k'_x) Z/K}}{(1 + k_0 Z/K)(e^{\pm Z} - 1)}$$

$$= \frac{e^{(k_x - k'_x)/k_0}}{k_0} \begin{cases} (1 - e^{-K/k_0})^{-1} & k_x < k'_x \\ \\ (e^{K/k_0} - 1)^{-1} & k_x > k'_x \ . \end{cases}$$

$$(5)$$

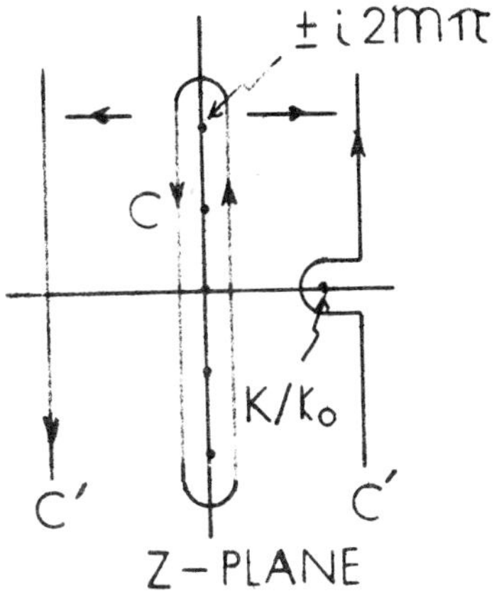

Fig.1. Deformation of
       contour from c
       to c' in the
       complex z-plane

Taking

$$f_0(k_\perp, k_x) = \theta(k_\perp - k_{\perp m})\{ \theta(k_x + k_{xm}(k_\perp)) - \theta(k_x - k_{xm}(k_\perp)) \}$$

with θ being the unit step functions, and k_{xm} defined by

$$E_x(k_{xm}) = E_f, \quad \text{and} \quad E(k_m) = E_f$$

at $k_\perp = 0$, and $k_x = 0$ respectively, $f(k_x)$ is given by

$$f(k_x) = \exp(-(k_x + K/2)/k_0) \frac{\sinh(k_{xm}/k_0)}{\sinh(K/2k_0)}, \qquad k_x < -k_{xm}$$

$$= 1 + \exp(-k_x/k_0) \frac{\sinh((k_{xm}-K/2)/k_0)}{\sinh(K/2k_0)}, \qquad |k_x| < k_{xm} \tag{6}$$

$$= \exp(-(k_x - K/2)/k_0) \frac{\sinh(k_{xm}/k_0)}{\sinh(K/2k_0)}, \qquad k_x > K_{xm}$$

and $f(\mathbf{k}) = \Theta(k_\perp - k_{\perp m}) f(k_x)$.

From

$$j_x = \frac{1}{4\pi^3\hbar} \int f(\vec{k}) \frac{\partial E}{\partial k_x} d\vec{k}$$

with

$$E_x = E_0 - E_1 \cos(2\pi k_x/K), \tag{6a}$$

we obtained for the final result

$$j_x = \frac{neE_1 d}{\hbar} \frac{\omega_B \tau}{(\omega_B \tau)^2 + 1} H \tag{7}$$

with

$$H \equiv \frac{1}{2} [(Q - \sin Q \cos Q)/(\sin Q - Q \cos Q)] \qquad E_0 - E \le E_f \le E_0 + E_1$$

$$\frac{1}{2} [E_1/(E_f - E_0)] \qquad E_f > E_0 + E_1 \tag{8}$$

where the superlattice period d is defined by $K \equiv 2\pi/d$, and $Q \equiv \cos^{-1}((E_0 - E_f)/E_1)$.

Next we shall show that the Chamber's path integral method for the Fermi-Dirac distribution leads to the same result in eq.(8). According to Chamber (1952), the distribution function

$$f = \frac{1}{\tau} \int_{-\infty}^{t'} f_0 (E - \Delta E) \; e^{-(t'-t)/\tau} \; dt' \;,$$ (9)

with

$$v_x = \frac{1}{\hbar} \frac{\partial E}{\partial k_x} \qquad\qquad \Delta E = \int_t^{t'} \vec{F} \cdot \vec{v} dt'',$$

$$\langle v_x \rangle = \frac{1}{4\pi^3} \int_k f v_x d\vec{k} = \frac{1}{4\pi^3} \int f_0 d\vec{k} \int_{-\infty}^{t'} \frac{1}{\hbar} \left(\frac{\partial E(t')}{\partial k_x} - \frac{\partial E(t)}{\partial k_x} \right) e^{-(t-t')/\tau} \frac{dt}{\tau} \;.$$

(10)

Let us take a more general field because we need this expression later in the response function,

$$F = F_0 + \sum_n F_n e^{-i\omega_n t}$$

with $eF = \hbar \dot{k}_x$,

$$k_x(t) = k_x(t') + \frac{eF_0}{\hbar}(t-t') + \sum_n \frac{ieF_n}{\hbar\omega_n}(e^{-i\omega_n t} - e^{-i\omega_n t'}) \;.$$ (11)

Substituting eq.(11) in (10), and integration by parts with $E(k_x)=E(-k_x)$, the expectation value of the velocity becomes

$$\langle v_x \rangle = nH v_0 \int_{-\infty}^{t} \sin(g(t,t') d) \; e^{(t'-t)/\tau} dt'$$ (12)

where $g(t,t')$ is defined by the sum of the second and third terms on the right side of eq.(11). Using eq.(6a) for E_x, $v_0 = E_1 d/\hbar$, it is trivial to show that the result can be obtained from eq.(12) because,

$$nH = \frac{1}{4\pi^3} \int \cos(k_x d) \; f_0 \; d\vec{k} \;.$$

Equation (12) will be used as a starting point of deriving the response of a superlattice to a combined constant field F_0 and time varying field given by F_n.

Next, we shall take $F = F_0 + 2F_1 \cos(\omega t)$, then

$$gd = \frac{eF_0 d}{\hbar}(t-t') + \frac{eF_1 d}{\hbar\omega_1}(\sin\omega_1 t - \sin\omega_1 t')$$

and

$$\langle v_x\rangle = v_0 H \sum_{m,n=-\infty}^{\infty} J_m(\frac{\omega_{B1}}{\omega_1}) J_n(\frac{\omega_{B1}}{\omega_1}) \frac{\sin(m-n)\omega_1 t + (\omega_B + n\omega_1)\tau\cos(m-n)\omega_1 t}{(\omega_B + n\omega_1)^2\tau^2 + 1}$$

(13)

where $\omega_B \equiv eF_0 d/\hbar$ and $\omega_{B1} \equiv eF_1 d/\hbar$. For $H = 1$, $\langle v_x\rangle$ in eq.(13) is same as that which first obtained by Tsu (1990).

For $\omega_{B1}/\omega_1 \ll 1$ (small F_1),

$$\langle v_x\rangle = \langle v_x\rangle_0 + Re\langle v_x\rangle_1 \cos\omega_1 t + Im\langle v_x\rangle_1 \sin\omega_1 t ,$$

(14)

where

$$\langle v_x\rangle_0 = v_0 H \frac{\omega_B\tau}{(\omega_B\tau)^2 + 1}$$

(15)

Accept the dependence of H in terms of E_f, for $H=1$ with low E_f, eq.(15) is identical to the expression first derived by Esaki and Tsu (1970), however, $2E_1$ in v_0 is the bandwidth of the energy band in question.

Let us discuss the case for $H=1$, then $v_0 H = E_1 d/\hbar$ and the maximum extent $\langle x\rangle_m = \langle v_x\rangle_m\tau = E_1/2eF_0$. Since length is measured by nd, the maximum period covered is given by

$$n = E_1/2eF_0 d$$

(16)

For $\omega_B\tau \gg 1$, $\langle v_x\rangle \to 0$,   and $\langle x\rangle = E_1/eF_0$,   (17)

the electrons will now oscillate with a period $T = 2\pi/\omega_B$, which was known to Bloch (1928), and discussed by Houston (1940).

Without collision, an electron will oscillate at a frequency of ω_B and covering a distance of E_1/eF_0. Note that the extent of an electron without collision is twice the maximum distance given by $\omega_1\tau = 1$. We shall make this point clearer with regards to the degree of localization. Rahinovich and Zak (1972) again casted doubt in the existence of this oscillation. In order to answer whether a Bloch electron can oscillate, we shall examine in more detail of eqs.(13) and (14). Before we do that, it is obvious without both F_0 and F_1, it is pointless to discuss whether a Bloch electron oscillates or not. Any coupled system should have oscillations, however, observations can be made only via a coupling to an external system. In this case, the $\mathbf{A \cdot P}$ term should be included in the Hamiltonian. We shall look at the linear system, the in-phase component with time goes as $\cos\omega t$ which we abbreviate by writing $\text{Re} < v_x >$; and the out of phase component with time goes as $\sin\omega t$ denoted by $\text{Im} < v_x >$. Thus, we sum all the terms in eq.(13) with n-m=1.

The equations describing the linear response

$$Re\langle v\rangle = \frac{Re\langle v_x\rangle_1}{v_0\,\cos\omega t}\left(\frac{2\omega}{\omega_{B1}}\right)$$

(18a)

and

$$Im\langle v\rangle \equiv \frac{Im\langle v_x\rangle_1}{v_0\,\sin\omega t}\left(\frac{2\omega}{\omega_{B1}}\right)$$

(18b)

are plotted for various $\omega_B\tau$ in Fig. 2.

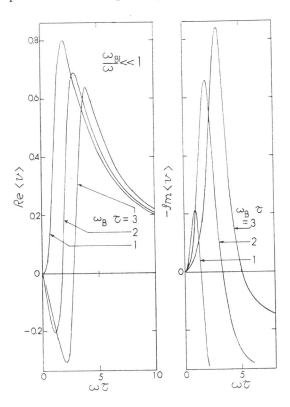

Fig.2. The in-phase, $\text{Re}<v>_1$ and out-of-phase $\text{Im}<v>_1$ components of the linear response function, for a superlattice with an applied electric field of $F=F_0+2F_1\cos\omega t$. $\omega_B\equiv eF_0d/\hbar$ and $\omega_{B1}\equiv eF_1d/\hbar$.

For $\omega_B\tau=1$, $\text{Re}<v>$ is always positive indicating the lack of gain, or self-oscillation. The $\text{Im}<v>$ has a maximum at $\omega=\omega_B$. For $\omega_B\tau=2$, $\text{Re}<v>$ has a minimum at $\omega=\omega_B/2$ and is negative, but Im$<v>$ has a peak at $\omega=\omega_B$. And for $\omega_B\tau=3$, Re$<v>$ has a maximum negative value at $\omega=2\omega_B/3$, and the Im$<v>$ has a peak at $\omega=\omega_B$. Thus the peak in Im part always appears at $\omega=\omega_B$, supposedly substantiating the intuitive understanding that the system is oscillating at the Bloch frequency. However, the Re$<v>$ always has a maximum negative value below, indicating that self-oscillation occurring at the maximum gain is never at the Bloch frequency. Only as $\omega_B\tau = \infty$ does the maximum gain coincide with the Bloch frequency. For $\omega_{B1} >> \omega$ and $\omega_B\tau >> 1$, Re$<v>_3$ can have a negative region indicating that in the regime of non-linear optics, an intense optical field is required.

Experimentally we need to arrange the polarization with a component in the superlattice direction. For $\tau=0.5ps$ and $\omega_B\tau=3$ give $eF_0d \sim$ 4mev corresponding to $F_0 \sim 4\times10^3$v/cm for d=100 Å. Therefore condition for self-oscillation at $\omega=2\omega_B/3$ or $\omega \sim 4\times10^{12}$Hz, should be reachable. Therefore we have answered the question raised whether a Bloch electron can oscillate. The result presented here shows that not only a Bloch electron can oscillate, it can serve as an optical oscillator and amplifier!

3.BAND MODEL AND HOPPING MODEL

Recently, there appeared numerous experimental confirmations regarding the NDC and electric field induced localization. Before we look into this aspect in detail, we shall review the hopping model of Tsu and Döhler (1975). Figure 3 shows possible transitions between electrons localized in a given cell to adjacent cells. Under the application of a field F_0, a constant potential energy difference exists between adjacent cells and a ladder structure for the energy states appears. If eF_0d is such that the level 1 coincides with level 2', electrons may tunnel resonantly from 1 to 2' marked by (a), followed by an inelastic scattering marked by (b) to level 1', in order to repeat the process onto the next cell. This process has been treated by Kazarinov and Suris (1972). Observations of NDC by Esaki and Chang (1974) is explained by this type of resonant tunneling processes, Tsu and Esaki (1973), Chang et al (1974).

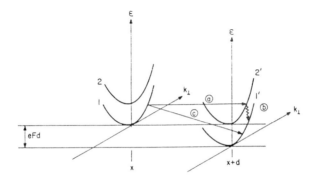

Fig.3. Energy states under an applied electric field F. Process (a) involve direct tunneling followed by an inelastic scattering (b), and process (c) involves inelastic scattering.

Here we shall take few essential derivations from Tsu and Döhler (1975) and discuss the relationship with experiments and several important theoretical studies. The hopping model of Tsu and Döhler, as in the case of the treatments of Saitoh (1972) and Fukuyama et al (1973), considers transitions between the energy levels of a Stark-ladder via phonon assisted transitions. The main motivation is to by pass the objections of using a band model for the case where the energy gain by an electron in an electric field in a period exceeds the energy bandwidth, i.e. $\hbar\omega_B \geq 2\,E_1$. Using the "Kane functions" (Kane 1959), Tsu and Döhler (1975) obtained the following expression for the energy

$$E_\nu(\vec{k}) = \frac{\hbar^2 k_\perp^2}{2m} + E_0 + eF_0 d(\nu + L/d) , \tag{19}$$

with L being a constant coming from the matrix X_{nn} (see Tsu and Döhler 1975); and the wavefunction

$$\psi_\nu(\vec{r}) = \left(\frac{d}{2\pi}\right)^{\frac{1}{2}} e^{ik_\perp \rho} \int_{-\pi/d}^{\pi/d} u(k_x, x) \exp\{i\,[k_x(x-\nu d-L) - \frac{E_1}{eF_0 d}\sin k_x d]\}\,dk_x. \tag{20}$$

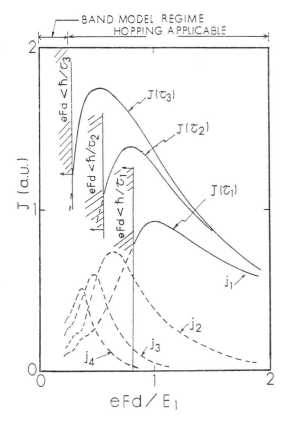

Fig.4. Current vs eFd/E_1, for three values of τ with F and E_1 being the applied field and the energy bandwidth of the miniband, and $\tau_3 > \tau_2 > \tau_1$. As τ increases, more terms representing hopping between adjacent cells and next nearest cells, etc. need to be summed. The lower case j represents hopping for $\nu-\nu' = 1,2,3,4$

For large F_0, the wavefunction is highly localized in a distance $\Delta x \sim E_1/eF_0$, and centered at $x = L + \nu d$. Using such a wavefunction to calculate the current due to an electron in the call ν (energy state ν) making a hop to cell ν' (energy state ν'), Tsu and Döhler (1975) calculated this current using the "golden rule". Their result is replotted in fig.4 taken from a recent article (Tsu and Esaki, 1991).

Before we discuss fig.4, we shall explain an important role of τ, the scattering time. First of all, the localization distance $\Delta x \sim E_1/eF_0$ agrees with eq.(17) which applies to the band model for $\omega_B \tau >> 1$. There should not be any surprise because all these treatments of Stark-ladder do not take into account of finite scattering time. We have clearly pointed out that within the model of a constant scattering time, the effect of scattering is to reduce this localization length to half the value without scattering as shown in eq.(16). There is a way to take scattering into account. The total current should be a sum of individual hopping between ν and ν', in other words,

$$J = \sum_{\gamma=1}^{n} j_\gamma \ , \qquad with \qquad \gamma = \nu' - \nu \tag{21}$$

How many terms we need to sum depends on n, and in turn depends on the applied electric field F_0 in eq.(16). Therefore only in the extreme localization case where hopping can only take place between adjacent cells, $n=1$, the Tsu and Esaki (1970) expression $\omega_B \tau \geq 1$ is not necessary because implicitly we already assumed that the free path is greater than the localization distance. This point may be made quite clear by an example. Taking a regime where $\hbar \omega_B \sim 2E_1$, such that we cannot confidently use the band model, but the localization distance $\Delta x = nd$, the sum in eq.(21) of course is required to take all these terms up to n. Figure 4 shows that the peak of the total current progressively shifts toward lower field, while the number of individual hopping currents increases. How many terms we take depends on τ which is not taken into account in any of these treatments on Stark-ladder. Thus we need the Esaki-Tsu (1970) condition that $\omega_B \tau \geq 1$, to tell us how many terms in eq.(21) should be taken. Since scattering is not accounted in hopping model, and the limit of sum is set by the band model, there is an intricate relationship between the band model and the Stark-ladder treatment. For $\omega_B \tau \geq 1$ and $\hbar \omega_B << E_1$, the band model of Esaki-Tsu applies. For $\hbar \omega_B \sim E_1$, the hopping model of Tsu-Döhler model applies, however, the Esaki-Tsu condition that $\omega_B \tau \geq 1$ is still needed to tell us how many terms are required in the sum of eq.(21). Only in the extreme localization case, the hopping model of Tsu-Döhler (1975) can be taken without the condition $\omega_B \tau \geq 1$, then NDC is always given by $eFd > 2 E_1$, the energy bandwidth of the band. As noted in Tsu and Esaki (1991), it is futile to apply the hopping model for $eFd << 2E_1$. There is lacking a quantitative theory in the regime where band model does not apply and without taking into account of scattering, hopping model also cannot apply. The situation is quite analogous to localization by magnetic field.

4. ANALOGY BETWEEN STARK LADDER AND CYCLOTRON LEVELS

I spent some time during the summer of 1991 at IBM Research working with Leo Esaki. He asked me to look into the confusion recently surfaced between the band model and hopping model which resulted in the paper Tsu and Esaki (1991). Although Esaki did ask me to make a comparison between cyclotron resonance and the Stark-ladder, because of the pressing need to clarify the role of scattering, only a passing remark was made in that work. Since the analogy is much more than casual, the situation will be discussed here in more detail.

Before the establishment of energy quantization in units of $\hbar\omega_c$ with $\omega_c = eH/mc$, the dynamics of electrons in a magnetic field was treated in the traditional equation $\dot{\mathbf{k}} = (e/\hbar c)$ $\mathbf{v} \times \mathbf{H}$. The situation is entirely analogous to our situation. The use of $\dot{\mathbf{k}} = (e/\hbar)\mathbf{F}$ for NDC by Esaki and Tsu (1970) established the condition for Bragg reflection. Without scattering, due to the oscillation nature, the electrons are entirely localized within a distance $<x>$ given by eq.(17). Similarly, without scattering, electrons are entirely localized within a cyclotron orbit. The quantization of cyclotron orbit is treated without scattering (see for example Ziman, (1988)), in much the same way as the treatment of stark-ladder. Experimentally, taking into account of scattering, the condition $\omega_c\tau \geq 1$ must be taken. In addition, as $\hbar\omega_c$ approached the band width of the energy band in question, the simple transport theory using the above equation of $\dot{\mathbf{k}}$ surely cannot apply. And yet, a formal quantization treatment by solving the Schrödinger equation with multiple bands is lacking the effect of scattering. Therefore we must realize the validity of the band model and hopping model. However, from experimental point of view, as in the case of cyclotron resonance, solution of Boltzmann equation is certainly the key to understandings of experimental results. As we shall discuss later that experiments on superlattice are exceeding the range of validity of the band model, therefore in many situations, the concept of transitions between Stark-ladder levels needs to be considered. There is one subtle point should be clarified, the role of scattering in localization. Again, we shall discuss magnetic quantization. Obviously when an electron undergoes a collision before completing one circular orbit, cyclotron resonance cannot be important. In our case, when an electron suffers a collision before Bragg reflection, Stark-ladder cannot be dominant. Intuitively, if a wavefunction is localized in one cell, scattering can couple wavefunction with wavefunctions in the adjacent cells, resulting in "delocalization". Similarly an electron in one cyclotron orbit can be scattered into another orbit. Then, why is the maximum excursion given by eq.(16) is reduced because of scattering? The answer of this paradox lies in the fact that localization in quantum mechanics is only defined in terns of stationary eigenstates, i.e., the extent of the eigenfunction, and scattering involves incoherent transitions among stationary states. Therefore scattering does reduce the coherent length. It is true that scattering "delocalize" the wavefunction causing it to spread. Since this spreading is incoherent, we should not consider that localization is reduced. As eq.(16) represents a reduction, therefore localization as defined by the extent of the coherent part of the wavefunction is increased. We can thus clearly state that localization is increased by scattering. This is certainly true in amorphous solids, as pointed out by the model of Tsu (1989) in the study of phase coherence in amorphous quantum wells. We shall close this discussion with a definite statement regarding the effects of scattering. Scattering increases localization. However, this increase in localization leads to a broadening of the Stark-ladder. Keeping the scattering fixed, increase in the applied electric field increases localization, and sharpens the SL levels. Keeping the field fixed, increase in scattering will also increase localization, but broadens the SL levels!

5. EFFECTS OF FINITE LENGTH

Some concerns about the existence of Stark-ladder in a periodic system of finite length led to the discussion by Schockly (1972) and Fukuyama (1973). Fukuyama considered the Hamiltonian

$$H = \sum_n H_n + H_n'$$

(22)

with $H_n = H_0 + V_n$, where $V_n \equiv eFdn \equiv nV_0$ and nearest neighbor interaction such that $<n'|\ H_n'\ |n> = 0$ for $n'=n$ and α for $n'=n+1$. With a wavefunction

$$\psi = \sum_{n=1}^{N} C_n a_n^+ |0>,$$

(23)

the expansion coefficient C_n satisfied the recursion relation

$$(\frac{E-E_0}{V_0} - n)\ C_n = \frac{\alpha}{V_0}(C_{n+1} + C_{n-1})\ .$$

(24)

Using a rigid wall boundary condition, i.e., $C_0 = C_{N+1} = 0$, an equation for the eigenstate E is obtained

$$J_{-\epsilon}(2\alpha/V_0)\ Y_{N+1-\epsilon}(2\alpha/V_0) = J_{N+1-\epsilon}(2\alpha/V_0)\ Y_{-\epsilon}(2\alpha/V_0)$$

(25)

in which $\epsilon \equiv (E-E_0)/V_0$.

Fukuyama showed that $\epsilon = n$ for $\alpha = 0$, which is a statement of the Stark- quantization. For $\alpha << V_0$, he found that only states near the band edges are effected $\sim N^{-1}$. We shall take only three coupled wells and examine the question of localization, whether the separations are given by V_0. With $V_0 \equiv eFd$, and nearest neighbor given by $<n|H'|n'> = \alpha$, the energy states are given by

$$\frac{E}{E_0} = 1 + (\frac{\alpha}{E_0})b, \qquad \psi_+ = \frac{1}{b}(\frac{|1>}{b-\beta} + |2> + \frac{|3>}{b+\beta})$$

$$\frac{E}{E_0} = 1, \qquad \psi_c = \frac{1}{b}(\ |1> - \beta|2> - |3>)$$

(26)

$$\frac{E}{E_0} = 1 - (\frac{\alpha}{E_0})b, \qquad \psi_- = \frac{1}{b}(\frac{-|1>}{b+\beta} + |2> - \frac{|3>}{b-\beta}),$$

where $b \equiv (2 + \beta^2)^{1/2}$ with $\beta \equiv V_0/\alpha$.

Figure 5 shows the plot of $|\Psi|^2$ for several values of β with $\alpha/E_0 = 0.1$, and Table I gives the values of the energy E/E_0.

Table I. Values of E/E_0 for various β and $\alpha = 0.1E_0$

	$\beta = 0$	$\beta = 1$	$\beta = 4$
E_+/E_0	1.141	1.173	1.424
E_c/E_0	1	1	1
E_-/E_0	0.859	0.827	0.576

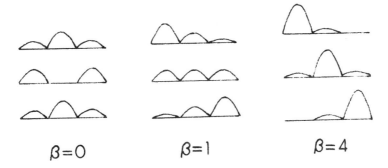

$\beta = 0$ $\beta = 1$ $\beta = 4$

Fig.5. The normalized wavefunctions $|\Psi|^2$ for three values of β. $\beta = 0$ represents no applied voltage. The coupling constant α between wells are taken as $0.1E_0$.

What is clear from these energies, only for $\beta >> 1$, does the separation approaches V_0, the applied voltage, indicating a true SL, and the corresponding wavefunction are truly localized. For example, taking the case for $\beta = 4$, i.e., $V_0 = 4\alpha$, $|\Psi|^2$ shows localization in the sense that the peak moves from the left to the right and the separation is dominated by the applied voltage.

There is much to be learned from a simple computation. Most of the features due to finite length is clearly demonstrated, i.e., localization as a function of the applied voltage at each cell for a given coupling between cells. It is difficult to avoid commenting that there appeared too much argument and counter argument as to the existence of Stark-ladder. Experimentalist generally have a pretty fair understanding of the subject. As pointed out in the original paper by Esaki and Tsu (1970): tunneling into other bands should be kept as low as possible, the scattering time τ should be such that $\omega_B \tau > 1$, and the mean free path should be at least several period. With all the subsequent theoretical investigations, the basic physics has not changed. As a matter of fact, the first theoretical treatment on the resonant tunneling in quantum well structures by Tsu and Esaki (1973) was motivated precisely by the finite length consideration. And in that paper, it was clearly pointed out the need to consider statistics, particularly in the case of, now famous, the so-called sequential tunneling discussed by Capasso et al (1986).

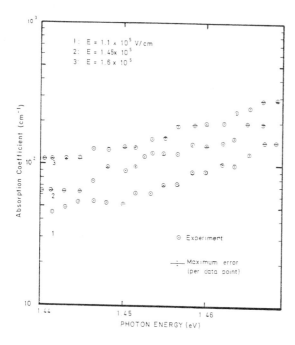

Fig.6. Absorption coefficient vs photon energy taken from Koss and Lambert (1972), is replotted with comparison to theory removed. The point is that stepwise increase at $\Delta\hbar\omega=neFa$ is not so obvious.

6.EXPERIMENT

Koss and Lambert (1972) reported the effect of a constant electric field on optical absorption in GaAs. Figure 6 is a re-plot of their data with lines removed. Without the guide from these calculated, the data do not clearly indicate jumps in $\hbar\omega=eFa$. The situation is quite different in superlattice. Mendez et al (1988) observed field dependent photo-luminescence and photo-current with results consistent with field induced localization. They observed increase in localization with increase in the applied electric field. Voisin et al (1988) observed the effect of Stark-quantization in a GaAs/AlGaAs superlattice in electro-reflectance. Leavitt (1990) put a quantum well in between superlattices serving as a notch for reference sake in their study of optical transition involving SL. It is interesting to note that Dignam and Sipe (1990) considered the treatment by Emin and Hart (1987) as the ultimate proof of the existence of the Stark-ladder. In this work, they take the constant electric field as a superposition of a saw-tooth ramp and a step. Actually this approach is no more valid a proof than previous work. This is a proper place to comment on the many theoretical investigations. Personally, the many body approach utilizing a linear chain model is basically pedagogical, Chen and Zhao (1991).

Recently, the negative differential conductance first predicted by Esaki and Tsu (1969) was observed by several teams: Beltram et al (1990), Sibille et al (1990). These are very important experimental confirmations of Stark-quantization. As we discussed earlier, for $\omega_B\tau >> 1$, an electron oscillates by virtual of repeated Bragg reflections, leading to localization. Again, an analogy to the magnetic quantization is in order. The observation of cyclotron resonance or de-Hass-van Alphen effect is a direct manifestation of magnetic quantization. The observation of NDC is a direct manifestation of electric field quantization. Beltram and Capasso (1990) were quite correct in pointing out that localization starts at $\omega_B\tau > 1$, the condition for NDC. SL theories predict that SL exists at any field because scattering is not considered. Obviously it is meaningless to consider quantized energy levels for $\omega_B\tau < 1$. This author tried to consider the effect of incoherent scattering on the energy states of a quantized system, Tsu (1989). It is common to first find the eigenstates, using perturbation calculations, scatterings are considered, an energy broadening is obtained from the scattering cross-section. When scattering is large, we know that not only levels are broadened, there should be an accompanied level shift. These considerations together with the discussion on the effects of scattering on localization should induce some theorists to formulate a realistic theory instead of redoing what has been done many times. The reason is very obvious: most semiconductors do not have sufficiently long τ, or mean free path, the effects of scattering is very dominant. In other words experiments with real solids instead of superlattices have a marginal chance of success. For this reason, scattering must be taken into account in any realistic theory. This is also why this author took the Boltzmann transport approach.

Apart from what has been discussed, there is another important difference between superlattice and real solids. In superlattices, the applied electric field mis-oriented with respect to the periodic axis has little consequences. However, in real solids, there is a build-in smearing effect. If the applied electric field is slightly off the major symmetry axes, they predicts no Stark-quantization, and yet, intuitively, no major problem should arise because there is always some spreading in the field axis. Perhaps a wave packet consisting of a solid angle of **k** vector centered about the field axis should be taken.

The coherent oscillations of a wave packet in a double-well structure have been recently observed by Leo et al (1991). Furthermore Bloch oscillation in a semiconductor superlattice has also been observed by Feldmann et al (1992). actually, superlattice and coupled quantum wells are not that all different theoretically and experimentally. As discussed earlier, electric field induces localization, in a periodic system, as well as coupled wells as shown in Fig.5, by the same mechanism, separating the energies between adjacent cells resulting in a decoupling of wavefunctions. Therefore these experimental observations confirm the localization induced by the application of a constant electric field.

7.CONCLUSION

The relationship between cyclotron resonance and magnetic quantization is identical to the Bloch oscillation and electric field induced quantization. At present, before scattering are included in the quantization schemes, Boltzmann transport equation provides better guide to experiments. At extreme high field where $\hbar\omega_B \sim E_1$, scattering considerations are less important in hopping models because localization is so complete that wavefunctions are almost decoupled between adjacent cells. This extreme quantization is unlikely observable in real solids except in molecular crystals. However, this extrem localization is actually easier to observe in superlattices and quantum well structures. Since the density of states obtained from Stark-ladder formulation is not obtainable from a semiclassical band model, effects such as optical transitions involving these Stark-levels, resonant tunneling when these levels are lined up between adjacent cells, etc., require a quantum mechanical approach. This aspect is of course apparent in the case of magnetic quantization: transitions involving Landau levels must be treated with the full Hamiltonian. Unlike the magnetic quantization, $\omega_c \tau > > 1$ is generally achievable, $\omega_B \tau > > 1$ is usually difficult to obtain. Theorists willing to tackle the scattering in SL can indeed provide important contribution to the field of electric field quantization.

From device point of view which was what provided the rationale in this field, with the advent of hyper-mobility materials (Pfeiffer 1990), many new effects such as NDC; material for non-linear optics (Tsu and Esaki 1971); parametric light amplifier (Tsu 1990), (Monsivais 1990); and ultimately a tera-Hz Bloch oscillator, are either already proven experimentally or readily realizable.

Acknowledgement: The support of Army Research Office and Office of Naval Research is gratefully acknowledged.

LIST OF REFERENCES

Beltram F., Capasso F., Sivco D., Hatchinson A.L., Chu S.N., and Cho A.Y., 1990, Phys. Rev. Lett. 64, 3167.
Bloch F., 1928, Zeits. f Physik 52, 555.
Budd H., 1963, J. Phys. Soc. Japan, 18, 142.
Capasso F., Mohammed K., and Cho A.Y., 1986, QE 22, 1853.
Chambers R.G., 1952, Proc. phys. Soc. (London), A65, 458.
Chang L.L., Esaki L., and Tsu R., 1974, Appl. Phys. Lett., 24, 593.
Chen S.G., Zhao X.G., 1991, Phys. Lett. A 155, 303.
Dignam M.M., and Sipe J.E., 1991, Phys. Rev. Lett. 64, 1797.
Emin D., Hart C.F., 1987, Phys. Rev. B 36, 7353.
Esaki L., and Chang L.L., 1974, Phys. Rev. Lett. 33, 495.
Esaki L., and Tsu R., 1969, IBM Research Note RC-2418.
Esaki L., and Tsu R., 1970, IBM J. Res Develop. 14, 61.
Felfmann J. et al, 1992, to be published.
Fukuyama H., Bari R.A., and Fogedby H.C., 1973, Phys. Rev. B 8, 5579.
Houston W.V., 1940, Phys. Rev. 57, 184.
James H.M., 1949, Phys. Rev. 76, 1611.
Kane E.O., 1959, J. Phys, Chem. Solids 12, 181.
Kazarinov R.F., and Suris R.A., 1971, Sov. Phys.-Semicond. 5, 707.
                        1972, Sov. Phys.-Semicond. 6, 120.
Koss R.W., and Lambert L.M., 1972, Phys. Rev. B 5, 1479.
Krieger J.B., and Iafrate G.J., 1986, Phys. Rev. B 33, 5494.
Leavitt R.P., and Little J.W., 1990, Phys. Rev. B 41, 5174.

Lebwohl P.A., and Tsu R., 1970, J. Appl. Phys. 41, 2664.

Leo K., Shah J., Gobel E.O., Damen T.C., Schmitt-Rink S., and Schafer W., 1991, Phys Rev. Lett. 66, 201.

Mendez E.E., Agullo-Rueda F., and Hong J.M., 1988, Phys. Rev. Lett 60, 2426.

Monsivais G., Castillo-Mussot M., and Claro F., 1990, Phys. Rev. Lett. 64, 1433.

Pfeifer L., West K.W., Starmer H.L. and Baldwin K.W., 1989, Appl. Phys. Lett. 55, 1888

Pippard A.B., 1965, "Dynamics of Conduction Electrons" (Gordon and Breach, Science Publication Inc., New York)

Rabinovitch A., snd Zak J., 1971, Phys. Rev. B 4, 2358.

Saitoh M., 1972, J. Phys. C. Solid State, 5, 914.

Shockley W., 1972, Phys. Rev. Lett. 28, 349.

Sibille A., Palmier J.F., Wang H., and Mokot F., 1990, Phys. Rev. Lett. 64, 52.

Tsu R., 1989, J. Non-Cryst. Solids 114, 708.

Tsu R., 1990, SPIE, 1361, 231.

Tsu R., and Döhler G., 1975, Phys. Rev. B 12, 680.

Tsu R., and Esaki L., 1971, Appl. Phys. Lett. 19, 246.

Tsu R., and Esaki L., 1973, Appl. Phys. Lett. 22, 562.

Tsu R., and Esaki L., 1991, Phys. Rev. B 43, 5204, and Erratum Phys. Rev. B 44,3495

Wannier G.H., 1960, Phys. Rev. 117, 432. 1961, Rev. Mod. Phys. 34, 645.

Ziman J.M., 1988, "Principles of the Theory of Solids", (Cambridge Univ. Press.) P. 313.

Chapter 2

Intersubband transitions and quantum well infrared photodetectors

K. K. Choi

U. S. Army Electronics Technology and Devices Laboratory
Fort Monmouth, New Jersey 07703, U.S.A.

ABSTRACT: Infrared detection based on intersubband absorption has become increasingly mature in the last several years. Besides their impact on infrared technology, infrared sensitive quantum well structures also help advancing fundamental device physics by providing a fertile testing ground for hot-electron phenomena and novel concepts of band-gap engineering. In this chapter, we will review the basic physics and the operating principles of quantum well infrared photodetectors, and discuss various design issues.

1. INTRODUCTION

Traditionally, infrared detection has been based on electron transitions from the valence band to the conduction band as in an intrinsic photoconductor, or from impurity states to the conduction band as in an extrinsic photoconductor. After Esaki and Tsu (1969,1970) proposed synthesized semiconductor superlattices, a new possibility of infrared detection emerged. Electrons trapped in a multiple quantum well (MQW) structure can be freed by photoexcitation and conduct through electron states above the quantum barriers, similar to an extrinsic photoconductor, except that the "impurity" states can now be engineered.

Chiu et al. (Chiu et al. 1983a; Smith et al. 1983; Chiu et al. 1983b) was the first group to propose a GaAs/AlGaAs MQW structure, in which electrons resided in the quantum wells absorb incoming radiation, and promote to the free traveling states above the barriers. The electrons can then be swept out of the MQW structure by applying an electric field across the structure. Even in these early studies, some of the potential advantages of MQW structures have already been anticipated, which include the flexibility of tailoring detection wavelengths and the possibility of large scale detector integration. However, in these studies, only free electron absorption process was considered, which turns out to be relatively weak.

On the other hand, if the electric vector of the incident light is made polarized parallel to the superlattice axis, the light absorption process can be much stronger due to the large dipole moment between the envelope wave functions of the quantum well states. Coon et al. (1984, 1985) applied intersubband transition mechanism to a single quantum well structure, and proposed a new mode of infrared detection which involves repeatedly charging and discharging the quantum well. Goossen et al. (1985, 1989) and Hasnain et al. (1989) proposed using a diffraction grating for light coupling to enhance the electric component parallel to the superlattice axis.

In 1985, West and Eglash (1985a) made a first experimental observation of intersubband absorption in GaAs/AlGaAs MQW structures. Subsequently, Choi et al. (1987a,b,c) and Levine et al. (1987a,b,c) made a detailed investigation on the infrared and transport properties of MQW structures, based on which the first sensitive MQW infrared photoconductor was demonstrated. The same detector structure since then has been under constant optimization (Levine et al. 1988a, 1989, 1990), and is the basic building block of the current state-of-the-art quantum well infrared photodetector (QWIP). Other than photoconductors, MQW photovoltaic detectors (Kastalsky et al. 1988, Goossen et al. 1988a) and phototransistors (Choi et al. 1990a,b,d, 1991a,b) also exist, although they will not be covered in this chapter.

In fact, intersubband absorption and its use in infrared detection are not completely new. Before its observation in MQW structures, a large amount of literature on intersubband absorption has already been accumulated in the Si/SiO_2 system (Kamgar et al. 1974, 1976; Kneschaurek et al. 1976). A n-channel metal-oxide-semiconductor field-effect-transistor as a photodetector, which is based on intersubband absorption, has also been made (Wheeler and Goldberg 1975). However, in this structure, photocurrent can only be measured along the channel between the source and the drain contacts. A large background current exists even without light illumination, which limits the detector sensitivity. Moreover, the separation of the subband energies in this system is relatively small, which is unsuitable for mid-infrared detection.

In the following sections, the optical and transport properties of GaAs/AlGaAs multiple quantum well structures, as well as their performance in infrared detection, will be reviewed.

2. BLACKBODY RADIATION

Before discussing the properties of an infrared detector, it is useful to mention some of the important features of blackbody radiation.

At a given temperature T, the spectral radiance $F(T,\lambda)$ of an ideal blackbody can be expressed as

$$F(T,\lambda) = \frac{2hc^2}{\lambda^5} \frac{1}{e^{(hc/\lambda kT)}-1} \quad \frac{W}{cm^2} \frac{1}{sr} \frac{1}{\mu m} \quad , \tag{2.1}$$

where T is the temperature and λ is the wavelength. To find the wavelength λ_{max} at which the spectral radiance is a maximum, we set $dF/d\lambda = 0$, which gives Wien's displacement law:

$$\lambda_{max} T = 2898 \quad \mu m\,K \quad . \tag{2.2}$$

At T = 300 K, λ_{max} is 9.7 μm. Therefore, a 10 μm thermal detector will be advantageous in detecting room temperature objects. Similarly, for those detectors which depend on the number of incident photons instead of the incident power, λ_{max} is given by

$$\lambda_{max} T = 3670 \quad \mu m\,K \quad , \tag{2.3}$$

which is 12.2 μm at 300 K. The power radiated per unit area between λ_1 and λ_2, denoted by $I_B(T,\lambda_1,\lambda_2)$, is given by integrating Eq. (2.1):

$$I_B(T,\lambda_1,\lambda_2) = \frac{2\pi h}{c^2}\left(\frac{kT}{h}\right)^4 \int_{x_1}^{x_2} \frac{x^3}{e^x-1}\,dx \quad,$$

$$= \frac{15}{\pi^4}\,\sigma\,T^4 \int_{x_1}^{x_2} \frac{x^3}{e^x-1}\,dx \quad, \tag{2.4}$$

where $\sigma = 5.67\times10^{-12}$ Wcm^{-2}K^{-4} is Stefan's constant and x is defined as hc/kTλ. Thus, I_B(300 K, 8 μm, 12 μm) $= 12$ mW/cm^2, and I_B(300 K, 3 μm, 5 μm) $= 0.58$ mW/cm^2. Since the average photon energy in 8 to 12 μm range is approximate 2x10^{-20} J, the number of photon N(300 K, 8 μm, 12 μm) emitted by a 300 K blackbody is then 6x10^{17} ph/s/cm^2 in this range. Likewise, the average photon energy of 3 to 5 μm radiation is 5x10^{-20} J, resulting a photon flux N(300 K, 3 μm, 5 μm) of 1.2x10^{16} ph/s/cm^2, 50 times less than the 10 μm range. Therefore, a 300 K blackbody will appear much brighter to a 10 μm detector than a 4 μm detector.

However, it can be shown that when x $>> 1$,

$$\frac{1}{N(T,\lambda_1,\lambda_2)}\,\frac{dN(T,\lambda_1,\lambda_2)}{dT} = \frac{hc}{kT^2\,\overline{\lambda}} \quad, \tag{2.5}$$

where $\overline{\lambda}$ is the average wavelength. Eq. (2.5) shows that the percentage change of the photon flux is inversely proportional to $\overline{\lambda}$. For example, if ΔT is 1 K, $\Delta N/N$ is 1.6 % in the 10 μm range and 4 % in the 4 μm range. Therefore, although a blackbody appears to be brighter in the 10 μm range, the contrast of the object against other possible background objects with slightly different temperatures will be greater in the 4 μm range.

In the following discussion, we will concentrate on 10 μm detectors.

3. INTERSUBBAND TRANSITIONS

Infrared absorption in n-type MQW structures is initiated by dipole transitions between two quantum well states within the conduction band. A typical MQW detector described here consists of 50 periods of GaAs quantum wells and AlGaAs barriers. The thickness of the quantum wells and the Al molar fraction of the barriers are designed such that the energy spacing of the quantum well states coincides with the photon energy. The barriers are thick enough to suppress tunneling dark current, and the quantum wells are doped to obtain a optimum sensitivity. The barriers are usually undoped to prevent a possible leakage current. In order to make ohmic contacts to the structure, heavily doped n-type GaAs layers are grown on the top and the bottom of the structure. The substrate material is usually semi-insulating GaAs. Four possible band structures of the detector are shown in Fig. 3.1.

In this section, the theoretical and the experimental results of intersubband transitions will be reviewed. The transition is usually initiated from the first subband E_1 of thick barrier structures or from the first miniband M_1 of the thin barrier structures, where the doped electrons reside. The optical transition to the first excited state E_2 is usually the strongest, so that $E_2 - E_1$ is usually designed to match the desired infrared photon energy.

The optical properties of MQW structures can be characterized by the absorption coefficient. Corresponding to 10 μm radiation, the energy of the quantum well states differs only by 0.12 eV, the overlap integral between the Bloch functions in the same conduction band is nearly unity, therefore only the envelope states are considered in the calculation of intersubband transitions. If we denote the envelope wave function by ψ_n for the nth subband

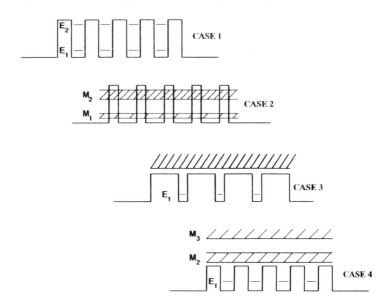

FIG. 3.1 Four possible subband structures of multiple quantum wells with different device parameters.

state E_n, or miniband M_n, N_q to be the photon occupation number inside the sample with the wave vector q, the transition rate W_q is given by

$$W_q = \frac{2\pi}{\hbar} \sum_f |\langle \psi_f, N_q - 1 | \frac{e}{m^*} A \cdot P | \psi_i, N_q \rangle|^2 \delta(E_f - E_i - \hbar\omega_q) , \qquad (3.1)$$

where A and P are the vector potential operator and the momentum operator respectively. Following the rules of second quantization, and expressing P as $m^*[z, H_0]/i\hbar$, the square of the matrix element $|M|^2$ can be obtained.

$$|M|^2 = \frac{e^2 \hbar \omega_q N_q}{2eV} z_{12}^2 \sin^2\theta , \qquad (3.2)$$

where z_{12}^2 is the dipole moment between the lowest two subband states, and θ is the angle between the light propagation direction and the superlattice axis. Substituting Eq. (3.2) into Eq. (3.1), W_q reads

$$W_q = \frac{2\pi}{\hbar} \frac{e^2 \hbar \omega_q N_q}{2eV} z_{12}^2 \sin^2\theta \int \rho(E) \delta(E - E_i - \hbar\omega_q) dE , \qquad (3.3)$$

where ρ is the density of the final states, depending on the MQW structural parameters. Here, we have assumed the width of E_i is negligible compared with that of E_f. If we further define the oscillator strength f to be $2m^* \omega_q z_{12}^2/\hbar$, and identify N_q/V to be $nI_i/c\hbar\omega_q$, I_i is the optical power inside the sample and n is the refractive index of the sample, W_q is given by

$$W_q = \frac{\pi e^2 \hbar}{2\epsilon_o m^* c} \frac{I_i}{\hbar \omega_q} f \frac{\sin^2 \theta}{n} \rho(\hbar \omega_q + E_i) \ , \tag{3.4}$$

Having found W_q, the number of transitions per unit volume, G, is given by $\rho_s W_q / L_p$, where ρ_s is the two-dimensional electron density per well, L_p is the thickness of one period of the MQW structure. Finally, the absorption coefficient α_q is related to G by $\hbar \omega_q G / I_i$, and is given by

$$\alpha_q = \frac{\rho_s}{L_p} \frac{\pi e^2 \hbar}{2\epsilon_o m^* c} f \frac{\sin^2 \theta}{n} \rho(\hbar \omega_q + E_i) \ . \tag{3.5}$$

With this definition of α_q, one can directly use the detector area A instead of $A\cos\theta$ as the optical area in calculating the detector response. In the following, we will consider four different cases, depending on the relative positions of the first excited state E_2 and the barrier height ΔE_c as shown in Fig. 3.1.

Case 1. Structures with thick barriers, E_2 below ΔE_c

In this case, the wells can be considered to be isolated. The scattering lifetime broadening of E_2 usually dominates the absorption width. The density of the final states can then be expressed as

$$\rho(E) = \frac{1}{\pi} \frac{\Gamma/2}{(E - E_2)^2 + (\Gamma/2)^2} \ , \tag{3.6}$$

where Γ is the full width at half maximum and is equal to $\hbar/\tau$, τ is the lifetime of the electron in E_2. Hence, $\rho(\hbar \omega_q + E_i)$ in Eq. (3.5) is given by

$$\rho(\hbar \omega_q + E_i) = \frac{1}{\pi} \frac{\Gamma/2}{(\hbar \omega_q - E_r)^2 + (\Gamma/2)^2} \ , \tag{3.7}$$

where E_r is the resonant energy $E_2 - E_1$.
  In addition to the lifetime broadening, other factors may also contribute to the absorption width. They include the band nonparabolicity (West 1985b) of the material, the nonparabolicity caused by different effective mass between the quantum well material and the barrier material (West 1985b; Ikonić et al. 1988), and the well width inhomogeneity. Since the first two factors do not give a symmetrical lineshape, the absorption peak will be shifted from E_1 - E_2 calculated at the transversal electron wave vector equal to zero.
  Other than the effects of nonparabolicity, many-body interactions will also shift the absorption energy. They include direct and exchange-correlation of electron screening of the external excitation field (Ando, Fowler, and Stern 1982), and the lowering of the ground state energy through Coulomb and exchange interactions (Bandara et al. 1988). Manasreh et al. (1991) examined a set of MQW structures, and concluded that many-body effects increases the absorption energy by 10 meV at the usual doping density.
  Fig. 3.2 shows an example of intersubband absorption of a MQW structure. The well width W of the structure is 70 Å, the barrier width B is 150 Å, the Al molar ratio x is 0.27,

and the doping density N_d is 2×10^{18} cm^{-3}. The absorption is measured by a Fourier transform infrared spectrometer at room temperature. The angle of incidence is at Brewster's angle to avoid Fabry-Perot oscillations. The absorption of a blank GaAs substrate is used for background subtraction. In Fig. 3.2, the data have been converted to 45° coupling angle by multiplying a scaling factor of 5.6.

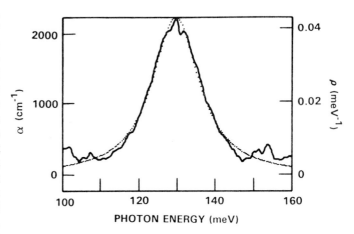

FIG. 3.2 The solid curve is the measured absorption coefficient α of a MQW at 45° light coupling angle and at room temperature. The dotted curve is the fitted joint density of states ρ assuming Lorentzian distribution.

The lineshape fits the Lorentzian distribution, indicating lifetime broadening dominant in this case with τ equal to 0.18 ps. Based on the nominal values of the sample parameters, the calculated subband energy E_1 is 47 meV, E_2 is 171 meV, while ΔE_c is 202 meV. Experimentally, the resonant energy E_r is measured to be 129 meV, 5 meV larger than E_2 - E_1 in this case. From the absorption coefficient at the peak, α_{peak}, the oscillator strength f is estimated to be 0.44 from Eq. (3.5).

Case 2. Structures with thin barriers, M_2 below ΔE_c

If level broadening is dominated by coupling among different wells, the wave functions will be extended and k_z is conserved in the light absorption process. One can use the tight binding model (Coon et al. 1985; Helm et al. 1991) to calculate the joint density of states with the same k_z. In this model, the dispersion relations for the first two minibands are:

$$E_1(k_z) = M_1 + \Delta_1[1 - \cos(k_z L_p)] \ ,$$
$$E_2(k_z) = M_2 + \Delta_2[1 + \cos(k_z L_p)] \ ,$$

(3.8)

where M_1 and M_2 are the bottom band edges of the minibands, and $2\Delta_1$ and $2\Delta_2$ are the respective miniband widths. Using Eq. (3.8), the joint density of states is

$$\rho_{joint}(\hbar\omega_q) = \frac{1}{\pi}\frac{1}{\sqrt{F}} \ ,$$

(3.9)

where $F = (\Delta_1 + \Delta_2)^2 - (\hbar\omega_q - M_2 + M_1 - \Delta_2 + \Delta_1)^2 \ .$

The absorption in this case is peaked at the photon energies equal to $M_2 - M_1 - 2\Delta_1$ and $M_2 - M_1 + 2\Delta_2$, which correspond to the transitions at the center and at the edge of the Brillouin mini zone, where $k_z = 0$ and $k_z = \pi/L_p$ respectively. In Fig. 3.3, ρ is plotted for a structure with W = 80 Å, B = 30 Å, and x = 0.27.

However, at present, there is no experimental evidence for such an energy dependence. To probe the miniband density of states, Helm et al. (1991) designed strongly coupled superlattices ($\Delta_2 \simeq 10$ meV) with energy spacing less than the optical phonon energy in order to avoid phonon lifetime broadening. Instead of an expected double-peaked structure in the absorption curve, they found a single absorption peak very similar to that found in case 1. They attributed the discrepancy to electron localization caused by the fluctuations in the barrier width.

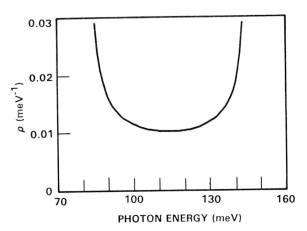

FIG. 3.3 The theoretical joint density of states of a superlattice assuming a miniband dispersion relation.

More significantly, not only Helm et al.'s result is negative, in fact, all the infrared absorption data reported on the bound-to-extended states experiments (Levine et al. 1988a,b, 1990; Gunapala et al. 1991) show single-peaked structure exclusively, regardless the miniband width of the excited state. These results are in direct contradiction to Eq. (3.9) since the double-peaked structure is predicted even in the case when Δ_1 approaches zero. When the miniband is above the barriers, miniband width can be much larger than other broadening effects. In this case, the miniband dispersion relation should be easier to observe.

The fact is when Δ_1 is small, well width and barrier width fluctuations and impurity scattering tend to localize the ground state electrons within each individual well. In this case, an optical transition can be described as a spatially direct transition between the localized ground state and the first excited state E_2 determined by the potential of a single quantum well, analogous to the first case. The difference in this case is E_2 being strongly coupled to the states of the adjacent wells through tunneling, and hence the lifetime width of E_2 is mainly contributed by tunneling rather than scattering. The absorption lineshape is therefore Lorentzian and the single-peaked absorption can be explained. In the following, we will describe the structures with only one bound state in the well, where the dispersion relations of the extended states can be important.

Case 3. Structures with thick barriers, E_2 above ΔE_c

In this regime, one can consider each well individually. The intersubband transition is from the localized ground state to one of the continuum states above the barriers. In principle, one can construct appropriate wave functions above the barriers and calculate the absorption coefficient directly (Ikonić et al. 1989). However, the calculation is rather lengthy and the result is not convenient to use. Here, we will follow a more intuitive approach in which the physical picture of case 1 and case 2 is preserved. We first consider the plane wave states above the barriers as the eigenstates, which form a continuum. Next, a resonant quantum well state E_2 in the continuum is introduced. Infrared transition is from the localized ground state to this resonant state, with an oscillator strength f. Since E_2 is strongly coupled to the plane wave states, the spectral function of E_2 is mainly determined by the coupling rather than scattering as in case 1. The task is to find the spectral function of E_2 for different band structures.

Before calculating the lineshape, we should discuss the nature of the resonant state. If the exact eigenstates are constructed above the barriers by matching the boundary conditions, it turns out that the amplitude of the wave function in the well region depends on the energy of the state and reaches a maximum when $k_2 W = \pi$, where k_2 is the wave vector of the state in the well region and W is the width of the well. This condition can be considered as a resonance. However, it is not the final state with which the absorption is maximum. The reason is that, for the usual device parameters, the wave function of the ground state penetrates substantially into the barriers, consequently the oscillator strength depends on the magnitude of the final wave function in the well region as well as that in the barrier region which is adjacent to the well. For example, when $x = 0.3$ and $W = 40$ Å, E_1 is equal to 91 meV, at which the decay length $1/\kappa$ inside each barrier on both sides is 18 Å, comparable to the well width. According to the calculation of Ikonić et al. (1989) for the present device parameters, maximum absorption occurs at the energy where $k_2 W = 0.78\pi$. One can then regard this energy as the position of the resonant state E_2. The decay length turns out to be insensitive to device parameters, for example, if $x = 0.25$ and $W = 40$ Å, $1/\kappa$ is equal to 21 Å; if $x = 0.20$ and $W = 45$ Å, $1/\kappa = 24$ Å. As a result, the resonant condition will not vary appreciably with different device parameters.

The spectral function $A(E)$ of a resonant state can be calculated using a Green's function approach (Mahan 1981). A Hamiltonian describing a resonant state of fixed energy E_2 in a material with thickness L_p has been introduced by Anderson (1961):

$$H = E_2 b^+ b + \sum_k E_k c_k^+ c_k + \sum_k D(c_k^+ b + b^+ c_k) , \tag{3.10}$$

where D is a coupling parameter. The retarded self-energy Σ_{ret} is given by

$$\Sigma_{ret} = \sum_k \frac{D^2}{E - E_k + i\delta} . \tag{3.11}$$

Using a free particle dispersion relation, the real part of Σ_{ret} is zero, and the imaginary part is given by

$$-Im\Sigma_{ret} = \frac{1}{C\sqrt{E}} ,$$

where $\qquad C = \frac{\hbar}{D^2 L_p} \sqrt{\frac{2}{m_b^*}} .$ $\qquad\qquad$ (3.12)

m_b^* is the effective mass of the barriers. Hence, the spectral function $A(E)$ is

$$A(E) = \frac{-2\,Im\Sigma}{(E - E_2)^2 + (Im\Sigma)^2} ,$$

$$= 2C \frac{\sqrt{E}}{1 + C^2 E(E - E_2)^2} . \tag{3.13}$$

Here, E is measured from the top of the barrier. If the origin of E is shifted to the bottom of the well, and $\rho(E)$ is equated with $A(E)/2\pi$, then

$$\rho(\hbar\omega_q + E_i) = \frac{C}{\pi} \frac{(\hbar\omega_q + E_1 - \Delta E_c)^{1/2}}{1 + C^2(\hbar\omega_q + E_1 - \Delta E_c)(\hbar\omega_q + E_1 - E_2)^2} . \tag{3.14}$$

Eq. (3.14) is valid when E_2 is well above ΔE_c. Caution has to be taken however in the case when E_2 approaches ΔE_c and their separation is less than the energy broadening. Since scattering effects are not considered in the present calculation, $\rho(E)$ is zero at E less than ΔE_c, and hence $\rho(E)$ needs to be renormalized to satisfy the condition $\int \rho(E)dE = 1$. Fig. 3.4 shows the normalized joint density of states of two MQW structures. For the first structure, $x = 0.19$, $W = 46$ Å with which $E_1 = 68$ meV, $E_2 = 160$ meV assuming $k_2W = 0.78\pi$, and $\Delta E_c = 145$ meV. Assuming the coupling parameter $D = 21$ meV, the result is plotted as curve (a). Note that the line shape is asymmetrical, which reflects the one-dimensional density of the plane wave states. If E_2 is moved further away from ΔE_c by decreasing the well width, ρ becomes more symmetrical, as indicated by curve (b) for a

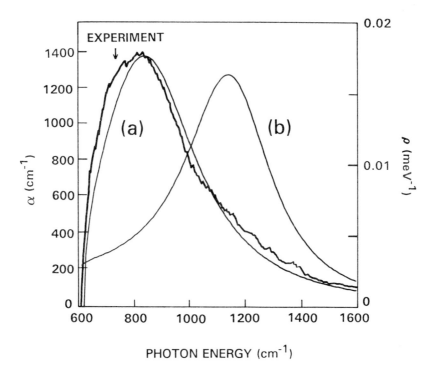

FIG. 3.4 Curve (a) shows the joint density of states ρ of a MQW with $W = 46$ Å, $x = 0.19$ and the coupling parameter $D = 21$ meV. Curve (b) shows ρ of a MQW with the same parameters except that $W = 40$ Å. The experimental absorption curve is for a sample with $W = 45$ Å and $x = 0.2$.

structure with W = 40 Å. In this case, E_1 = 82 meV, E_2 = 214 meV. Both of these lineshapes have been observed (Levine et al. 1988b, 1990). The experimental result obtained from a structure with W = 45 Å, B = 140 Å and x = 0.2 is shown in Fig. 3.4, which agrees with the theory. From the peak absorption, f is deduced to be 0.90.

Note that the width of the barriers need not be thick. As long as the miniband dispersion is not prominent, which can be a result of material inhomogeneity, Eq. 3.14 will be applicable in such case.

Case 4. Structures with thin barriers, E_2 above ΔE_c

When the miniband structure above the barriers is prominent, one can use the tight binding modal as in case 2 to obtain the dispersion relation. Note that the energy E_2 is determined by the well width only, while the locations of minibands can be changed by changing the barrier width. As a result, the positions of the resonant states and the minibands are uncorrelated, contrary to case 2. If we for simplicity assume the resonant state E_2 is at the center of one of the miniband e.g. M_3 in this example, and take $M_3 + \Delta_3$ as the origin, A(E) is then given by (Mahan 1981)

$$A(E) = 2\pi Z [\delta(E - \xi_c) + \delta(E + \xi_c)] \Theta(E^2 - \Delta_3^2)$$

$$+ 2D^2 \frac{\Theta(\Delta_3^2 - E^2)}{(\Delta_3^2 - E^2)^{1/2} \left(E^2 + \dfrac{D^2}{(\Delta_3^2 - E^2)} \right)} , \qquad (3.15)$$

where
$$\xi_c = \frac{1}{\sqrt{2}} \left[\Delta_3^2 + (\Delta_3^4 + 4D^2)^{1/2} \right]^{1/2} ,$$

and
$$Z = \frac{\xi_c^2 - \Delta_3^2}{2\xi_c^2 - \Delta_3^2} .$$

Eq. (3.15) indicates that the spectral function of the resonant state E_2 is continuous within the miniband but becomes discrete outside the miniband. Since the additional states are localized states, they are undetected within the Konig-Penney model. These essential features are preserved even if E_2 is not centrally located inside the miniband, and can even be outside the miniband. In reality, the delta functions are broadened by scattering and can be represented by a Lorentizan distribution ρ_l of the form of Eq. (3.6), hence ρ is given by

$$\rho(\hbar\omega_q + E_i) = \Theta(E^2 - \Delta_3^2) Z [\rho_l(E - \xi_c) + \rho_l(E + \xi_c)]$$

$$+ \Theta(\Delta_3^2 - E^2) \frac{F}{\pi} \left\{ \frac{(\Delta_3^2 - E^2)^{1/2}}{1 + F^2 E^2 (\Delta_3^2 - E^2)} \right\} , \qquad (3.16)$$

where
$$E = \hbar\omega_q + E_1 - M_3 - \Delta_3 ,$$
$$F = D^{-2} .$$

Fig. 3.5 (a) is a plot of the theoretical $\rho(\hbar\omega_q + E_i)$ as a function of $\hbar\omega_q$. It is calculated using parameters suitable for a MQW structure with W = 40 Å, B = 200 Å, and x = 0.25. With these structural parameters, the calculated value of E_r = M_3 +Δ_3 - E_1 is 140 meV and that of Δ_3 is 8 meV. The experimental absorption is shown in Fig. 3.5 (b) (Choi et al. 1991b). It is measured at Brewster's angle and at room temperature. In order to fit the data, Γ of ρ_1 and D are set to be 9 meV and 11 meV respectively. For the present MQW structure, the observed E_2 (225 meV) approximately coincides with the center of the M_3 subband, the corresponding resonant condition based on the nominal well width is k_2W = 0.80π. The data show more than two subsidiary peaks, they are due to the interaction of the resonant state E_2 with the lower miniband M_2 and the upper miniband M_4 respectively. From the peak absorption, the oscillator strength f is deduced to be 0.37. Note that the parameter D determines both the relative magnitudes of the peaks and their separations. The fact

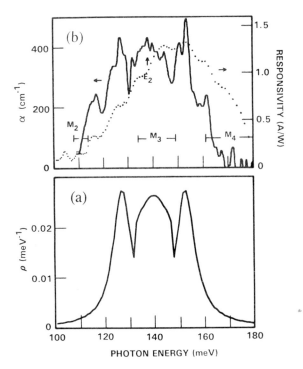

FIG. 3.5 (a) The curve shows theoretical joint density of states ρ of a MQW structure with W = 40 Å, B = 200 Å and x = 0.25. (b) The solid curve is the experimental absorption coefficient α, and the dotted curve is the measured responsivity which shows matching peaks as α.

that there exists a single value of D satisfying both conditions lends strong support to the present theory. Fig. 3.5 (b) also shows the measured responsivity of the MQW structure as a function of photon energy. The fact that the photocurrent peaks match the absorption peaks confirms that the absorption structures are due to electron transitions.

Incidentally, if the material quality is such that the dispersion relation of the unperturbed states lies between case 3 and case 4, the magnitude of the subsidiary peaks will be reduced. They can be easily dismissed as random fluctuations, and are lost in the curve smoothing process.

4. ELECTRON TRANSPORT IN MQW STRUCTURES

A determining factor for the detector sensitivity, in addition to optical absorption, is the amount of current flow in the detector even without the presence of radiation, which is usually known as the dark current. Since the quantum well of the structure is usually n-doped, there are always electrons in the conduction band. In order to optimize the detector performance, it is important to understand the electron transport mechanisms in these structures. In this section, the transport properties of the MQW structures will be reviewed.

The transport properties of a doped superlattice have been a subject of intense research for various device applications (Esaki and Chang 1974, Capasso et al. 1986). Here, we concentrate on MQW structures with relatively thick barriers. In this case, the coupling between adjacent wells is weak, and the ground state miniband width ($\Delta E_1 \simeq$ 0.4 meV) is much smaller than the impurity broadening ($\hbar/\tau \simeq$ 7 meV). Thus, the electrons are localized in each well, and the conduction is determined only by the local potential between the adjacent wells and not by the global structure. When the applied voltage across the MQW structure is small, the conduction can be attributed to three basic transport mechanisms: sequential resonant tunneling, thermionic emission, and phonon assisted tunneling.

In principle, due to the two-dimensional nature of the electron gas in the well, resonant tunneling is possible only when the energy levels in each well coincide, a condition generally not fulfilled in the presence of an electric field.

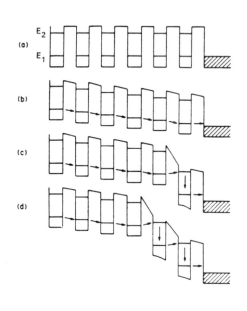

FIG. 4.1 The band diagrams show the energy level alignment at different biasing conditions. The arrows indicate the paths of the tunneling current.

However, Kazarinov and Suris (1972) showed that in the presence of acoustic phonon and impurity scattering within each well, conservation of energy and momentum can be relaxed and resonant tunneling is possible provided that $eV_p << \hbar/\tau_1$, where V_p is the potential difference between the adjacent wells and τ_1 is the ground-state scattering time. Therefore, at small bias, the electrons are able to conduct by resonant tunneling through the ground states of each well as shown in Fig. 4.1 (b).

The derivation of sequential resonant tunneling current (I_{st}) and thermionic emission current (I_{th}) is elementary (Choi et al. 1987a,b).

$$
\begin{aligned}
I_{st} &= \int_0^{e\Delta_1} Ae\frac{\rho_{2d}}{W}f(E)v_x\,D(E_1)\,dE \ , \\
&= \int_0^{e\Delta_1} Ae\frac{m^*}{\pi\hbar^2 W}f(E)\frac{\hbar\pi}{m^*W}D(E_1)\,dE \ , \\
&= \frac{eA}{\hbar W^2}D(E_1)kT\ln\left\{\frac{1+e^{E_F/kT}}{1+e^{(E_F-e\Delta_1)/kT}}\right\} \ ,
\end{aligned}
\tag{4.1}
$$

where W is the well width, Δ_1 is the potential drop across one barrier, ρ_{2d} is the two dimensional density of states, $f(E)$ is the Fermi distribution function, v_x is the electron velocity in the first subband along the superlattice axis, and $D(E_1)$ is the tunneling probability which can be obtained by Wentzel-Kramers-Brillouin (WKB) approximation:

$$D(E_1) = \exp\left\{ \frac{-4B}{3e\hbar\Delta_1} (2m_b^*)^{1/2} \left[(H-e\Delta_2-E_1)^{3/2} - (H-eV_p-E_1)^{3/2} \right] \right\}. \qquad (4.2)$$

Here, B is the thickness of the barrier, m_b^* is the effective mass of the barrier, Δ_2 is the potential drop within one well and $\Delta_1+\Delta_2 = V_p$. We estimated Δ_1 and Δ_2 to be $0.85V_p$ and $0.15V_p$ respectively, from the distortion of the wave function in a tilted well. Similarly, I_{th} is given by

$$I_{th} = \frac{e^2 m^*}{\pi\hbar^2} \frac{A v_d}{W} \Delta_1 \, e^{-(H-e\Delta_2-E_F-E_1)/kT},$$

$$(4.3)$$

where v_d is the drift velocity near the top of the barrier. In the above equations, E_F is temperature dependent,

$$E_F = kT \ln\left[\exp\left(\frac{\pi\hbar^2\rho_s}{m^*kT} \right) -1 \right], \qquad (4.4)$$

where ρ_s is the two-dimensional electron density. The optical phonon-assisted tunneling current I_{pt} is obtained by Calecki et al. (1984), which is not listed here. The relative contributions of each current component at different temperatures at $V_p = 1$ mV are shown in Fig. 4.2. The same figure also shows the experimental result of a MQW structure with $W = 70$ Å, and $B = 140$ Å, Al molar ratio $x = 0.36$ and doping density of 1.4×10^{18} cm^{-3}. From the fitting, one obtains $x = 0.38$, $B = 133$ Å, and $v_d = 2\times10^7$ cm/s, quite consistent with the nominal values. From this result, a general conclusion can be drawn that at low temperatures, tunneling current dominates, while at high temperatures,

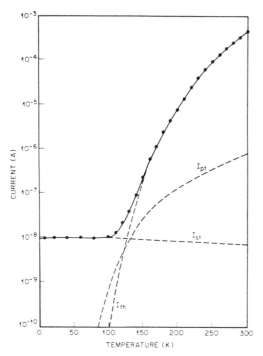

FIG. 4.2 The circles show the experimental temperature dependence of a MQW structure at low bias. The dashed curves show the relative current contributions from sequential tunneling (I_{st}), phonon-assisted tunneling (I_{pt}) and thermionic emission (I_{th}).

thermionic emission current becomes more important. Phonon-assisted tunneling is significant only in a very narrow range of temperatures.

When the applied bias increases such that V_p is larger than $\hbar/\tau_1$, resonant tunneling is disrupted, leading to a negative differential conductance (NDC). Under a bias, the potential drop across the structure is nonuniform due to the screening of the electric field by the space charge. Space charge exists in the sample because of electron accumulation near the anode, caused by the unbalanced quasi-Fermi levels and also because of the existence of the thermal electrons in the barrier region, the same mechanism which causes the nonuniform field in a vacuum tube. Consequently, the potential drop is largest at the anode and it progressively decreases toward the cathode. The resonant tunneling condition hence is disrupted first near the anode, and is restored only when the applied bias is increased further until E_1 of the adjacent well approximately aligns with the second state E_2. As a result, a high field domain is formed. The condition is depicted in Fig. 4.1b and 4.1c. If the increase of the applied bias continues, the process repeats itself and a series of NDC will be observed, as shown in Fig. 4.3 (a) for a structure with W = 76 Å, B = 90 Å, and x = 0.27. From the applied voltage where the first NDC occurs, τ_1 is estimated to be 0.1 ps. Since the spacing of the NDC is approximately equal to E_2-E_1-$\hbar/\tau_1$-$\hbar/\tau_2$, τ_2 is estimated to be 0.06 ps by substituting the calculated value of E_2-E_1 (105 meV) for this sample. For a given applied bias, the structure is generally divided into two regimes, a high field regime near the anode and a low field regime near the cathode.

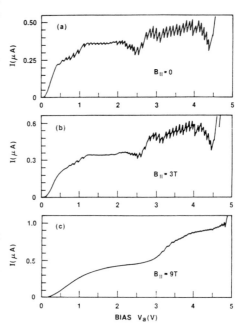

At higher temperatures, sequential resonant tunneling becomes less important, and there is no preferred level alignment for the thermionic emission process. Therefore, the NDC becomes less apparent as shown in Fig. 4.4 for a structure with W = 40 Å, B = 200 Å, and x = 0.25. However, the nonuniformity of the potential drop inside the sample still persists as the existence of space charge is independent of the level alignment. The NDC features at low temperature can also be suppressed by a strong magnetic field applied parallel to the material layers as shown in Fig. 4.3 (b) and (c) (Choi et al. 1988a, Grahn et al. 1991) due to the relative shift of the dispersion relation of two adjacent wells under the influence of a Lorentz force.

FIG. 4.3 show the characteristics of negative differential conductance (NDC) of a MQW structure with area 3.14×10^{-4} cm^{-2}. NDC structures can be suppressed by an applied magnetic field parallel to the material layers.

At large voltages, the thermionic emission current described by Eq. (4.3) turns out to be insufficient to account for the experimental voltage dependence for some MQW structures (Pelvé et al. 1989, Levine et al. 1990). In order to explain the discrepancy, a thermally assisted tunneling (TAT) mechanism is introduced, in which the ground state electrons with large thermal velocities can be scattered by the impurities and transfer most of their energy toward the direction of the current flow. As a result, the tunneling probability is determined by the total energy of the electrons instead of the parallel energy component. The larger voltage dependence is hence caused by the larger tunneling probability at higher voltages.

In order to better understand the transport mechanisms of MQW structures, it is advantageous to know the energy distribution of the conduction electrons at different temperatures. For this reason, Choi et al. (1989a,c 1990c) extended the work of Heiblum

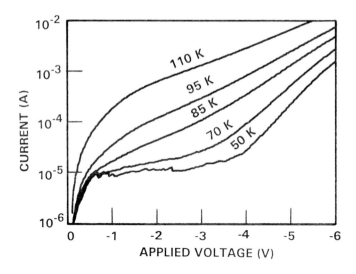

FIG. 4.4 shows the I-V characteristics of a MQW structure with area of 7.92×10^{-4} cm^2 at different temperatures.

et al. (1985) and Levi et al. (1985) in using a quantum barrier to analyze the energy of hot-electrons. Specifically, a quantum barrier as an electron energy analyzer is constructed next to the MQW structure as shown in Fig. 4.5 (a). The energy distribution of the hot-electrons injected from the MQW structure is examined at different temperatures. The technique is referred as thermally stimulated hot-electron spectroscopy (TSHES). The entire structure includes a MQW structure S under study, sandwiched between the emitter layer E and the base layer B. The barrier filter is then located next to the base, followed by a collector layer C.

Under an external emitter to base bias V_E, electron conduction occurs within each subband, either by sequential tunneling from well to well within the same subband in the low field regime, or by tunneling to the higher subband of the adjacent well with subsequent relaxation back to the original subband in the high field regime as indicated in Fig. 4.1. At low temperatures, the conduction electrons from the dopant reside primarily in the lowest subband E_1 and hence the electrical conduction occurs only in this subband. As the temperature increases, the electron population in the higher subbands increases. Although the population in the higher subbands may be still very small compared with the lowest subband when kT is less than the subband spacings, the increase in the tunneling probability at higher subbands greatly compensates this factor and leads to comparable current density through each subband even at relatively low temperatures.

In the present device configuration, conduction electrons in each subband inject into the base and form hot-electrons with different launching energies. The hot-electrons then travel through the base, and for those electrons with energy higher than the barrier height δE_c of the analyzer, they can overcome the analyzer and will be collected at the collector. Electrons with less energy will then contribute to the base current. In these experiments, V_E is varied at each temperature with the collector to base voltage V_C fixed at zero volts. The advantage of changing V_E instead of V_C is that it allows a larger electron energy range to be probed. The transfer ratio α, defined as I_C/I_E, is then measured as a function of V_E, from which the derivative r ($\equiv d\alpha/dV_E$) can be obtained. r can be shown to be proportional

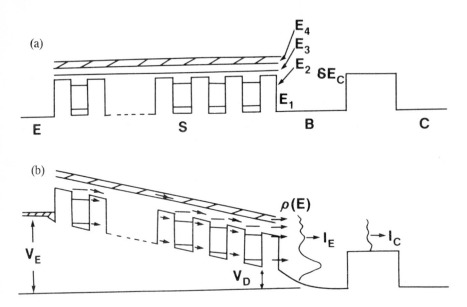

FIG. 4.5 shows the band diagram of an energy analyzer between B and C used to probe the electron energy distribution injected from the structure S located between E and B. V_E is the applied emitter bias and V_D is the resulting potential energy in the depletion layer.

to the normalized electron energy distribution $\rho(V_E, E)$ reaching the front boundary of the analyzer as follows. Let $n_o(E)$ be the total number of injected electrons from the MQW structure, then the emitter current I_E can be written as $en_o(V_E)v$, where v is the average velocity of the injected electrons. Likewise, the collector current I_C is given by

$$I_C(V_E) = e\,n_o(V_E)\,vf\int_{\delta E_c}^{\infty}\rho(V_E, E)\,dE \quad , \tag{4.5}$$

where f is the collection efficiency, accounting for the quantum mechanical reflection and the loss of the electrons with large momentum parallel to the interface. Thus,

$$\alpha = f\int_{\delta E_c}^{\infty}\rho(V_E, E)\,dE \quad . \tag{4.6}$$

The functional dependence of ρ on V_E can be separated into two parts, one accounting for the rigid shift without changing the shape of the distribution. It is due to the existence of the self-consistent voltage drop (V_D) across the depletion layer in the base plus the voltage drop at the first barrier as shown in Fig. 4.5 (b). Another V_E dependence accounts for the change of the shape of the distribution. If we assume $V_D = \beta V_E$ for some parameter β, then

$$r(V_E) = fe\beta\,\rho(V_E, \delta E_c) + f\int_{\delta E_c}^{\infty} \frac{\partial\rho}{\partial V_E}\,dE \quad , \tag{4.7}$$

In Eq. (4.7), the first term account for the rigid shift and the second term account for the change of the shape of the distribution. The second term is negligible since when V_E is small, the electron density in each subband is independent on V_E, while at large V_E, the majority of the hot-electrons are above δE_c, so that the integral of ρ is a constant by the normalization condition. Therefore, r is directly proportional to the electron distribution. By raising the temperature in steps, one can sequentially populate the higher subbands, from which the information of the dominant transport mechanism at each temperature and the subband structure can be obtained.

Fig. 4.6 shows the measured parameter r for a MQW structure with W = 74 Å, B = 150 Å, x = 0.32 and N_d = 2x 10^{18} cm^{-3}. The MQW is sandwiched between a heavily doped (n^+ = 2x10^{18} cm^{-3}) GaAs layer on the top as the emitter contact, and lightly doped (n = 3x10^{17} cm^{-3}, 1500 Å) GaAs layer at the bottom as the base contact. The analyzer is a 1000-Å-thick undoped $Al_{0.32}Ga_{0.68}As$ barrier, and the collector layer is a heavily doped (n^+ = 2x10^{18} cm^{-3}, 6000 Å) GaAs layer. The device is grown on a semi-insulating substrate. At 4.2 K, only one peak in r is observed at V_E equal to 5.9 V. This peak represents a group of electrons injected from the lowest subband E_1. At higher temperature around 100 K, another peak develops at V_E equal to 3.1 V due to the thermal excitation of conduction electrons from E_1 to E_2. Finally, at an even higher temperature of 130 K, the conduction through E_3 and E_4 subbands detected at 1.3 V and 0.3 V respectively becomes dominant. From Fig. 4.5 (b), it is apparent that the peak from the nth subband will occur when $eV_D + E_n - E_p = \delta E_c$, where E_p is

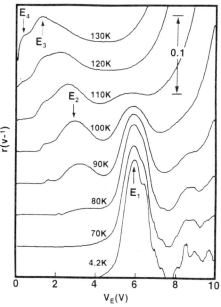

FIG. 4.6 shows the TSHES of a MQW structure having W = 74 Å, B = 150 Å and x = 0.32. The arrows indicate the subband locations detected at different temperatures. The origins of the curves have been shifted for clarity.

the amount of the hot-electron energy relaxation in the base. Substituting $V_D = \beta V_E$, we get

$$\begin{aligned}E_{nm} &= E_m - E_n \\ &= e\beta\,(V_n - V_m) \quad .\end{aligned} \tag{4.8}$$

where V_n is the emitter voltage for the nth peak to occur. From the value of V_n the ratio of the subband separation can be determined from Eq. (4.8). The subband separation E_{12} can be measured from the intersubband absorption experiment, from which E_{12} is found to be 0.136 eV. The value of β is then deduced to be 0.051. With this value of β, the separations between different subbands can be determined: $E_{23} = 0.084$ eV and $E_{34} = 0.049$ eV. These values can be compared with the calculated value from the nominal device parameters: $E_{12} = 0.128$ eV, $E_{23} = 0.082$ eV and $E_{34} = 0.044$ eV. To proceed further, if we use the calculated value of E_1 (= 0.046 eV), the value of E_p (= 0.083 eV) can be determined. This value of E_p is consistent with the value (0.062 eV) deduced using a single barrier at the emitter as the hot-electron injector (Choi et al. 1989a,c 1990c).

Since the hot-electron peaks are distinct, it is evident that direct thermionic emission from each higher subband dominates the temperature dependent process for the present MQW structure. On the other hand, TSHES of a MQW structure with only one bound state shows a slightly different picture. For a structure with $W = 40$ Å, $B = 200$ Å, and $x = 0.22$, the E_1 peak and the E_2 peak are less distinct as shown in Fig. 4.7. At temperatures around 80 K, a large portion of electrons are observed with energy between E_1 and E_2, indicating thermally assisted tunneling mechanism significant in this MQW structure. In general, it can be concluded that TAT current is important in MQW structures when E_2 is above the barrier height since the scattered ground state electron with energy slightly below E_2 do not need to tunnel at all.

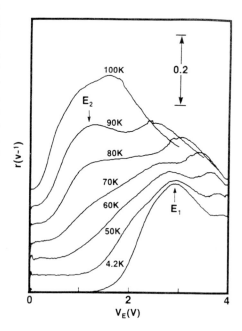

FIG. 4.7 shows the TSHES of a MQW structure having $W = 40$ Å, $B = 200$ Å and $x = 0.22$. The arrows indicate the subband locations detected at different temperatures. The origins of the curves have been shifted for clarity.

5. QUANTUM WELL INFRARED PHOTODETECTORS

Before discussing the properties of quantum well infrared detectors, let us first review some basic terminologies of infrared detection.

5.1 Background discussion

Let us consider a thin detector in which the electron motion can be described by diffusion with a diffusion constant D. Due to surface recombination, the spatial variation of photogenerated electron density n(x) can be approximated by

$$n(x) = N_o \cos\left[\xi\left(x - \frac{L}{2}\right)\right], \tag{5.1}$$

where L is the thickness of the sample along the x direction. The boundary conditions are

$$D \frac{\partial n}{\partial x} = v_s n , \quad at \ x = 0 ,$$

$$-D \frac{\partial n}{\partial x} = v_s n , \quad at \ x = L , \tag{5.2}$$

where v_s is the surface recombination velocity. When v_s is small, $\xi^2 = 2v_s/DL$. If we consider also the time dependence, after a light pulse, $n(x,t)$ satisfies

$$\frac{\partial n}{\partial t} = D \frac{\partial^2 n}{\partial x^2} - \frac{n}{\tau_o} , \tag{5.3}$$

where τ_0 is the bulk recombination lifetime of the electrons. The solution of Eq. (5.3) is

$$n(x,t) = N_o \exp\left(-\frac{t}{\tau_{eff}}\right) \cos\left[\xi\left(x - \frac{L}{2}\right)\right],$$

where
$$\frac{1}{\tau_{eff}} = D \xi^2 + \frac{1}{\tau_o} , \tag{5.4}$$

$$= \frac{2v_s}{L} + \frac{1}{\tau_o} .$$

τ_{eff} is the effective lifetime of the electrons in the presence of surface recombination. Under a steady optical power, the spatially averaged electron density n can be shown to be

$$n = \frac{1}{AL} \eta \frac{P}{hv} \tau_{eff} , \tag{5.5}$$

where A is the optical area, η is the internal quantum efficiency equal to $(1 - \exp(-\alpha L))$, P is the optical power on the area A. Here, we have assumed the material coated with antireflection coating. In the presence of a bias along the x direction, the photocurrent I_p is given by

$$I_p = A n e v_d ,$$

$$= \frac{\eta P}{hv} e \frac{v_d \tau_{eff}}{L} ,$$

$$= \frac{\eta P}{hv} e g , \tag{5.6}$$

where
$$g = \frac{L_{eff}}{L} \quad or \quad \frac{\tau_{eff}}{\tau_{tr}} .$$

v_d is the drift velocity. g is the photoconductive gain of the detector, τ_{tr} and L_{eff} are the transit time and the effective mean free path of the photoelectrons respectively. The responsivity R is then defined as I_p/P.

The noise of a photoconductor comes from several sources, which include 1/f noise, Johnson noise and generation-recombination (g-r) noise. g-r noise is given by $(4egIB)^{1/2}$, where B is the bandwidth and I is the total current flow caused by all physical mechanisms. In this chapter, only the noise caused by the detector dark current will be considered. The sensitivity of a detector is specified by the noise equivalent power (NEP) which is the optical power to give a signal to noise ratio of unity. In order to compare the performance of different detectors, a convenient figure of merit is the detectivity D^* defined as $(AB)^{1/2}$/NEP which turns out to be independent of the detector area and the measuring bandwidth. Another useful figure of merit is the noise equivalent temperature difference (NEΔT) which is defined as NEP/dP/dT, and is given by

$$NE\Delta T = \left(\frac{B}{A}\right)^{1/2}\frac{1}{D^*\sin^2(\theta/2)}\frac{T}{I_B}\frac{kT}{h\nu} \quad , \tag{5.7}$$

where θ is equal to 28° for f/2 optics, and I_B is given by Eq. (2.4).

5.2 DETECTIVITY OF QWIP

In sections 3 and 4, we have discussed the optical and transport properties of MQW structures. In this section, we will discuss their optoelectronic properties.

The photoconductivity of a MQW structure is very similar to that of an extrinsic photoconductor. An electron located in the lowest state E_1 absorbs an incoming photon and is excited to the higher state E_2 where the conductivity along the superlattice axis is higher. If a bias is applied in this direction, an increase in current flow will occur and the responsivity R is equal to

$$R = \frac{e}{h\nu}\eta\,p\,g \quad , \tag{5.8}$$

where p is the tunneling probability of the E_2 state. If E_2 is above the barrier height, p is equal to unity and the responsivity will be the same as a usual photoconductor.

On the other hand, the dark current of the detector in general can be divided into tunneling current and thermally activated current. Since the tunneling current can be practically eliminated by increasing the thickness of the barriers, it is not a determining factor for the detector sensitivity. If only the thermal current from the E_2 state is considered, D^* can be expressed as

$$D^* = \frac{\eta\,p}{2h\nu}\left(\frac{g}{n_{th}p\,v_d}\right)^{1/2} \quad , \tag{5.9}$$

where n_{th} is the thermally activated 3-dimensional electron density in the E_2 subband, and v_d is the drift velocity. From this expression, it is apparent that D^* can be maximized when E_2 is above the barriers where the tunneling probability p is maximum. This type of detector is referred as extended state (ES) detector here, otherwise it is referred as bound state (BS) detector. For the ES detector, since the photoconductive gain g and the drift velocity v_d are approximately fixed for a given material, the available parameter to optimize D^* is the doping density N_d, which directly determines the quantum efficiency η as well as the thermal electron density n_{th}. For a given doping density, α at the peak of absorption is given by

$$\alpha_{peak} = \frac{N_d W}{L_p} \frac{e^2 \hbar}{2 \epsilon_o m^* c} f \frac{\sin^2\theta}{n} \frac{2}{\Gamma} \quad , \tag{5.10}$$

where W is the well width, and η is equal to $0.5(1 - \exp(-2\alpha L))$ for unpolarized light with a double pass, L is the total thickness. n_{th} is given by

$$n_{th} = \frac{m^*}{\pi \hbar^2 L_p} k T \exp\left(- \frac{E_{12} - E_F}{kT} \right) \quad , \tag{5.11}$$

where $\quad E_F = \frac{\pi \hbar^2}{m^*} N_d W \quad ,$

and E_{12} is the subband spacing. For W = 50 Å, L_p = 500 Å, E_{12} = 124 mev, θ = 45°, Γ = 30 mev, and T = 77 K, the optimum doping density is 6.8×10^{17} cm^{-3}. The corresponding D_{max}^* is $1.47 \times 10^{10} \sqrt{g}$ cm$\sqrt{Hz}$/W, if the saturated velocity = 1×10^7 cm/s is assumed for v_d. The quantum efficiency in this case is 7.4 %. By using a reflection grating light coupler (Andersson et al. 1991a, 1991b), the quantum efficiency can be greatly improved and can be close to unity. The D^* can then be in the 10^{11} cm$\sqrt{Hz}$/W range since the measured g is order of unity.

It should be emphasized that the above analysis is done in an ideal case where the thermal current is carried in the second subband alone. In practice, there are two complications. First, it has been shown in section 3 that the resonant state for ES detectors may not coincide with the second miniband, and can span across many higher minibands for thick barriers. Thermal current from all of these minibands must be included. Second, besides the direct thermionic emission current, thermally assisted tunneling (TAT) current described in section 4 actually dominates the current flow and limit the detectivity at 77 K due to the lower activation energy for this current component. Kinch and Yariv (1989) considered only part of the TAT current with energy above the barrier height and concluded that the D^* is limited to 6.1×10^9 cm$\sqrt{Hz}$/W. Therefore, the estimated D^* can be greatly reduced.

In comparing the performance between an ES detector and a BS detector, a different conclusion may be reached if one considered NEΔT as the figure of merit, since NEΔT not only depends on the D^* but also the bandwidth of a measuring system, which in turn is inversely proportional to the signal integration time. For a given charging capacitor, the smaller dark current will allow a longer integration time and hence gives a smaller NEΔT. As the thermal current carried by E_2 subband in a BS detector is a factor of p smaller than that in an ES detector, NEΔT turns out to be independent of p and hence the type of the detector. If one considers the additional nonideal current contribution in an ES detector and the larger power consumption, the BS detector can be more advantageous especially at elevated temperatures.

5.3 PHOTOCONDUCTIVE GAIN OF A QWIP

In section 5.1, we have discussed the gain g of a typical photoconductor. Here, we provide a more detailed discussion specific to the QWIPs. While the photoexcitation mechanism and the band structure of a QWIP are very similar to an impurity doped extrinsic photoconductor, there are two significant differences. First, the total thickness of a MQW structure is very small ($\simeq 2~\mu$m), the typical applied electric field ($\simeq 25$ kV/cm) can be much higher in the QWIPs. Thus the electron motion may not be described by diffusion. Second, the contact layer of a MQW structure is made of quantum well material and hence its conduction band is equivalent to the impurity states of an extrinsic photoconductor, recombination of photoelectrons can occur at the contact interfaces; while the ohmic contacts of an extrinsic photoconductor are made at the conduction band of the host material, which is equivalent to the E_2 subband of the QWIP, consequently, there is no carrier recombination at the extrinsic photoconductor contacts, and the gain characteristics are different.

In the early detector development (Levine et al. 1987b, Choi et al. 1987b), the electron transport is described as quasi-ballistic: an electron travels ballistically within the MQW structure until it emits a phonon, and it thermalizes immediately after reaching the MQW-contact interface. With these assumptions, it turns out that in order to explain the experimental photocurrent data (Levine et al. 1987b, Choi et al. 1987b,c 1988b), the energy loss of a photoelectron will always balance out the energy gain from the applied electric field in order to maintain a finite bulk lifetime; and the maximum gain of the detector is necessarily less than 0.5.

After the introduction of the TSHES, it becomes clear that the hot-electron distribution at any given temperature is mainly determined by the thermal population in each subband and not by the hot-electron dynamics within the MQW structure. This observation indicates that without light, the electrons are always in thermal equilibrium with the lattice, and no appreciable amount of electrons in an applied field acquires significant amount of energy above the barriers. The last conclusion is consistent with that from the photocurrent fittings. On the other hand, the TSHES also shows that although the hot-electrons relax some of their energy when traveling across the base, the hot-electrons do not completely thermalize and hence the number of hot-electrons with energy E_2 is nonzero at the MQW-contact interface. Therefore, the original boundary condition needs to be modified to include a contact recombination velocity that is finite. Consequently, the gain can be larger than 0.5, as observed by Levine et al. (1990) and Hasnain et al. (1990).

It needs to be clarified why quasi-ballistic transport is observed in the base region but not in the MQW region. The main reason is due to the difference in the applied field. The MQW structure is usually under bias, the electrons above the barriers which travel ballistically initially will be accelerated by the electric field and be scattered into the L valleys, where energy and momentum relaxations are extremely efficient due to the increase of the density of states by 94 times (Sze 1981). The dispersion relation of the miniband structure of a MQW will also prevents a Bloch electron to acquire large energy (Lei et al. 1991). On the other hand, the electric field in the base is small, the ballistic electrons will not be accelerated to exceed the energy of the L valleys, and hence energy relaxation is small.

In the following, we will describe the optoelectronic properties of a specific QWIP as an example (Choi et al. 1991c). It is grown by molecular beam epitaxy on a (100) semi-insulating substrate. The first layer grown is a 6000-Å-thick n^+-GaAs layer as the emitter layer. Next, an infrared sensitive MQW structure is grown. It consists of 50 periods of GaAs quantum wells and $Al_{0.25}Ga_{0.75}As$ barriers. The well width and the barrier width are 40 Å and 200 Å respectively. On the top of the MQW structure, a thin (300 Å) pseudomorphic $In_{0.15}Ga_{0.85}As$ is grown as the base layer. In order to probe the hot-electron distribution injected from the MQW structure, a 2000-Å-thick $Al_{0.25}Ga_{0.75}As$ barrier is grown next to the base as an electron energy analyzer, followed by a 1000-Å-thick GaAs as the collector layer. All the layers except the barriers are doped to n = 1.2×10^{18} cm^{-3}. According to the nominal material parameters, the ground state E_1 is inside the wells and is 85 meV above the conduction band edge. The higher energy states are outside the wells and form minibands denoted by M_n in Fig. 5.1. The infrared absorption has been shown in Fig. 3.5 (b).

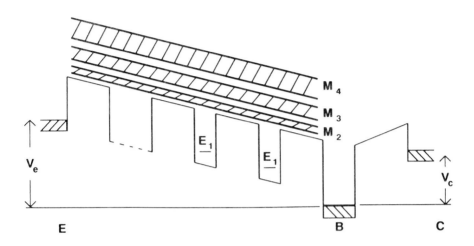

Fig. 5.1 shows the band diagram of a device used to probe the hot-electron energy distribution generated in a MQW structure with W = 40 Å, B = 200 Å and x = 0.25.

     The detector is fabricated using standard photolithography with the emitter area equal 7.92×10^{-4} cm^2 and the collector area equal to 2.25×10^{-4} cm^2. Infrared radiation is coupled into the device through a 45° angle as shown in the insert of Fig. 5.2. The figure also shows the measured emitter dark current I_E (normalized to the collector area) and the collector dark current I_C as a function of emitter bias V_E at 77 K, with the collector voltage equal to zero volts. The sign of the applied voltages follows the common base configuration. From the measured current at different temperatures, the thermally stimulated hot-electron spectra can be constructed and represented in Fig. 5.3 by the curves (a), (b), and (c) for temperatures equal to 100 K, 77 K, and 50 K respectively. In this figure, the large fluctuations of r at low temperatures are reproducible structures caused by the strong negative differential conductance observed at these temperatures. As discussed before, the data show that the dark current at 77 K is dominated by the thermally assisted tunneling since the hot-electrons are distributed between the ground state E_1 and the resonant state E_2, and is lower in energy than the photoelectron distribution indicated by the curve (d). Curve (d) is a plot of $r_p \equiv d\alpha_p/dV_E$, where α_p is the ratio of collector photocurrent to the emitter photocurrent, measured with a 9.25 μm monochromatic light source at temperature equal to 90 K. The photocurrent peak is found to be independent on temperature. From the separation of E_1 and E_2, and the width of r_p, the width of the photoelectron distribution Γ_ρ is deduced to be 17 meV.

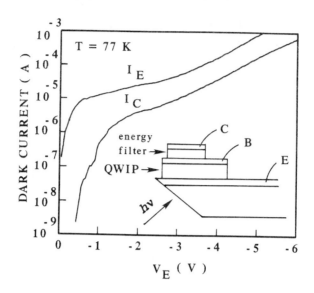

FIG. 5.2 shows the emitter current I_E and the filtered collector current I_C as a function of emitter voltage V_E. The insert shows the device configuration.

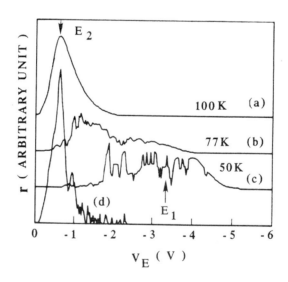

FIG. 5.3 Curves (a), (b), and (c) show the thermal stimulated hot-electron spectra at T = 100 K, 77 K and 50 K respectively. Curve (d) shows the optical stimulated hot-electron spectrum at T = 90 K. The curves have been shifted vertically for clarity.

Γ_ρ can also be measured by fixing V_E and V_C, equal to -1.2 V and 0 V respectively in this example, but changing the incident photon energy. The result plotted in Fig. 5.4 shows the increase of α_p as a function of photon energy. The solid curve is a fitting to the experimental data assuming that the photoelectron distribution is Lorentzian and the photoelectron injection energy increases with the incident energy. The width is again deduced to be 17 meV.

In order to probe the photoelectron distribution at different V_E, dI_{pc}/dV_C is measured as a function of V_C at different V_E, where I_{pc} is the collector photocurrent. As V_C becomes more negative, the barrier height of the analyzer becomes higher as shown in Fig. 5.1, and less photocurrent will be collected. The rate of change of I_{pc} is then a measure of the photoelectron distribution ρ. The photocurrent is generated by a CO_2 laser with photon energy equal to 134 meV modulated at 200 Hz. The result is plotted in Fig. 5.5. It shows that the photoelectron distribution is indeed Lorentzian and the width is insensitive to the applied

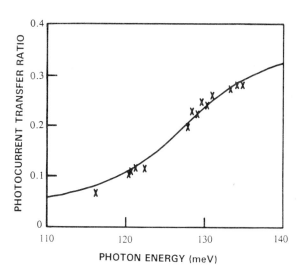

FIG. 5.4 The crosses are the experimental photocurrent transfer ratio as a function of incident photon energy. The curve are the theoretical fitting assuming that the photoelectron distribution is Lorentzian with a width of 17 meV.

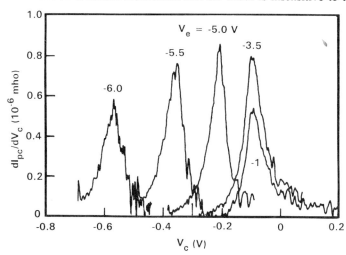

FIG. 5.5 is a plot of dI_{pc}/dV_C as a function of V_C at different emitter voltage V_e at temperature equal to 10 K.

voltage. However, Γ_ρ probed by this method is found to be 76 meV. The difference is due to the fact that the present method uses the far edge of the barrier to filter the photocurrent whereas the previous two methods use the front boundary, and there is substantial energy broadening ($\simeq$ 59 meV) in the analyzer. When the detector is under 1000 K blackbody radiation, the present method yields 90 meV for Γ_ρ. Subtracting the broadening occurred in the analyzer, Γ_ρ is then equal to 31 meV at the front boundary of the analyzer, consistent with the absorption width of 35 meV shown in Fig. 3.5(b).

The above results indicate that the photoelectron distribution in the MQW structure is extremely narrow, consistent with the diffusion picture, and the position of the photoelectron peak is independent on V_E below -4V. Hence, the standard theory of Eq. (5.6) for short photoconductors is applicable to the QWIPs. In order to apply Eq. (5.6) to the present example, it is necessary to know the specific characteristics of the device. For the present structure, high field domain (HFD) formation is observed up to 77 K. Below a critical emitter voltage (V_k) equal to -3 V, an increase in V_E results in an increase in the number of HFDs, but keeping the electric field in the low field regime and the high field regime constant. Since the major contribution of photocurrent is from the HFDs, the effective η is then proportional to V_E below V_k and constant above V_k. Under a constant electric field below V_k, v_d is also constant and actually equal to the saturated drift velocity since the electric field ($\simeq$ 60 meV per period) is much larger than the critical field (7 meV per period) for maximum drift velocity to occur (Lei et al. 1991). Since V_d is constant, τ_{tr}, being the time to travel across the high field regime, is proportional to V_E below V_k and constant above V_k. If we further assume v_s and τ_0 to be constant, the dc current responsivity can be fitted to Eq. (5.7):

$$R = \frac{e}{h\nu} \eta(V_E) g(V_E)$$
$$= \frac{e}{h\nu} \frac{\eta(V_E)}{\dfrac{2v_s}{v_d} + \dfrac{\tau_{tr}(V_E)}{\tau_o}} \quad, \tag{5.7}$$

as shown in Fig. 5.6 for $|V_E| < 4$ V with $\eta = 0.15$, $v_s/v_d = 0.13$ consistent with the assumption that $v_s << v_d$, and $\tau_0/\tau_{tr} = 2.5$ at V_k.

For $|V_E| > 4$ V, R is larger than the fitting predicted. In order to understand this discrepancy, it is necessary to examine R at the lower bias. When V_E is small, τ_{tr}/τ_0 is less than $2v_s/v_d$, therefore, R is proportional to η and hence V_E. At larger bias, the reverse is true, R is then proportional to $\eta\tau_0/\tau_{tr}$, which is independent of V_E as observed. The rising of R after -4 V then indicates that τ_0 increases at large bias. This conclusion is in fact consistent with the increasing photoelectron peak position at large V_E shown in Fig. 5.5. As the average energy of the photoelectrons above the quantum barrier is increased, τ_0 is also expected to increase. In Fig. 5.6, we plot the peak position eV_m above the analyzer barrier, which show strong correlation with R. We should note that eV_m is not the position of the photoelectrons above the MQW barriers, but is equal to $E_h + eV_D - E_p$, where E_h, eV_D, and E_p are the peak position above the MQW structure, the depletion potential energy in the base and the energy loss in the base and the analyzer. An estimation shows that E_h is approximately equal to 40 % of eV_m.

We emphasize that the contact recombination term is crucial to the theory. It explains the voltage dependence at low voltages. At high voltage when τ_0 tends to be infinite, Eq. (5.7) predicts a maximum gain of $v_d/2v_s$, independent on the applied voltage. Such a

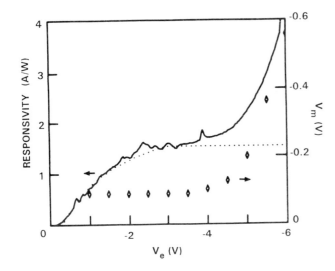

FIG. 5.6 A plot of experimental (solid curve) and theoretical (dotted curve) dc current responsivity R at photon energy equal to 150 meV as a function of emitter bias V_e. The diamonds are the position of the photoelectron peak eV_m shown in Fig. 5.5. R and V_m show strong correlation.

plateau region at high voltage has been observed in InGaAs/InP material system by Gunapala et al. (1991). Since the nonuniformity of the potential distribution in MQW structures is due to the screening of space charge, and is not caused by the high field domain formation, the present theory is also applicable to the MQWs even when the HFD formation is not apparent.

At 77 K and V_E equal to -0.85 V, the noise current i_n at the emitter is measured to be 4.1 pA/$\sqrt{\text{Hz}}$. Combined with the measured responsivity shown in Fig. 5.6, D^* is evaluated to be 4×10^9 cm$\sqrt{\text{Hz}}$/W at 8.2 μm. Likewise, D^* is equal to 5.5×10^9 cm$\sqrt{\text{Hz}}$/W and 6.0×10^9 cm$\sqrt{\text{Hz}}$/W at V_E = -1.33 V and -1.8 V respectively. From the noise measurement, the gain g is deduced to be 0.66 at -0.85 V, from the relation $i_n = (4egI_d)^{1/2}$, differing from the value 3.0 we deduced from Eq. (5.7) at that voltage. The difference is due to the fact that the gain determined in the noise measurement is the average gain of the entire MQW, including the high field regime and the low field regime, while the average gain in Eq. (5.7) is the average gain at the high field regime only. The quantum efficiency of the MQW obtained in the photocurrent measurement is 15 %, while the value obtained from the infrared absorption measurement is 5 % and the theoretical value assuming f = 1 is 11 %. In general, photocurrent measurement gives a more absolute value of intersubband absorption because it is relatively free from other optical interference, such as plasmon absorption in the contact layers and Fabry-Pérot oscillations between the epilayers, as demonstrated in the next section.

6. PHOTOEXCITED COHERENT TUNNELING AND MULTI-COLOR PHOTODETECTOR

In some applications, it is advantageous to expand the spectral response of the detector to cover both the 3 to 5 μm and 8 to 12 μm ranges. One simple method is to stack two different MQW structures together, each of which covers a specific range of radiation. Here, we described an alternative approach of using coupled quantum wells. Coupled quantum well structures not only provide multi-color detection capability, but also provide a wavelength tunable feature, and a dark current reduction mechanism (Choi et al. 1987c, 1989b).

The first sample we described was grown with 50 periods of 65-Å GaAs quantum well (doped at n = 1.0×10^{18} cm^{-3}), 40-Å undoped $Al_{0.25}Ga_{0.75}As$, 14-Å undoped GaAs, and 150-Å undoped $Al_{0.25}Ga_{0.75}As$. The MQW are sandwiched between the top (0.5 μm) and the bottom (1 μm) GaAs contact layers in which n = 1×10^{18} cm^{-3}. The band diagram of the structure is shown in Fig. 6.1. If the wells are isolated, there would be two bound states (E_1 = 50 meV, E_2 = 181 meV) in the thicker well and one bound state (E_1' = 163 meV) in the thinner well. In the actual structure E_2 and E_1' are coupled, thus shifting the energy levels to the values E_1 = 50 meV, E_2 = 154 meV, and E_3 = 188 meV. The resulting wave functions for these bound states, together with that for the first continuum state (ψ_4) are depicted in Fig. 6.2.

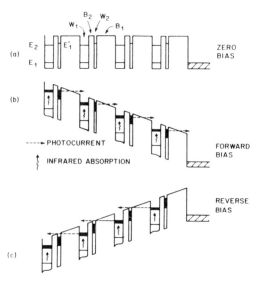

FIG. 6.1 (a) The band diagram of the coupled quantum-well structure. (b) The band diagram under forward bias. The shaded area indicates the lifetime broadening of the state. (c) The band diagram under reverse bias.

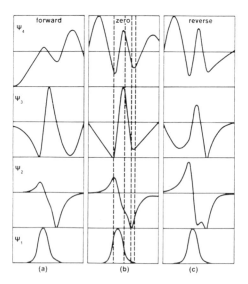

FIG. 6.2 The shape of the wave functions in the coupled quantum-well structure under (a) forward, (b) zero, and (c) reverse biases. The theoretical oscillator strength under bias is based on these wave functions. The dashed lines indicate the locations of the band-edge discontinuity.

Based on these wave functions, the oscillator strength (f_n) between the ground state and the excited state n can be calculated, and the results are $f_2 = 0.46$, $f_3 = 0.45$, and $f_4 = 0.13$. The approximate equality between f_2 and f_3 is consistent with the equal absorption amplitudes at ν_2^0 and ν_3^0 shown in Fig. 6.3, where ν_n^0 denotes the fitted wave number of the absorption peak due to the transition between the ground state and the nth state at zero bias. The expected weaker absorption at ν_4^0 is not observed due to the rapidly rising background on both sides.

Fig. 6.3 also shows the photocurrent spectra at two biasing conditions. The solid curve shows the spectrum when the device is under reverse bias with the potential drop per period $V_p = 63.5$ mV, whereas the dashed curve shows the spectrum under forward bias with a similar V_p (58.4 mV). Under reverse bias, three peaks in the spectrum are observed, in contrast with the forward-biasing case where only one peak with a left shoulder is found. These peaks can be explained by the subband transitions, which can be shifted under the influence of an electric field. The arrows indicate the calculated positions of the transitions. The magnitude of the shift of the peak position under different biases shows that the wells are strongly coupled. For example, if the wells were isolated the maximum upward shift of ν_3^r could only be 16 cm^{-1} (Stark shift) by assuming all the potential drop within the thicker well, contrary to the experimental shift (100 cm^{-1}).

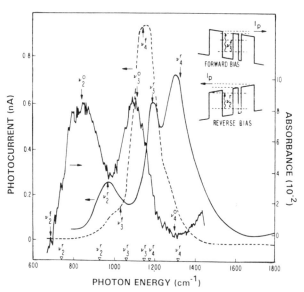

FIG. 6.3 The absorption spectrum for sample having the waveguide geometry and the photocurrent tunneling spectra for forward (dashed curve) and reverse (solid curve) biases. The inverted triangles indicate the transition energies assuming linear potential drop. The arrows indicate the fitted transition energies with nonlinear potential drop in a unit cell. The insert indicate the device under different biasing conditions.

In order to understand the general shape of the spectrum, we calculated, for example, the oscillator strengths under reverse bias: $f_2 = 0.76$, $f_3 = 0.11$, and $f_4 = 0.14$. The fact that $f_2 >> f_3$ under this bias greatly compensates the difference in the tunneling probabilities ($p_2 = 0.31$ and $p_3 = 0.91$) and leads to comparable I_p.

Next, we describe the absorption and photocurrent spectra of another MQW structure that shows additional interesting features. The device consists of 50 periods of 72-Å GaAs (n = 1×10^{18} cm^{-3}), 39-Å undoped Al$_{0.31}$Ga$_{0.69}$As, 20-Å undoped GaAs, and 154-Å undoped Al$_{0.31}$Ga$_{0.69}$As. The barrier height of this device is higher than the former device, and hence the coupling is weaker. The calculated transition energies for this structure indicated in Fig. 6.4 are in satisfactory agreement with the absorption spectrum. However, a calculation of the oscillator strengths ($f_2 = 0.27$, $f_3 = 0.75$, and $f_4 = 0.01$) shows that the apparent strengths at ν_2^0 and ν_4^0 could have been modified by the background absorption due to the small oscillator strengths. On the other hand, the photocurrent spectrum show all the expected features. For example, the E_2 transition, almost unobservable in the absorption curve, is evident in the I_p spectrum under reverse bias. The enhanced f_2 ($f_2 = 0.74$, $f_3 = 0.26$) under reverse bias again leads to comparable I_p between E_2 and E_3 states. Another striking feature in this spectrum is the broad I_p peak associated with the ν_4^r observed around 1800 cm^{-1} ($\lambda = 5.6$ μm). It indicates that although this level is significantly (40 meV) above the barriers, the

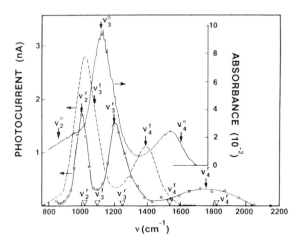

FIG. 6.4 The absorption spectrum for sample having the waveguide geometry and the photocurrent tunneling spectra for forward (dashed curve) and reverse (solid curve) biases. The inverted triangles indicate the transition energies assuming linear potential drop. The arrows indicate the fitted transition energies with nonlinear potential drop in a unit cell. ν_2^f (= 548 cm^{-1}) is out of the scale and is not shown.

level is still relative well defined, confirming the existence of the Stark ladder even for the extended continuum states as found in the tunneling characteristics (Choi et al. 1987b).

The large f_3 under forward bias ($f_2 = 0.19$, $f_3 = 0.85$) contributes to the large I_p near ν_3^f. The fact that the photocurrent peak near ν_3^f does not coincide with the calculated absorption peak is a noteworthy feature of the I_p spectrum. Since I_p is proportional to both the absorption strength and the tunneling probability at a particular energy, I_p is peaked at the maximum absorption only if p is a monotonic function of energy. For the present device, it is designed such that p is peaked near $\nu = 1000$ cm^{-1} using the coherent tunneling mechanism; the resulting I_p is peaked toward the energy where p is a maximum, and the combined factors make this peak dominant over all other peaks. The detailed mechanism is described as follows.

When an electric field is applied to the MQW structure, the MQW is broken up into high-field domains and low-field domains. At low temperatures, electron transport out of E_1 is via sequential tunneling in both regimes. However, if the electrons from E_1 are photoexcited to the excited state with energy close to E_1', the conduction mechanism can be quite different. In the high field regime under forward bias where E_1 is approximately aligned with E_2 of the adjacent period [Fig. (6.1)], the tunneling lifetime width of the state E_1' in the thinner well can be large (> 50 meV) compared with impurity broadening ($\approx$ 7.4 meV) because of the large $t_1(E_1')$ and $t_2(E_1')$ at high field, where t_1 and t_2 are the transmission coefficients of the thicker barrier B_1 and thinner barrier B_2 in Fig. 6.1

respectively. In this case, coherent tunneling of the photoexcited electron from the thicker well W_1 through the state E_1' and out of the barriers is possible. In addition, t_1, originally smaller than t_2 at zero bias, increases more rapidly under an applied field, finally becoming greater than t_2 at large bias. Therefore, in certain range of applied bias, $t_1 \simeq t_2$ and a large coherent enhancement is expected. This is in contrast to the case of reverse bias shown in Fig. 6.1c, in which t_1 is always less than t_2. Using WKB approximation, the coherent transmission coefficient can be shown to be

$$T_{coh}(E) = 4\left[\left(\frac{t_1}{t_2} + \frac{t_2}{t_1}\right)\cos^2\theta + \left(\frac{t_1 t_2}{16} + \frac{16}{t_1 t_2}\right)\sin^2\theta + 2\right]^{-1}, \qquad (6.1)$$

where $\theta = (2m^*)^{1/2}W_2[E^{1/2}-(E_1')^{1/2}]/\hbar$ is the phase angle measured relative to the angle at the resonant energy. t_1 and t_2 can be obtained by WKB approximation. In Fig. 6.5, T_{coh} for forward (T_{coh}^f) and reverse (T_{coh}^r) bias at $E = 170$ meV measured from the center of W_1 is plotted as a function of voltage drop per period V_p. Their ratio starts from unity at small bias and rapidly increase to $\simeq 100$ at $V_p = 100$ meV, and then falls back to $\simeq 8$ at large bias. In contrast, if sequential tunneling were the dominant mechanism, then $T_{seq} = t_1 t_2$, the maximum forward-to-reverse transmission coefficient ratio would only be 10.

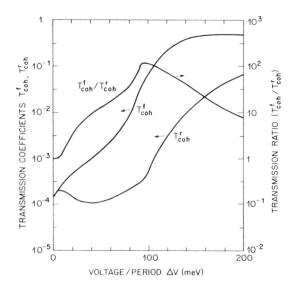

FIG. 6.5 The theoretical transmission coefficients as a function of voltage drop per period ΔV under forward bias (T_{coh}^f) and reverse bias (T_{coh}^r) and their ratio based on the assumption of coherent tunneling.

Fig. 6.6 shows the responsivity R of the device at $T = 15$ K with infrared radiation of $\lambda = 10.3~\mu$m, corresponding to the energy used in calculating Fig. 6.5. Below 5 V, the dark current characteristics show that the increase of the applied voltage increases the number of high field domains but keeps the voltage drop across each high field domain constant at 93 meV. Hence, according to Fig. 6.5, R in the forward bias should be 2 orders of magnitude larger than R in the reverse bias, which is shown to be the case in Fig. 6.6. At higher bias, they approach each other, as predicted by the theory of coherent tunneling. Therefore, it can be concluded that the photoexcited electrons tunnel out of the wells coherently even though the dark current is transported via sequential tunneling.

In this section, we have described two MQW structures which can be used for multi-color detection. The asymmetrical unit cell breaks the parity symmetry, leading to nonzero oscillator strengths among the energy levels and the plurality of the detection wavelengths. The shifts of the energy levels by an applied bias has been demonstrated to be much larger for coupled wells. This feature can be used for wavelength tuning. With the coherent tunneling mechanism, the tunneling probability can be made close to 1 even though the excited state is still

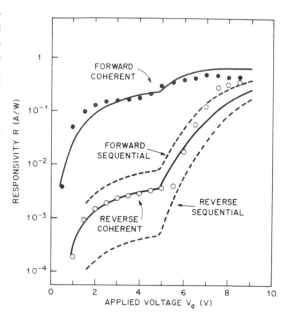

FIG. 6.6 The experimental result of responsivity R in forward bias (filled circles) and reverse bias (open circles), and theory with the assumption of coherent tunneling (solid curves) under forward bias T_{coh}^{f} and reverse bias T_{coh}^{r} and of sequential tunneling (dashed curves) under forward bias T_{seq}^{f} and reverse bias T_{seq}^{r}.

below the barrier height. This means that the coupled well structure will have the same responsivity as the extended state detector without the additional nonideal current. Hence, the present structure is potentially more sensitive than the extended state detector. By adding more thin barriers to each period, the E_1' state forms miniband (Borenstain et al. 1991; Yu et al. 1991), which can conduct photoelectrons preferentially. Such a structure can also have the same advantages over the extended state detector.

7. FINAL REMARKS

In this chapter, we have discussed the electrical and infrared properties of multiple quantum well structures. With the observation of intersubband absorption and photocurrent in these structures, a new infrared technology emerges that can be very competitive with the existing technologies. By generating hot-electrons with infrared radiation, the transport properties of a quantum well structure can be studied in much greater detail, based on which detectors with greater sensitivity and new functionality can be designed. Several techniques mentioned in this chapter, such as infrared absorption spectroscopy, thermally stimulated hot-electron spectroscopy and infrared photoelectron tunneling spectroscopy, are shown to be very useful in studying quantum well structures. The same techniques should be applicable to other material systems and different types of dopants. Hence, these studies not only contributes to infrared technology but also to several branches of science including basic quantum mechanics, band structures of materials and hot-electron phenomena. With the versatility of band-gap engineering, the present detector design and material system may not represent the full potential of quantum well infrared technology. The invention and demonstration of a better detector poses an interesting and challenging problem to all solid-state scientists.

REFERENCES

Anderson P W 1961 Phy. Rev. **124** 41

Andersson J Y, Lundqvist L and Paska Z F 1991a Appl. Phys. Lett. **58** 2264

Andersson J Y and Lundqvist L 1991b Appl. Phys. Lett. **59** 857

Ando T, Fowler A B and Stern F 1982 Rev. Mod. Phys. **54** 437

Bandara, K M S V, Coon D D, Byungsung O, Lin Y F and Francombe M H 1988 Appl. Phys. Lett. **53** 1931

Borenstain et al. 1991 Phys. Rev. B **43** 9320

Capasso F, Mohammed K and Cho A Y 1986 IEEE J. Quan. Elect. QE-22 1853

Calecki D, Palmier J F and Chomette A 1984 J. Phys. C **17** 5017

Chiu L C, Smith J S, Margalit S, Yariv Y and Cho A Y 1983a Infra. Phys. 23 93

Chiu L C, Smith J S, Margalit S and Yariv Y 1983b Appl. Phys. Lett. **43** 331

Choi K K, Levine B F, Malik R J, Walker J and Bethea C G 1987a Phys. Rev. B **35** 4172

Choi K K, Levine B F, Bethea C G, Walker J and Malik R J 1987b Appl. Phys. Lett. **50** 1814

Choi K K, Levine B F, Bethea C G, Walker J and Malik R J 1987c Phys. Rev. Lett. **59** 2459

Choi K K, Levine B F, Jarosik N, Walker J and Malik R J 1988a Phys. Rev B **38** 12362

Choi K K, Levine B F, Bethea C G, Walker J and Malik R J 1988b Appl. Phys. Lett. **52** 1979

Choi K K, Newman P G, Folkes P A and Iafrate G J 1989a Appl. Phys. Lett. **54** 359

Choi K K, Levine B F, Bethea C G, Walker J and Malik R J 1989b Phys. Rev B **39** 8029

Choi K K, Newman P G, Folkes P A and Iafrate G J 1989c Phys. Rev. B **40** 8006

Choi K K, Dutta M, Newman P G, Saunders M.-L. and Iafrate G J 1990a Appl. Phys. Lett. **57** 1348

Choi K K, Dutta M, Newman P G, Calderon L, Chang W and Iafrate G J 1990b Phys. Rev B **42** 9166

Choi K K, Newman P G and Iafrate G J 1990c Phys. Rev B **41** 10250

Choi K K, Fotiadis L, Newman P G and Iafrate G J 1990d Appl. Phys. Lett. **57** 76

Choi K K, Dutta M, Moekirk R P, Kuan C H and Iafrate G J 1991a Appl. Phys. Lett. **58** 1533

Choi K K, Taysing-Lara M, Fotiadis L, Chang W and Iafrate G J 1991b Appl. Phys. Lett. **59** 1614

Choi K K, Taysing-Lara M, Chang W and Iafrate G J 1991c Tech. Disgest IEEE IEDM

Coon D D and Karunasiri R P G 1984 Appl. Phys. Lett. **45** 649

Coon D D, Karunasiri R P G and Liu L Z 1985 Appl. Phys. Lett. **47** 289

Easki L and Tsu R 1969 IBM Research Note RC-2418

Easki L and Tsu R 1970 IBM J. Res. Dev. **14** 61

Esaki L and Chang L L 1974 Phys. Rev. B **33** 495

Goossen K W and Lyon S A 1985 Appl. Phys. Lett. **47** 1257

Goossen K W, Lyon S A and Alavi K 1988a Appl. Phys. Lett. **52** 1701

Goossen K W, Lyon S A and Alavi K 1988b Appl. Phys. Lett. **53** 1027

Grahn H T, Haug R J, Müller W and Ploog K 1991 Phys. Rev. Lett. **67** 1618

Gunapala S D, Levine B F, Ritter D, Hamm and Panish M B 1991 Appl. Phys. Lett. **58** 2024

Hasnain G, Levine B F, Bethea C G, Logan R A, Walker J and Malik R J 1989 Appl. Phys. Lett. **54** 2515

Hasnain G, Levine B F, Gunapala S and Chand N 1990 Appl. Phys. Lett. **57** 608

Helm M, Peeters F M, DeRosa F, Colas E, Harbison J P and Florez L T 1991 Phys. Rew. B **43** 13983

Heiblum M, Nathan M I, Thomas D C and Knoedler C M 1985 Appl. Phys. Lett. **55** 2200

Ikonić Z, Milanović V, Tjapkin D and Pajević S 1988 Phys. Rev. B **37** 3097

Ikonić Z, Milanović V and Tjapkin D 1989 Appl. Phys. Lett. **54** 247

Kamgar A, Kneschaurek P, Dorda G and Koch J F 1974 Phys. Rev. Lett. **32** 1251

Kamgar A and Kneschaurek P 1976 Surf. Sci. **58** 135

Kararinov R F and Suris R A 1972 Sov. Phys. Semicond. **6** 120

Kastalsky A, Duffield, Allen S J and Harbison J 1988 Appl. Phys. Lett. **52** 1320

Kinch M A and Yariv A 1989 Appl. Phys. Lett. **55** 2093

Kneschaurek P, Kamgar A and Koch F 1976 Phys. Rev. B **14** 1610

Lei X L, Horing N J M and Cui H L 1991 Phys. Rev. Lett. **66** 3277

Levi A F J, Hayes T R, Platzman P M and Weigmann 1985 Phys. Rev. Lett. **55** 2071

Levine B F, Malik R J, Walker J, Choi K K, Bethea C G, Kleinman D A and Vandenberg 1987a Appl. Phys. Lett. **50** 273

Levine B F, Choi K K, Bethea C G, Walker J and Malik R J 1987b Appl. Phys. Lett. **51** 934

Levine B F, Choi K K, Bethea C G, Walker J and Malik R J 1987c Appl. Phys. Lett. **50** 1029

Levine B F, Bethea C G, Hasnain G, Walker J and Malik R J 1988a Appl. Phys. Lett. **53** 296

Levine B F, Bethea C G, Choi K K, Walker J and Malik R J 1988b J. Appl. Phys. **64** 1591

Levine B F, Hasnain G, Bethea C G and Chand N 1989 Appl. Phys. Lett. **54** 2704

Levine B F, Bethea C G, Hasnain G, Shen V O, Pelve E, Abbott and Hsieh S J 1990 Appl. Phys. Lett. **56** 851

Mahan G D 1981 Many-Particle Physics (New York: Plenum Press) pp 256

Manasreh M O, Szmulowicz F Vaughan T, Evans K R, Stutz C E and Fisher D W 1991 Phys. Rew. B **43** 9996

Pelvé E, Beltram F, Bethea C G, Levive B F, Shen V O, Hsieh S J and Bbott R R 1989 J. Appl. Phys. **66** 5656

Smith J S, Chiu L C, Margalit S, Yariv A and Cho A Y 1983 J. Vac. Sci. Technol. B **1** 376

Sze 1981 Physics of Semiconductor Devices (New York: John Wiley and Sons) pp 647

West L C and Eglash S J 1985a Appl. Phys. Lett. **46** 1156

West L C 1985b Ph. D. Thesis

Wheeler R G and Goldberg H S 1975 IEEE Trans. Elect. Dev. ED-**22** 1001

Yu L S and Li S S 1991 Appl. Phys. Lett. **59** 1332

Chapter 3

Real-time spectroscopic ellipsometry monitoring of semiconductor growth and etching

R. W. Collins, Ilsin An, A. R. Heyd, Y. M. Li, and C. R. Wronski

The Pennsylvania State University, Materials Research Laboratory, University Park, PA 16802, USA.

ABSTRACT: With the development of a novel multichannel ellipsometer, a powerful spectroscopy for real time monitoring of semiconductor preparation and processing has been realized. The new instrument provides ellipsometric spectra over the photon energy range of $1.5 \leq h\nu \leq 4.5$ eV with a minimum time resolution of 40 ms. Here multichannel ellipsometry is described in detail and compared to real time single-photon energy approaches. An overview of recent applications of real time spectroscopic ellipsometry for monolayer-level characterization is presented, including plasma-enhanced chemical vapor deposition of amorphous semiconductors, and ion beam etching of III-V semiconductors.

1. INTRODUCTION AND MOTIVATION

Because of the complexity of modern semiconductor device structures and the stringent demands on their specifications, techniques for real time monitoring of semiconductor characteristics during preparation and processing have attracted considerable attention. Although the information desired is a function of the device structure and application, the following characteristics are of general interest: the mode of growth, microstructural uniformity, surface morphology, layer thickness, band structure, band gap, composition, doping, defects, surface temperature, etc. Obtaining such information in real time is a tall order considering that the probe should be non-invasive, non-perturbing, and hence, contactless. If the data collected in real time can also be interpreted in real time, then it becomes possible to control material characteristics through closed-loop adjustment of process variables. Even if real time data interpretation is not possible, the information deduced in post-process analyses of real time measurements is especially important for understanding the process. In addition, reproducibility can be assessed, problems identified, and new processing procedures designed with expeditious arrival at the appropriate process variables. In contrast, the fundamental mechanisms underlying processing-property relationships may be hidden from post-process measurements because of the complexity of the final structure.

Electron diffraction techniques have been used widely to provide feedback on the growth mode, surface morphology, and growth rate during semiconductor preparation under the ultrahigh vacuum conditions of molecular beam epitaxy (MBE). Surface probes involving either electrons or ions, however, cannot be used in the high pressure, reactive environment of chemical vapor deposition (CVD). Probes of semiconductor preparation based on photons, on the other hand, enjoy wider applicability from MBE to CVD environments. At first glance, it may appear that any optical probe will not be sufficiently powerful to extract the semiconductor characteristics of general interest listed above. It is the purpose of this article to demonstrate the full range of capabilities of real time spectroscopic ellipsometry, the most powerful optical technique yet developed for in-process characterization of semiconductors.

The power of ellipsometry over other optical techniques derives from its ability to obtain measures of both the amplitude (ψ) and phase (Δ) changes that the electric field of a polarized monochromatic light wave incurs when it reflects from a specular surface (Azzam and Bashara, 1977). These changes of course lead to a change in the polarization state of the light upon reflection. When the surface of an isotropic material displays an atomically-smooth, abrupt termination of the bulk stucture, then the measured ellipsometry parameters, (ψ,Δ), provide a complete description of the linear optical response of the material; i.e. the real and imaginary parts of the bulk dielectric function, ($\varepsilon_1,\varepsilon_2$). The problem that arises in most practical situations, however, is that of the three medium sample (substrate/film/ambient). If the substrate dielectric function is known from prior measurement, then there are three unknown parameters: ($\varepsilon_{1f},\varepsilon_{2f}$) of the film and the film thickness, d. Thus, a single measurement of (ψ,Δ) is insufficient to provide useful information, and multiple measurements are required, for example, versus photon energy (Aspnes, 1976) or film thickness during growth (Hottier and Theeten, 1980). Owing to the high precision phase measurement, ellipsometry is sensitive to film thickness at the sub-monolayer level. This sensitivity is an advantage in film growth studies, but a disadvantage in attempts to extract accurate bulk dielectric functions free of artifacts arising from unintentional surface layers.

Although the ellipsometry measurement originates in the work of Drude over a century ago (Drude, 1889), its capabilities for repetitive measurement were limited owing to the time-consuming nature of manual polarization state analysis. It was not viable as a continuous spectroscopic or real time probe until widespread use of laboratory computers in the mid-1960's led to the development of automatic, high speed ellipsometers [time resolution from 10 μs to >1 s per (ψ,Δ) pair depending on instrument design (Muller, 1976)]. Since then, a majority of the real time studies of semiconductor preparation have been performed at a fixed photon energy and have been used to determine film thickness, the dielectric function at the single photon energy, and rudimentary microstructural information (Hottier and Theeten, 1980). There are two main limitations to real time ellipsometry at a single photon energy.

The first limitation concerns characterization of the electronic structure of the material. It is difficult to identify materials based on a measurement of the optical properties at a single photon energy. Hence, information such as the band gap, composition, doping, and surface temperature are not generally accessible. A recent exception is the case of epitaxial growth; for example, the crystalline perfection of $Al_xGa_{1-x}As/GaAs$ heterostructures allows one to determine Al composition, x, based on $\{\psi(t),\Delta(t)\}$ measurements at one photon energy (Aspnes et al., 1990a). This has been achieved through calibrations that determine ($\varepsilon_1,\varepsilon_2$) for $Al_xGa_{1-x}As$ as a function of x and are valid for the selected photon energy and growth temperature. Any changes in temperature or surface or interface smoothness, or any inaccuracy in measurement generate errors in the deduced composition. The errors could be eliminated or at least minimized with a real time approach at multiple photon energies. This subject will be discussed further in Sec. 4.

The other limitation of real time single-photon energy ellipsometry concerns the microstructural evolution of the material, which may be reflected in the deviations of $\{\psi(t),\Delta(t)\}$ from a perfect layer-by-layer growth model. These deviations are generally associated with growth complexities such as interfacial reactions or initial nucleation phenomena, the latter characterized by cluster formation, coalescence, and surface roughness evolution. These are of great interest because they provide insights into the mechanisms underlying process-property relationships. For example, the simplest model to characterize nucleation requires two parameters, the layer thickness (d) and void volume fraction in the layer (f_v), in addition to the bulk dielectric function of the material (Hottier and Theeten, 1980). If the bulk dielectric function can be estimated (for example, from measurements of the layer after it reaches opacity), then the evolution of d and f_v can be extracted from $\{\psi(t),\Delta(t)\}$ by exact numerical inversion. (In the inversion an effective medium approximation is used to calculate the dielectric function of the density deficient layer, modeling it as a sub-microscopic mixture of bulk material and voids; see Sec. 3.1.) Unfortunately, experimental errors, errors in the bulk dielectric function, and deviations from the assumed model can lead to large errors in d and f_v. More often, it is possible to fit $\{\psi(t),\Delta(t)\}$ to a nucleation model which assumes a geometry (e.g. hemispherical clusters), allowing f_v to be expressed as a function of d. It is difficult, however, to ensure that the geometric model is unique. Here again, the real time spectroscopic capability can resolve such ambiguities. This subject will be discussed at length in Sec. 3.

Spectroscopic ellipsometry, as perfected with the automatic rotating analyzer instrument, typically requires ~30 min to collect a continuous spectrum in (ψ,Δ), covering the near-infrared to near-ultraviolet photon energy range (Aspnes and Studna, 1975). This instrument consists of the following sequence of elements: Xe source, monochromator, polarizer, compensator (optional), sample, continuously rotating analyzer, and photomultiplier tube (PMT) detector. The exact acquisition time depends on the desired signal/noise ratio, the number of spectral points collected (usually ~100), and the stepping speed of the monochromator used for scanning. Thus, the instrument is restricted to studies of static materials. More recently, however, Muller and Farmer (1984) have developed a rapid-scanning spectroscopic ellipsometer based on the principle of self-nulling and intended for applications in electrochemistry. With this instrument, a single (ψ,Δ) pair can be collected in ~1 ms using two high speed Faraday cells, one mounted between the polarizing and compensating elements of the incident beam path, and the other mounted between the sample and a fixed analyzer in the reflecting beam path. Spectra consisting of 400 such pairs can be acquired from 1.8 to 3.1 eV in 3 s using a continuously rotating interference filter to monochromate the Xe source output before it enters the polarizer. The detection system employs a PMT for sequential nulling versus photon energy. For a 100-point spectrum the ultimate speed of this instrument is ~200 ms, but this limitation depends on the nature of the surface and the rate of change of the optical characteristics. The main disadvantage of the self-nulling ellipsometer in semiconductor applications is the high energy limitation in the spectrum, imposed by the Faraday cells.

A completely different rapid-scanning spectroscopic ellipsometer has been newly developed based on the rotating polarizer principle and intended for semiconductor film growth and processing applications (Kim et al., 1990). With this instrument, spectra consisting of 128 (ψ,Δ) pairs over the energy range from 1.5 to 4.5 eV likewise can be collected in ~3 s. The instrument differs from the slower rotating analyzer ellipsometer in that the monochromator is removed from the source side of the instrument, and the PMT is replaced with a combination spectrograph and photodiode array detector. In addition, the polarizer is designed to rotate continuously and the analyzer element is fixed. The primary advantage of the integrating multichannel detector is that the spectra are acquired in parallel rather than sequentially. The ultimate speed of the rotating polarizer multichannel ellipsometer is linked to the technology of commercially-available optical multichannel analyzers. With the current state of the art, 128 point (ψ,Δ) spectra in theory can be collected every ~5 ms, albeit at relatively low precision. Until the present work, the fastest acquisition time, while retaining monolayer precision, is 320 ms achieved in a study of Al growth by evaporation onto SiO_2 (Collins et al., 1991).

In Section 2, the developmental details of this rotating polarizer multichannel ellipsometer will be discussed. This section will begin with a description of the ideal, error-free operation of the instrument, and then overview the system errors that must be corrected in order to obtain high accuracy data in semiconductor studies. In Section 3, applications of the multichannel ellipsometer will be described for monitoring the growth of hydrogenated amorphous silicon (a-Si:H) by plasma-enhanced CVD (PECVD) and by physical vapor deposition. Because the properties of a-Si:H depend on the preparation and processing parameters, the real time probe is a very important tool for characterizing not only the evolution of the physical structure of the material, but also the optoelectronic properties. In Section 4, applications of the multichannel ellipsometer will be described for monitoring III-V semiconductor structures during ion beam etching. The purpose of the study is to assess the capabilities of the new techniques for characterizing III-V semiconductor materials, surfaces, and interfaces during preparation and processing. In this study, semiconductor layer thicknesses and composition, surface damage layer thickness and composition, and surface temperature are deduced through linear regression analyses of the (ψ, Δ) spectra collected during etching. The unique outcome of the real time observations is the time evolution of many of these quantities; instantaneous etch rates and the extent of ion-induced intermixing at interfaces can be determined.

2.   INSTRUMENTATION

2.1   Description of the Ideal Rotating Polarizer Multichannel Ellipsometer

Figure 1 shows a schematic of the multichannel ellipsometer configured to monitor the growth of a-Si:H by PECVD. It consists of a collimated Xe arc source, a continuously rotating polarizer assembly with quartz Rochon element (P_{rot}), a vacuum preparation and processing system (S), a stepping-motor controlled analyzer with a calcite Glan-Taylor element (A), a single Littrow prism spectrograph, and a 1024-pixel silicon photodiode array detector. The vacuum chamber is fitted with strain free windows, aligned for optical access to the sample surface at a 70° angle of incidence. The detector is part of a commercially-available optical multichannel analyzer that includes a detector controller for versatile scanning capabilities. In the studies reported below, spectral resolution has been sacrificed for signal/noise ratio, and the pixel-grouping capability of the detector controller has been employed to collect 128 spectral positions from the near-infrared to near-ultraviolet. At 1.5 eV and 3.2 eV, the eight-pixel groups intercept photon energy bands 0.05 and 0.025 eV in width, respectively.

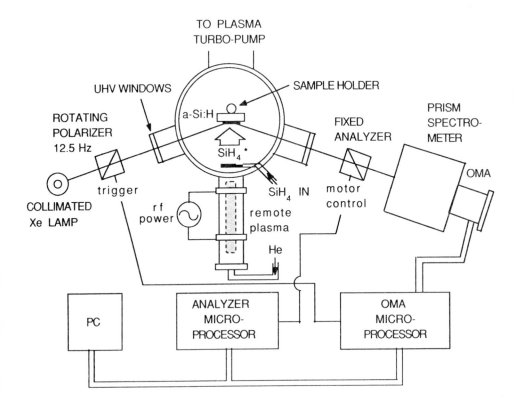

Fig. 1   Rotating polarizer multichannel ellipsometer designed for real time spectroscopic studies of semiconductor film growth and etching. Here the deposition system is configured for remote He plasma-enhanced CVD of hydrogenated amorphous silicon.

An error-free ellipsometer in the $P_{rot}SA$ configuration of optical elements can be shown to exhibit a time-dependent output irradiance at the detector of the following theoretical form (Collins, 1990):

$$I_{th}(h\nu,t) = I_{0,th} \{1 + \alpha(h\nu) \cos2(P-P_S) + \beta(h\nu) \sin2(P-P_S)\}. \tag{1}$$

Here, α and β are the 2ω Fourier coefficients which depend on photon energy $h\nu$, and $P=\omega t$, where ω is the mechanical rotation frequency of the polarizer and t is time. The phase factor P_S is defined so that $P-P_S$ represents the true polarizer angle measured in a counterclockwise sense with respect to the plane of incidence. P_S accounts for the fact that at time zero (defined experimentally below), the polarizer is not necessarily aligned along the plane of incidence. In the absence of system errors, α and β are related to the sample parameters, ψ and Δ, by:

$$\alpha = [\tan^2\psi \cos^2(A-A_S) - \sin^2(A-A_S)]/[\tan^2\psi \cos^2(A-A_S) + \sin^2(A-A_S)] \tag{2a}$$

$$\beta = \tan\psi \cos\Delta \sin2(A-A_S)/[\tan^2\psi \cos^2(A-A_S) + \sin^2(A-A_S)]. \tag{2b}$$

The angles ψ and Δ are defined by $\tan\psi\exp(i\Delta) = r_p/r_s$, where r_p and r_s denote the sample's complex amplitude reflection coefficients for light polarized parallel (p) and perpendicular (s) to the plane of incidence. The reflection coefficients in turn are defined for single or multiple interfaces in terms of the electric field amplitudes, E_o, of the incident (i) and combined reflected (r) waves for the p- and s-polarizations: $r_p = (E_{or}/E_{oi})_p$ and $r_s = (E_{or}/E_{oi})_s$. In Eqs. (2), the quantity $A-A_S$ is the true angle of the analyzer measured in a counterclockwise sense with respect to the plane of incidence. The angle A represents the experimental scale reading and A_S is a correction to the reading that provides the true angle. Thus, in a system free of errors, (ψ,Δ) can be calculated from the Fourier coefficients by inversion of Eqs. (2):

$$\tan\psi = [(1+\alpha)/(1-\alpha)]^{1/2} \tan(A-A_S) \tag{3a}$$

$$\cos\Delta = \beta[1-\alpha^2]^{-1/2}. \tag{3b}$$

In the experiment, however, one measures the coefficients α' and β' given by the simple waveform that does not include phase correction by P_S:

$$I_{ex}(h\nu,t) = I_{0,ex} \{1 + \alpha'(h\nu) \cos2\omega t + \beta'(h\nu) \sin2\omega t\}. \tag{4}$$

α' and β' can be employed to calculate α and β of the theoretical version, Eq. (1), from a simple $2P_S$ rotation transformation:

$$\alpha = \alpha' \cos2P_S + \beta' \sin2P_S \tag{5}$$

$$\beta = -\alpha' \sin2P_S + \beta' \cos2P_S. \tag{6}$$

In summary then, an experimental determination of the Fourier coefficients α' and β' at each photon energy can provide spectra in ψ and Δ through Eqs. (5-6) and (3) if the polarizer phase and analyzer correction angles, P_S and A_S, are known. The discussion to this point is valid for any rotating polarizer ellipsometer (or rotating analyzer ellipsometer upon interchange of the angular variables A and P), irrespective of whether a single-channel PMT for sequential acquisition or a photodiode array for parallel acquisition is used. In the following paragraphs, the procedure for extracting α' and β' from the raw output of the multichannel detector is described. Then, a calibration procedure that determines P_S and A_S will be outlined briefly.

For an instrument based on a PMT, the detector output is sampled as a function of polarizer angle using an analog-to-digital converter with sample-and-hold which is triggered by the output of an optical encoder attached to the polarizer shaft (Aspnes and Studna, 1975). Typically 60 data points are collected per optical cycle of the polarizer (1/2 mechanical cycle), and these data are Fourier analyzed to determine α' and β' of Eq. (4) at a single photon energy. Because the photodiode array is an integrating detector, an approach based on Fourier analysis cannot be used. On the other hand, if the array is read out N times per optical cycle (again triggered by an optical encoder), the resulting signal output collected from the pixel group designated k (k=1,128) can be written:

$$S_{jk} = \int_{(j-1)\pi/N\omega}^{j\pi/N\omega} I_{0,ex,k}\,\{1 + \alpha_k'\cos2\omega t + \beta_k'\sin2\omega t\}\,dt, \tag{7}$$

where j=1,N represents the read-out index, and the subscript k on $I_{0,ex}$, α', and β' denotes the pixel group-dependence via the photon energy. For the instrument of Fig. 1, the array is read out four times per optical cycle (N=4). The resulting Eqs. (7) for j=1,4 can be solved for the three unknowns, $I_{0,ex,k}$, α_k', and β_k', and one consistency check is provided (Laurence et al., 1981). The latter three results are of interest here:

$$\alpha_k' = (\pi/2)(S_{1k}-S_{2k}-S_{3k}+S_{4k})/(S_{1k}+S_{2k}+S_{3k}+S_{4k}) \tag{8a}$$

$$\beta_k' = (\pi/2)(S_{1k}+S_{2k}-S_{3k}-S_{4k})/(S_{1k}+S_{2k}+S_{3k}+S_{4k}) \tag{8b}$$

$$0 = (S_{1k}-S_{2k}+S_{3k}-S_{4k}). \tag{8c}$$

Thus from (α_k',β_k'), a 128-point (ψ,Δ) spectrum can be computed from Eqs. (5-6) and (3) as long as a calibration of the spectrograph/detector is available that relates pixel group number, k, to photon energy, $h\nu_k$.

For the instrument of Fig. 1, a single 8-pixel group read time is 35 μs, and the full array read time is 4.5 ms. Thus, when the polarizer rotation frequency is set at 12.5 Hz, the exposure time of the array is 10 ms (with N=4), and 5.5 ms elapses between reading the final pixel group of read-out j [see Eq. (7)] and the first pixel group of read-out (j+1). In this case, the minimum acquisition time for one (ψ,Δ) spectrum is 40 ms. In the relatively slow semiconductor growth and etching applications described below, S-values [see Eq. (7)] are averaged over 80 consecutive optical cycles acquired in 3.2 s. This provides submonolayer precision [standard deviation in (ψ,Δ) of $(0.01°, 0.02°)$ for Au at 2.0 eV] as well as monolayer resolution. Advances in the development of multichannel detection systems, can provide an acquisition time of ~5 ms when operated at maximum speed, albeit at a reduced precision.

To conclude the description of the ideal rotating polarizer multichannel ellipsometer, the calibration procedure used to obtain P_S in Eqs. (5-6) and A_S in Eq. (3a) will be described (Aspnes, 1974). This procedure is performed on the static sample mounted in the $P_{rot}SA$ configuration prior to real time measurements of deposition or etching. It relies on the fact that when linearly polarized light reflects from an isotropic surface with $\Delta\neq0°$ or 180°, the light remains linearly polarized if and only if the incident electric field is oriented along the p- or s-directions. As the incident field is rotated away from the p- or s-directions, the reflected polarization state becomes increasingly elliptical. For linear polarization, $\alpha_k'^2+\beta_k'^2=1$, and the deviation from linear polarization is given by the residual function, $R_k=1-(\alpha_k'^2+\beta_k'^2)$. Thus, in the p- and s-directions, where $A=A_S$ and $A=\pi/2+A_S$, R_k exhibits minima as a function of A. In the absence of errors the residual function should vanish at the two minima. It can also be shown that the phase function, given by $\Theta_k=[\tan^{-1}(\beta_k'/\alpha_k')]/2$, evaluated at the minima ($A=A_S$ or $A=\pi/2+A_S$) equals P_S. Experimentally then, spectra in R_k and Θ_k are obtained as a function of fixed analyzer reading A in the vicinity of $A=A_S$ or $A=\pi/2+A_S$, using the computer-controlled stepping motor attached to the analyzer. Each of $R_k(A)$ and $\Theta_k(A)$ is fit to a quadratic function of A, and the fits are evaluated to obtain precise values of A_S and P_S. For a system free of errors, A_S should be independent of pixel group k. On the other hand, P_S should be a linear function of pixel group number and is thus written P_{Sk}. The reason is that t=0 for the integral in Eq. (7) is defined by the time at which the first of the four exposures begins for pixel k. Thus, the time origin for successive pixel groups is shifted by the group read-out time of 35 µs, and over this duration the polarizer rotates by 0.16°.

2.2 Errors in the Rotating Polarizer Multichannel Ellipsometer

Next we will describe various errors encountered during the development of the real time spectroscopic ellipsometer that require extensions beyond the data reduction procedures of Sec. 2.1 (Nguyen et al., 1991; An and Collins, 1991). In general, the errors are small when the sample converts the incident rotating, linearly polarized light to circularly polarized light (i.e., with $\alpha\sim0$ and $\beta\sim0$; $\psi\sim|A-A_S|$). Thus, if the flexibility exists to choose a substrate in film growth studies, e.g. in investigations of amorphous semiconductors, then a metal is best if high accuracy during the nucleation stage is desired. But because the bulk optical functions of metals are not well characterized, it is sometimes difficult to verify that their surfaces are near-atomically smooth, and this is a requirement for studying monolayer-scale nucleation phenomena with confidence. For this reason, reproducibly smooth c-Si wafer substrates have been used in most of the amorphous semiconductor studies. In the low energy region (1.5 to 2.5 eV), the reflected light from c-Si remains nearly linearly polarized and errors outlined below can be considerable. In this Section, the sources of the errors will be summarized.

The first error to be described is detector non-linearity, a concern for all photometric ellipsometers, i.e. those in which (ψ,Δ) are determined through an analysis of detector irradiance vs. time. For the photodiode array, non-linearity was characterized by measuring the observed count rate (counts integrated over the exposure time per unit exposure time) registered by the different pixel groups using a cw source at different irradiance levels. For a given irradiance level and pixel group (or photon energy), the observed count rate was found to be a function of exposure time rather than being constant as required for true linearity. By plotting the observed count rate as a function of the observed counts for different pixel groups and irradiance levels, a correction factor was derived which could be used to divide the observed counts and obtain values strictly proportional to the incident irradiance integrated over the exposure time. The correction factor was found to depend *only* on the observed counts, and not explicitly on the count rate, exposure time, or photon energy. Figure 2 shows the correction factor for an array detector with a dynamic range of 2^{16}. In order to correct the ellipsometry data for non-linearity, the following relationship is applied to the S_{jk} of Eqs. (8): $S_{jk,c}=S_{jk,r}/C_{NL}(S_{jk,r})$, where the 'c' and 'r' denote corrected and raw counts, and $C_{NL}(S_{jk,r})$ is a functional relationship specific to the detector such as that in Fig. 2.

A second detector error of importance is an apparent image persistence effect. This effect can be observed by directing a cw light beam onto the detector and blocking the beam with a fast shutter during a series of sequential read-outs at the same rate used in the ellipsometry experiments (100 Hz). After appropriate correction for background and dark counts, the first pixel group exposures commencing after the shutter is completely closed should read zero counts. In reality, however, such exposures give a low count level found to be 0.55% of the observed counts registered in the previous read-out. Such a persistence effect leads to a correction to the raw S_{jk} values of Eqs. (8), having the form: $S_{jk,c} = S_{jk,r} + C_{IP}(S_{jk,r}) - C_{IP}(S_{(j-1)k,r})$, where $C_{IP}(z)=0.0055z$. The first correction term on the right arises from the unread counts from the j-read-out which remain to effect the (j+1)-read-out; the second term arises from the unread counts from the (j-1)-read-out which remain to effect the j-read-out. A correction of this form vanishes under dc light conditions (e.g. incident circularly polarized light in the ellipsometry experiment), in agreement with observations.

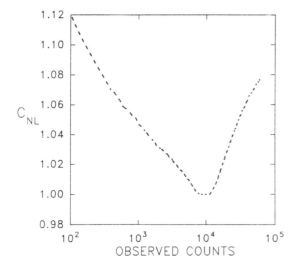

Fig. 2 Non-linearity correction factor, C_{NL}, as a function of the observed count level. This correction factor is used to divide the observed counts to obtain values proportional to the true count level (after An and Collins, 1991).

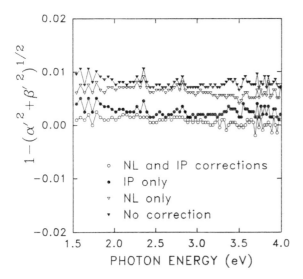

Fig. 3 Effective ac/dc gain deviation, $[1-(\alpha_k'^2+\beta_k'^2)^{1/2}]$, measured in the straight-through ellipsometer configuration, demonstrating the efficacy of the non-linearity (NL) and apparent image persistence (IP) corrections. In the absence of these errors, the gain deviation should vanish (after An and Collins, 1991).

Figure 3 shows the effect of non-linearity and apparent image persistence corrections on the experimentally-determined quantity $[1-(\alpha_k'^2+\beta_k'^2)^{1/2}]$, collected in the straight-through $P_{rot}A$ configuration with the sample removed and the analyzer fixed at $A-A_S=30°$, the value used in all real time experiments. The quantity $[1-(\alpha_k'^2+\beta_k'^2)^{1/2}]$ is a measure of the deviation from unity of the ratio of the 2ω gain to the dc gain. This should vanish for a system which is linear and free of errors. Note then from Fig. 3 that the effects of both non-linearity and image persistence are to suppress the ac signal component with respect to the dc component. The non-linearity correction provides a ~0.002 reduction in the gain deviation whereas image persistence provides a more significant reduction of ~0.005. As a result, an order of magnitude improvement in effective linearity is achieved by applying the corrections.

A third error associated with the detection system is stray light, which is light detected at pixel group k having an energy outside the usual bandwidth centered at $h\nu_k$. Stray light occurs at three different levels, classified according to the distance from the point of generation to the detector surface: (1) light scattering from the optical elements within the spectrograph; (2) multiple reflections between the detector and its window; and (3) crosstalk between detector elements which can be enhanced inadvertently when a protective coating (SiO) is applied to the detector surface. The effects of stray light are greatest for pixel groups of the array that receive the lowest irradiance levels. For all samples, this occurs for the high energy pixel groups of the spectrum (3.5 to 4.5 eV) because of the weak output of the Xe source over this range. Thus, it is the stray light error which limits the usable spectral range of the multichannel ellipsometer. Low irradiance levels are also associated with the minima in the interference patterns of weakly absorbing films (thicknesses > 500Å), and larger errors from stray light occur there. Stray light sources (2) and (3) are identified by their characteristic spectral dependences observed when a high energy cut-off filter is inserted at the entrance to the spectrograph. They lead to spectral replicas or to weak tails both extending into the blocked region of the spectrum. At present, these stray light sources have been minimized by system design; for example source (2) is eliminated by removing the detector window and maintaining the detector under a N_2 purge. An improvement planned for the future is to correct error (3) through appropriate software. In contrast, stray light source (1) is nearly uniform over the detector and is proportional to the observed counts summed over the entire array for the same exposure period. As a result, it has been corrected at the raw data level, but the detailed procedure is too complex to describe here.

In addition to the detection system, the source and polarization systems also introduce errors in the ellipsometry measurement (Collins, 1990; Nguyen et al., 1991). First, Eqs. (1)-(3) are valid only if the source is completely unpolarized. Any residual source polarization will be modulated by the rotating polarizer, influencing the 2ω Fourier coefficients in Eqs. (2), and generating weak 4ω coefficients as well. Because the right side of Eq. (8c) is proportional to the $\sin4\omega t$ coefficient, one indication of residual source polarization is a non-zero consistency relationship. Other indications are non-zero minima in the residual function $R_k(A)$ for A near A_S and $A_S+\pi/2$. (For a pure source polarization error, the deviations from zero in the minima of $R_k(A)$ for the p- and s-directions are equal in magnitude but opposite in sign.) Second, Eqs. (2-3) are valid only if the polarizing elements are ideal. Quartz Rochon polarizers are often used to extend the spectral range of the ellipsometer above 4.5 eV or to provide pseudo-depolarization of incident partially polarized light. However, the normal modes of the Rochon element are not linear polarization states, but exhibit a slight ellipticity (Aspnes, 1974). This behavior is related to the optical activity of quartz, and it also influences the 2ω Fourier coefficients in Eqs. (2). The effects of both source polarization and polarizer optical activity can be included through corrections to α and β in Eqs. (2), expressed to first order in parameters that characterize the two errors. Because such corrections also enter into the residual and phase functions (see Sec. 2.1), A_S is no longer given by the minimum in R_k, and P_{Sk} is no longer given by Θ_k evaluated at the minimum. The parameters that characterize the two errors, in fact, can be determined experimentally in the calibration procedure of Sec. 2.1 when it is performed in both the p- and s-directions. Once the error parameters are known, then error-corrected values of A_S and P_{Sk} can be obtained. Finally, the effects of the errors can be included in exact equations analogous to Eqs. (3) that express (ψ,Δ) in terms of (α,β).

As an example of the influence of errors on the calibration, Figure 4 shows results obtained at an angle of incidence of 70°, from a c-Si substrate covered with a 590 Å thermal oxide. The experimental apparatus used here (see Fig. 1) employs a quartz Rochon polarizer and a calcite Glan-Taylor analyzer. The latter does not introduce significant error as calcite is not optically active. In Fig. 4(a), the analyzer offset angle, denoted A_1, obtained by neglecting all errors is given versus photon energy. The accompanying results, denoted A_S, are corrected to first order for polarizer optical activity and source polarization. It is clear that the spurious photon energy dependence in A_1 is eliminated, demonstrating the validity of the first-order corrections. The largest correction to A_1 arises from polarizer optical activity, and the magnitude of the associated error parameter, the optical activity coefficient, is a linearly increasing function of hv. This gives rise to the increasing discrepancy between A_1 and A_S with photon energy. As noted in Sec. 2.1, the corresponding polarizer calibration angle P_{Sk}

should be a linear function of k. Figure 4(b) shows the deviation from linearity in P_{Sk} with and without correction for the same errors, denoted δP_S and δP_1, respectively. The oscillations in δP_1 appear to result from a photon energy-dependent rotation of the residual source polarization by the Rochon element. The oscillations are damped by a factor of 3-4 in the correction procedure, again providing support for its validity. The source of the oscillations in δP_S remaining after correction ($\pm 0.02°$, maximum) is unclear at present. A linearized P_{Sk} function is used in Eqs. (5-6) and so these oscillations tend to be averaged, giving calibration values believed to be accurate to within $\pm 0.01°$ over the optimum photon energy range of 1.8 to 3.5 eV.

3.   APPLICATIONS IN PLASMA-ENHANCED CHEMICAL VAPOR DEPOSITION OF HYDROGENATED AMORPHOUS SILICON

3.1   Overview

The microstructural development of PECVD a-Si:H has been studied in the greatest detail by electron microscopy (Knights, 1980; Ross et al., 1984) and single-photon enegy ellipsometry (Collins, 1988; Drevillon, 1989). Owing to the uniformity of the optimized PECVD process, only the ellipsometry studies have provided information on the monolayer-level nucleation and coalescence patterns under these conditions. As an example of the ellipsometry results, Figure 5 shows the experimental pseudo-dielectric function trajectory at 3.4 eV that is swept out versus time during the growth of an a-Si:H thin film by remote He PECVD (Lucovsky et al., 1991) on a native oxide-covered crystalline Si substrate held at 250°C (Collins and Yang, 1989). A single-channel rotating analyzer ellipsometer was used, and Fourier analysis was performed on sampled data to deduce (α',β') vs. time. The pseudo-dielectric function in Fig. 5 is calculated from $\rho = \tan\psi \exp(i\Delta)$ through the following equation:

$$\langle\varepsilon\rangle = \langle\varepsilon_1\rangle + i\langle\varepsilon_2\rangle = \varepsilon_a \sin^2\theta \left\{1 + [(\rho-1)/(\rho+1)]^2 \tan^2\theta\right\}, \qquad (9)$$

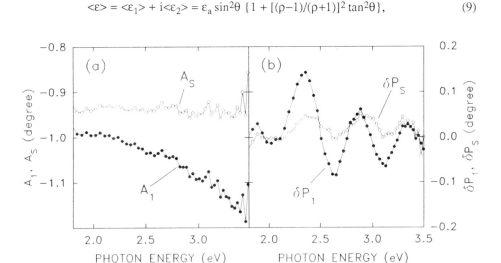

Fig. 4   Calibration data obtained as a function of photon energy for a c-Si substrate covered with a 590 Å thermally-grown oxide. (a) Analyzer offset uncorrected (A_1) and corrected (A_S) for source polarization and optical activity errors; (b) Deviation of the polarizer phase angle from linearity with pixel group number uncorrected (δP_1) and corrected (δP_S) for the same errors. In the absence of errors, A_S should be independent of photon energy, and δP_S should vanish (after Nguyen et al., 1991).

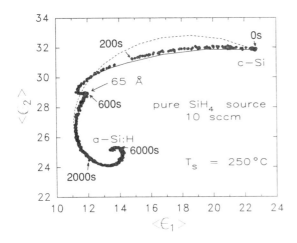

Fig. 5 Experimental pseudo-dielectric function trajectory at 3.4 eV for the growth of optimized a-Si:H on oxide-covered c-Si at 250°C by remote He plasma-enhanced CVD (points). Broken and solid curves were calculated assuming a layer-by-layer growth model and a nucleation and coalescence model, respectively. The latter nearly overlaps the data and uses a structural evolution model similar to those presented in Fig. 6(b) (after Collins and Yang, 1989).

where ε_a is the dielectric function of the ambient and θ is the angle of incidence (70°). This is the same mathematical relationship that allows one to calculate ε from ρ for a perfect interface between an ambient of known dielectric function ε_a and an opaque material of dielectric function $\varepsilon = \langle \varepsilon \rangle$ (Azzam and Bashara, 1977). Along the trajectory in Fig. 5, as will be seen, there is no point at which this ideal two-medium model is strictly valid [hence the left side of Eq. (9) is called the pseudo-dielectric function]. The points along the trajectory in the earliest stages in Fig. 5 were collected at 0.9 s intervals, and the a-Si:H deposition rate (time-averaged over the full run) was ~5 Å/min estimated on the basis of the real time ellipsometry data analysis to be described shortly. The relatively slow deposition rate is characteristic of the remote PECVD process; however, nearly identical experimental data are obtained for growth rates at least up to 150 Å/min, using the more conventional PECVD process in which pure SiH_4 gas is decomposed in an rf plasma established between two parallel-plates. Two static positions on the trajectory in Fig. 5 are noted: (1) the starting point corresponding to the c-Si substrate with its 18 Å native oxide and (2) the ending point corresponding to opaque a-Si:H (~ 600 Å thick). The oxide thickness is determined from a spectroscopic measurement of the room temperature substrate, requiring ~20 min with the single-channel instrument.

Before discussing further the results of Figure 5, it is helpful to provide an overview of the linear regression analysis (LRA) procedure applied to obtain the oxide thickness from (ψ, Δ) spectra obtained on the static substrate (Aspnes, 1981). This same procedure is also used extensively to analyze the real time spectroscopic data sets on more complex structures presented later in this Section, as well the data sets presented in Sec. 4. In order to apply LRA, all unknown quantities in the problem must be wavelength-independent. Thus for the sample under study, the dielectric function of each component material must either be known or expressible itself in terms of wavelength-independent parameters. To start the LRA procedure, an educated choice is made for the structure of the sample, including the number of layers and the identity of the component materials for each of the layers, which may in fact be microscopic composites. The dielectric functions of any composite layers are determined from the dielectric functions and volume fractions of the component materials though an effective medium theory (EMT; see next paragraph). Thus, in the simplest case in which all the dielectric functions are known, the wavelength-independent free parameters are simply the layer thicknesses and any required volume fractions. For the particular choice of sample structure, initial guesses are made for all the free parameters. Once these guesses are in place, (ψ, Δ) can be obtained for the simulated sample structure using conventional multilayer optical analysis to calculate first r_p and r_s (Azzam and Bashara, 1977). As the core of the analysis procedure, LRA determines the set of free parameters that minimize the deviations between the experimental and calculated values of $(\tan\psi, \cos\Delta)$. The quality of the final fit is quantified by the unbiased estimator of the mean square deviation:

$$\sigma = (N-p-1)^{-1/2} \left\{ \sum_{i=1}^{N} (| \tan\psi_e(hv_i) - \tan\psi_c(hv_i) |^2 + | \cos\Delta_e(hv_i) - \cos\Delta_c(hv_i) |^2) \right\}^{1/2}, \quad (10)$$

where N is the number of spectral points, p is the number of independent free parameters for the fit, and the subscripts 'e' and 'c' denote the experimental and best fit calculated (ψ,Δ) spectra. As necessary, a series of alternative choices for the structure of the sample are made in attempt to obtain a global minimum in σ, and explain all the features of the experimental data. One must be cognizant of the limits on the accuracy of the experimental data and avoid overcomplication of the structure which leads only to marginal improvements in σ. As a result, it is helpful to consider both the final confidence limits on all free parameters as well as the elements of the correlation matrix to prevent overparameterization of the model. The characterization of the c-Si/SiO$_2$ structure used for the substrate in a-Si:H growth is perhaps a trivial example of the LRA in which a single free parameter, oxide thickness is determined using as input the known room temperature dielectric functions of SiO$_2$ (Malitson, 1965) and c-Si (Aspnes and Studna, 1983). Real time studies presented later in this section, demonstrate beautifully the power of the LRA procedure in much more complex situations.

Next, a brief description of EMT's is in order. The EMT of Bruggeman (1935) has been applied most extensively, and among simple alternatives it has been demonstrated to be most appropriate for a-Si materials (Aspnes et al., 1979). The Bruggeman EMT is expressed by:

$$0 = \sum_{i=1}^{N_c} f_i \; \frac{\varepsilon_i - \varepsilon}{\varepsilon_i + 2\varepsilon} \; , \quad (11)$$

where ε is the effective dielectric function of the composite to be determined; ε_i and f_i are the dielectric function and volume fraction of component i; and N_c is the number of components, no more than two in the following studies. The assumptions implicit in Eq. (11) are that the geometry of the aggregate structure is spherical and that only dipole interactions are involved. Furthermore, the application of an EMT in general is based on the assumption that the microstructural inhomogeneity is much smaller than the wavelength of light but large enough so that the component materials retain their bulk dielectric properties. For amorphous semiconductors, the former assumption is reasonable for the thin films (<1000 Å) and uniform microstructures explored here; the latter is reasonable given the disordered atomic structure of the materials and their associated short electron mean free paths.

Now, we are in a much better position to consider the modeling of Fig. 5. The broken line in the Figure shows the trajectory expected from a perfect layer-by-layer growth model with a thickness independent dielectric function given by the convergence point of the experimental spiral where the a-Si:H is opaque. Conventional multilayer optical analysis for a four-medium structure (c-Si/SiO$_2$/a-Si:H/ambient) was used in the calculation. Although this model fits the experimental data well in the later stages of growth after about 125 Å, the fit in the earlier stages is poor and the lobe/cusp feature cannot be explained. In an attempt to understand the latter behavior, nucleation models were proposed in a similar vein to that of Theeten and Hottier (1980). Figure 6(a) shows the expected behavior if hemispherical nuclei were to form on a square grid, increase in radius and coalesce (Collins and Yang, 1989). After coalescence, the surface is assumed to form from interconnecting spherical segments with a radius of curvature equal to the film thickness. For simplicity a one-layer model for the a-Si:H was assumed in the calculation, and the Bruggeman EMT was applied to determine the dielectric function as a mixture of bulk a-Si:H and void with volume fractions determined

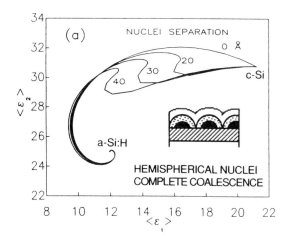

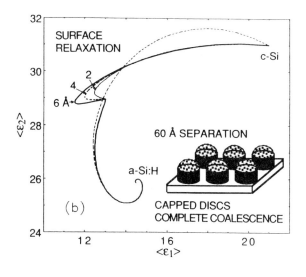

Fig. 6 Simulated pseudo-dielectric function trajectories at 3.4 eV for the growth of a-Si:H on oxide-covered Si. In (a) a Volmer-Weber mode of nucleation is assumed whereby hemispherical nuclei form on a square grid, increase in radius, and coalesce. The values given are the grid size, and 0 Å indicates layer-by-layer growth. In (b) a more complex process in which capped disks 23 Å in radius with a spacing of 60 Å increase in radius and coalesce. The lobe feature is generated by a surface relaxation process once nuclei make contact. The values in (b) indicate the magnitude of smoothening of the surface (after Collins and Yang, 1989).

from the chosen hemispherical geometry. It is clear that although such a model generates a lobe feature, a comparison of Figs. 5 and 6(a) indicates that it cannot explain the experimental behavior. After consideration of a number of geometries, the model of Figure 6(b) appears to incorporate the required elements. Disc-shaped structures are assumed to form in the first monolayer with a substrate coverage of about 70%, and these increase in size both in-plane and out-of-plane and coalesce. At the same time, spherical segments serve as caps on the disks; a reduction in the height of the caps, occurring upon coalescence of the discs is the necessary element for reproducing the lobe/cusp behavior observed in the experiment. Thus, this model suggests that once the separate a-Si:H clusters make contact in the early stages of growth, surface diffusion of newly-adsorbed film precursors results in a smoothening or relaxation of the nucleation-related surface structure.

The lobe/cusp feature is suppressed in the parallel-plate PECVD process at substrate temperatures less than 200°C, with SiH_4 diluted in inert gas or H_2 at flow ratios < 1:10, or for doped films with $B_2H_6:SiH_4$ flow ratios $\geq 5:10^4$, or for a-Si:H prepared by ion beam sputtering whatever the conditions (Collins, 1988). These results indicate two possibilities for such films: (1) coalescence is incomplete leading to columnar-type microstructure or (2) the surface diffusion is lower, leading to a weaker surface relaxation effect. The lobe/cusp feature is absent for optimized sputtered films comparable in bulk void fraction to the film of Fig. 5, as deduced from LRAs of (ψ,Δ) spectra for the final static films using reference dielectric function data of Aspnes et al. (1984) for a-Si. This observation suggests that the surface diffusion length is the key factor controlling the appearance of the lobe/cusp feature in the ellipsometry data at 3.4 eV. This feature also appears to presage good photoelectronic properties for the resulting films in studies performed to date.

Although the single-photon energy data have provided significant physical insight into the fundamental origins of process/property relationships for PECVD a-Si:H, there are too many free parameters in the model of Fig. 6(b) to obtain a unique fit. Thus, quantitative sample-to-sample comparison is not possible. More importantly, even if the model of Figure 6(b) was reduced to the simplest one possible, two layer thicknesses (to simulate the bulk and surface roughness) are required to describe surface relaxation. Although in principle (ψ,Δ) could uniquely establish the two thicknesses as a function of time over the trajectory, it would be impossible to ascertain whether the two-layer model was in fact valid. In particular, in some films coalescence is only partial, and voids are trapped at the interface (Collins, 1988). In such cases, the a-Si:H film may need to be modeled as three layers; a substrate interface layer of high void fraction, a bulk layer of low void fraction, and a surface layer of high void fraction. In single-photon energy ellipsometry, it is impossible to separate out the effects of the interface and surface layers, and thus to know whether the two layer model is appropriate. As noted previously in Sec. 1, the real time spectroscopic approach is needed for a quantitative description of the nucleation process with a validity that can be assured.

3.2   Real Time Spectroscopic Ellipsometry: Determination of Optical Functions

Figure 7 shows typical real time spectroscopic ellipsometry data expressed as pseudo-dielectric functions, $(<\varepsilon_1>,<\varepsilon_2>)$, for a-Si:H PECVD onto a native oxide-covered c-Si substrate at 250°C. For these data a parallel-plate configuration was used and the time-averaged deposition rate was 80 Å/min, as determined from an analysis of the final pair of ellipsometric spectra as described below. Each spectrum in Fig. 7 was constructed from 106 data points from 1.5 to 4.5 eV. The acquisition time for one pair of spectra was 160 ms, representing an average over 2 consecutive rotations of the polarizer (see Fig. 1); pairs of spectra were collected repetitively every 1 s. With this approach, sub-monolayer sensitivity and resolution were achieved. In Fig. 7 at the front of the plot, the optical properties of the c-Si are apparent with critical point structure appearing near 3.3 and 4.2 eV. At the rear of the plot where the a-Si:H is 337 Å thick, the 3.3 eV feature is damped and broadened, but still visible, indicating that the film is semitransparent throughout much of the spectral range at the end of data collection.

Before modeling pseudo-dielectric functions such as those in Fig. 7, the optical functions and structure of the substrate are required. For c-Si/oxide, LRA is performed on spectra collected at room temperature prior to heating the substrate to the deposition temperature. This procedure provides the native oxide thickness as described earlier (Sec. 3.1). It is then assumed that the oxide thickness and optical properties remain unchanged upon heating the substrate under vacuum. Thus, with its thickness known, the effect of the oxide can be analytically eliminated from the spectra obtained at the deposition temperature. The outcome is the true dielectric function of the c-Si at the elevated temperature, which is needed in subsequent LRAs of pseudo-dielectric functions during a-Si:H growth. In some studies, thin film chromium substrates were used. The Cr was deposited to opacity onto c-Si in an ion beam sputtering process that was optimized for surface smoothness, density, and uniformity. In this case, the Cr was chemically etched prior to insertion into the vacuum chamber, and the pseudo-dielectric function collected at the deposition temperature is assumed to be the true dielectric function of the Cr. Thus, with the optical functions of the substrate materials known at the deposition temperature, we can now discuss analyses to determine the optical functions of the growing a-Si:H from the real time spectra.

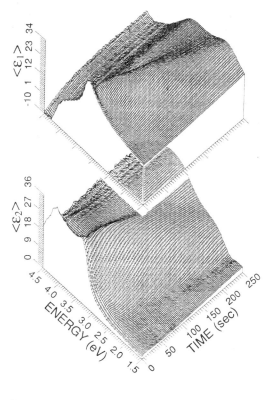

Fig. 7 Typical three-dimensional plot for the time dependence of the pseudo-dielectric function during the growth of a-Si:H onto oxidized c-Si at 250°C using conventional parallel-plate, radio frequency, plasma-enhanced CVD under optimized conditions. About 250 pairs of spectra were collected, each consisting of 106 points between 1.5 and 4.5 eV.

Fig. 8 The two-layer model and associated free parameters used to characterize the microstructural evolution for selected amorphous semiconductor depositions.

Because an analytical description of the dielectric function of an amorphous semiconductor requires at least 5 free parameters (Forouhi and Bloomer, 1986), then the simplest LRA of the film growth process (including film thickness and void fraction) would involve at least 7 wavelength-independent parameters. Because of this, such an approach is not expected to be successful in analyzing spectra collected in the earliest stages of film growth. As an alternative approach, we exploit the information contained in the complete time dependence of the $(\langle\varepsilon_1\rangle,\langle\varepsilon_2\rangle)$ spectra to separate the dielectric function from the microstructural features. In this approach, we assume that over some time interval during growth, the bulk dielectric function is independent of thickness. As will be verified later in this Section, this appears to be a good assumption for amorphous semiconductors beyond the monolayer level presumably due to the relatively short mean free path. We also assume that the film can be characterized by a two-layer structure including a bulk layer and a surface roughness layer of variable void volume fraction (see Fig. 8). The unknown dielectric function is that of the bulk layer, and it is also used as a component (along with voids) in the surface layer. Thus, analysis requires determining not only the dielectric function of the a-Si:H, but also three microstructural parameters as shown in Fig. 8: d_s and f_{sv}, the surface layer thickness and void fraction, and d_b, the bulk layer thickness. Further support for the two-layer model is presented later in this Section, and a simple criterion is presented in Sec. 3.3 to determine whether the two layer model is valid under the particular conditions of growth. It is the latter criterion that is absent from analyses of single-photon energy ellipsometry data.

The analysis procedure begins by guessing $\{d_s,d_b,f_{sv}\}$ values associated with a particular pair of $(\langle\varepsilon_1\rangle,\langle\varepsilon_2\rangle)$ spectra of interest, say at $t=t_0$, and inverting the $(\langle\varepsilon_1\rangle,\langle\varepsilon_2\rangle)$ spectra to obtain a trial dielectric function appropriate for the bulk material at $t=t_0$ (Cong et al., 1991). As usual, the Bruggeman EMT is used to calculate the optical functions of the density-deficient surface layer in the numerical inversion routine. The trial dielectric function for the bulk layer

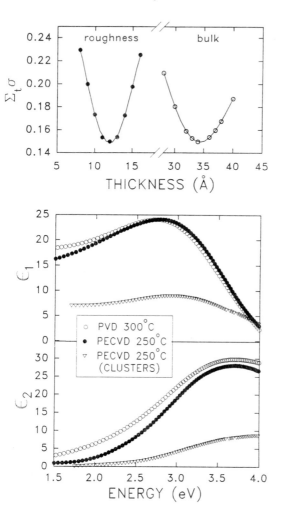

Fig. 9   A plot of the two-layer analysis used to determine the surface roughness and bulk layer thicknesses as well as the bulk layer dielectric function for ultrathin remote PECVD a-Si:H from pseudo-dielectric function spectra vs. time. The ordinate depicts the sum of unbiased estimators of the mean square deviation obtained in linear regression analysis fits of the spectra collected from 240 to 1800 s. A trial dielectric function applied in the analysis was determined from the pair of spectra at 575 s using a guess for one thickness (plotted along the abscissa) and a value for the other at the minimum (12 Å or 34 Å).

Fig. 10   Optical functions collected in real time for a 34 Å thick bulk layer of remote PECVD a-Si:H prepared onto oxidized c-Si at 250°C (solid circles) and for a 36 Å bulk layer of sputtered a-Si prepared onto oxidized c-Si at 300°C (open circles). Also shown are the optical functions for a 21 Å thick cluster film of remote PECVD a-Si:H measured in real time prior to bulk film formation on a Cr substrate at 250°C.

is then applied in LRA fits to all spectra within the time interval (including t_0) over which the dielectric function is assumed to be independent of thickness. The σ values obtained for such fits [see Eq. (10)] are summed over time; the correct guesses for $\{d_s, d_b, f_{sv}\}$ are those that provide a minimum in $\Sigma_t \sigma$. The trial dielectric function obtained with these choices is correct, and the time evolution of the three parameters is also correct. The evolution of the microstructural parameters will be discussed in Sec. 3.3.

Analysis results for a-Si:H growth by remote He PECVD on native oxide-covered a-Si:H held at 250°C are shown in Fig. 9. The time-averaged deposition rate obtained for remote PECVD was 5.0 Å/min, much lower than that for the parallel-plate method. The data in Fig. 9 were obtained by concentrating on a pair of ($\langle \varepsilon_1 \rangle, \langle \varepsilon_2 \rangle$) spectra collected at $t_0 = 575$ s (relative to plasma ignition). The left and right parabolic curves show $\Sigma_t \sigma$ for different choices of d_s and d_b at time t_0; for the curve on the left d_b was fixed at 34 Å, and for the curve on the right d_s was fixed at 12 Å. Spectra from 240 s to 1800 s were used in the summation over time; after 1800 s, $d_b \sim 150$ Å. Figure 9 suggests a global minimum in $\Sigma_t \sigma$ for the two choices, $d_s = 12$ Å and $d_b = 34$ Å. Finally, because of a relative insensitivity to f_{sv}, it was fixed at 0.50 in these analyses. This value led to consistency with the full microstructural evolution to be described

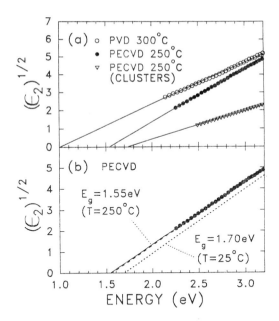

Fig. 11 (a) Optical gap determination from the real time dielectric functions of Figure 10; (b) optical gap determination for the 34 Å remote PECVD a-Si:H reproduced from (a) (solid circles), along with corresponding results obtained from data collected during the same deposition, but after 380 Å of growth at 250°C (dashed line). The optical gap determination for the final 380 Å film after cooling to 25°C is also included in (b) (dotted line).

in Sec. 3.3. Figure 10 shows the outcome of the analysis (solid points): the dielectric function of the 34 Å bulk a-Si:H layer at the deposition temperature of 250°C, appropriately corrected for the effects of surface roughness. The broad feature in ε_2 centered at 3.75 eV is typical of the results obtained on a-Si:H more than two orders of magnitude thicker (Collins, 1988). Figure 11(a) shows the analysis technique applied to determine the optical gap of the same 34 Å thick film. It involves plotting $\varepsilon_2^{1/2}$ vs. photon energy, E, and extrapolating the observed linear relationship to zero ordinate (Cody et al., 1982). This provides the difference between the onset energies of the valence and conduction bands, assuming parabolic densities of states and a constant dipole transition matrix element. Thus, Fig. 11(a) reveals an optical gap of 1.55 eV for remote PECVD a-Si:H at the deposition temperature of 250°C (solid points). The alternative Tauc optical gap expression (Tauc et al., 1966), which involves plotting $(E^2\varepsilon_2)^{1/2}$ versus E tends to show upward curvature. Thus, the Tauc optical gap is somewhat dependent on the energy range of the data and so is less suitable.

In addition, the $\Sigma_t\sigma$ minimization procedure was also used for the same remote PECVD a-Si:H to extract the optical functions independently from the final $(<\varepsilon_1>,<\varepsilon_2>)$ spectra where $d_s=10$ Å and $d_b=380$ Å. Thus, the resulting optical gap determination (dashed line) is compared with that for the 34 Å thickness (solid points) in Fig. 11(b). The gaps for the two thicknesses are the same within experimental error (1.55±0.01 eV), as are the optical functions themselves, demonstrating that a two-layer bulk/surface model is in fact appropriate and that the bulk layer optical properties do not evolve with thickness due to an evolution of the void fraction. Furthermore it can be concluded that there are no intrinsic size effects in the optical functions at least down to 34 Å. In order to compare these optical gap values with those in the literature, the final remote PECVD a-Si:H film was cooled to 25°C, and the bulk layer dielectric function was extracted under the assumption that $\{d_s,d_b,f_{sv}\}$ remain unchanged from their final values, $\{10$ Å, 380 Å, $0.50\}$. This gap determination appears in Figure 11(b) (dotted line), and the resulting value of 1.70 eV is in reasonable agreement with the literature results for conventional parallel-plate PECVD a-Si:H (1.64 eV), considering that the latter value was obtained by ex situ transmittance/reflectance measurements which are subject to errors due to surface oxides and roughness (Cody, 1984).

An accurate dielectric function for remote PECVD a-Si:H has been extracted in even thinner layers on metallic Cr substrates. When the total a-Si:H film thickness on Cr is less than 23 Å, however, d_b is less than one monolayer, as determined in an LRA that uses the a-Si:H dielectric function of Fig. 10 and the Cr substrate dielectric function (see also Sec. 3.3 for further details). Thus in this regime of thickness, d_b should be set to zero, meaning that the film consists of isolated clusters, and the one-layer model for the film is appropriate. Figure 10 shows the dielectric function of the cluster film, and Fig. 11(a) shows the optical gap determination (open triangles in each case). In the one-layer analysis, the cluster thickness of 21 Å was determined by trial-and-error under the condition that ε_2, the imaginary part of the dielectric function of the film, drop smoothly to zero for photon energies below the optical gap. The weak absorption strength in ε_2 compared with the 34 Å thick bulk film can be attributed to a high volume fraction of voids in the cluster film, estimated to be 0.48. A more interesting aspect is the blue-shifted peak energy for ε_2, 4.0 eV in comparison to 3.75 eV for the 34 Å thick bulk film. This shift cannot be explained in terms of voids. It most likely arises from a high fraction of SiH_n ($1 \leq n \leq 3$) which presumably covers the surfaces of the clusters. This proposal is also consistent with an optical gap which is wider by ~0.22 eV in comparison to the bulk a-Si:H. As a result of this study, we conclude that the dielectric function of the 21 Å thick cluster film is a better approximation to the optical properties of the surface layer on bulk a-Si:H, owing to the presence of Si-H_n bonding there. A refined analysis of the remote PECVD a-Si:H on c-Si/oxide using this improved surface layer dielectric function, however, leads to a bulk layer dielectric function and best fit microstructural parameters (d_s and d_b) that differ negligibly from that of the conventional bulk/void mixture used above (see Fig. 8).

For comparison with the remote PECVD results, Figs. 10 and 11(a) also include the dielectric function and optical gap (open circles) for a pure a-Si film (An et al., 1991). This film was prepared at a time-averaged rate of 1.8 Å/min using the magnetron sputter deposition method (PVD: physical vapor deposition) onto a c-Si substrate held at 300°C. The Ar sputtering gas pressure was set to the lowest possible value while maintaining a stable plasma, as this led to a-Si film deposition which could be described in terms of a simple two-layer model (see Fig. 8). For this deposition, an analysis similar to that of Fig. 9 concentrating on the spectra obtained at t_0=1420 s, provided surface roughness and bulk layer thicknesses of 18 Å and 36 Å, respectively. The differences in optical functions and gap between pure a-Si by PVD and a-Si:H by PECVD are consistent with the well-known role that H incorporation plays in relaxing strain in the growing Si-Si network, leading to an a-Si:H alloy with 5-15% H (Paul and Anderson, 1980). Thus, H-incorporation prevents strain-relief from occurring through less desirable coordination defects which are electronically active. H-incorporation tends to shift states from the conduction and valence band edges deeper into the bands; an effect also noted above for the hydrogenated cluster layer.

The effect of H incorporation can be observed clearly in another use of real time spectroscopic ellipsometry: characterization of the modification of the electronic structure of thin films. In this study, a-Si was sputtered to a bulk thickness of 250 Å onto c-Si at 300°C under low pressure conditions. Real time monitoring during growth was performed so that the bulk dielectric function and microstructure, characterized by $\{d_s, d_b, f_{sv}\}$, could be determined for the final film. Then the film was exposed at the temperature of deposition to an atmosphere rich in atomic H generated by a W filament in H_2 gas. Filament conditions were set to ensure that no etching occurred, as determined by exposing an ultrathin film of a-Si to the H atmosphere for an extended period. In the absence of etching, the same set of microstructural parameters can be used to analyze spectra obtained throughout the H exposure. Thus, it is straightforward to extract the dielectric functions and optical gap as a function of time throughout exposure by numerical inversion of ($<\varepsilon_1>,<\varepsilon_2>$). The evolution of the optical gap is shown in Figure 12. A 0.1 eV increase in gap after 30 min exposure is consistent with H incorporation which breaks Si-Si bonds, and removes electronic states preferentially from the band edges. For a 250 Å film, modification appears to occur uniformly throughout the film thickness, being limited by the H reaction kinetics rather than by diffusion. In separate investigations of a-Si:H prepared by post-hydrogenation, the dielectric function was

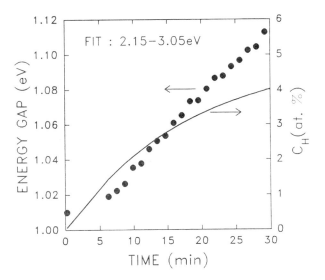

Fig. 12 Evolution of the optical gap of a 250 Å thick a-Si film sputtered onto oxidized c-Si, held at 300°C, during a 30 min exposure to atomic H generated by a heated filament (solid circles). For a film of this thickness, H-modification occurs uniformly throughout the bulk. The solid line is an estimate of the H-content based on deconvoluting the a-Si:H dielectric function into pure Si and Si:H components, and calibrating the Si:H component with an infrared transmission measurement.

decomposed into volume fractions associated with Si bonding and Si:H bonding contributions and the latter was calibrated with infrared transmittance measurements. Thus, the H-content estimated from the Si-H volume fraction is also plotted vs. time in Fig. 12. Possibly the Si and Si:H contributions are associated with Si_4 and Si_3H tetrahedra (Mui and Smith, 1988).

The unprecedented ability to determine the optical gap of semiconductor films as thin as 20 Å from real time observations is expected to be important for designing device structures that employ ultrathin layers. In addition, the ability to controllably modify materials based on real time observations of optical gap and chemical bonding also have important implications for tailoring device interfaces. Thus, utilizing the power of real time spectroscopic ellipsometry, the preparation and processing of semiconductor materials and devices can be guided by an understanding of physical phenomena rather than by trial-and-error.

3.3 Real Time Spectroscopic Ellipsometry: Determination of Microstructural Evolution

The microstructural evolution within a one-layer model for film growth is represented by the time dependences of two wavelength-independent parameters, the film thickness, d, and the void volume fraction, f_v. Within a two-layer model, the time dependences of three parameters are required, the surface and bulk layer thicknesses and the surface layer void volume fraction (relative to that of the bulk), i.e. $\{d_s, d_b, f_{sv}\}$ (see Fig. 8). Under ideal conditions of amorphous film growth, we might anticipate a transition from a one- to two-layer model, representing the transition from nucleation to bulk film growth through complete coalescence. Microstructures more complex than a two layer model are usually characterized not only by a surface roughness layer, which is a vestige of initial nucleation, but also by a layer trapped at the interface to the substrate having a higher void fraction than the final bulk film (Messier and Yehoda, 1985). In general, as noted earlier, this layer results from incomplete coalescence, and the void fraction in the layer is highest at the substrate interface and gradually decreases into the bulk (i.e. an inverted cone morphology). Thus, a three-layer model for such a structure is a only a crude approximation.

A criterion is needed for determining which models are valid in different thickness regimes and under different deposition conditions. One test for assessing the adequacy of the two-layer model is to determine the bulk optical functions at very different thicknesses as in Sec. 3.2 [see Fig. 11(b)]. If the optical functions are identical, then this suggests that a third layer is not required. Another approach that allows one to assess the continuous microstructural evolution is to compare the unbiased estimators of the mean square deviation [Eq. (10)] obtained from LRA best fits to the pseudo-dielectric function spectra assuming different models. In this analysis, the bulk optical functions deduced in the two-layer model of the final film are used as a first approximation. As an example, Fig. 13 shows $\sigma(t)$ in the early stages of growth for the remote PECVD a-Si:H of Figs. 9 and 11(b). For t<3.5 min in Fig. 13, the one- and two-layer models give nearly identical σ values, whereas for t>3.5 min, $\sigma(t)$

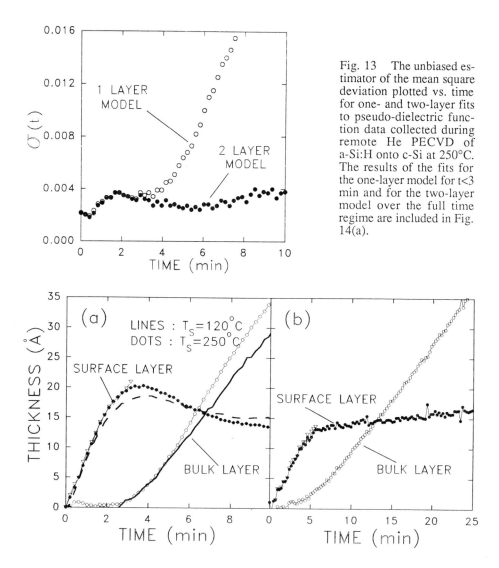

Fig. 13   The unbiased estimator of the mean square deviation plotted vs. time for one- and two-layer fits to pseudo-dielectric function data collected during remote He PECVD of a-Si:H onto c-Si at 250°C. The results of the fits for the one-layer model for t<3 min and for the two-layer model over the full time regime are included in Fig. 14(a).

Fig. 14   Surface and bulk layer thickness obtained in microstructural analyses of amorphous semiconductor growth. (a) Remote PECVD a-Si:H prepared on oxidized c-Si at 250°C and 120°C. (b) magnetron sputtered a-Si prepared on oxizided c-Si at 300°C. The open triangles in (a) and (b) were obtained from a one-layer model over its regime of validity; all other results were obtained from a two-layer model.

for the one-layer model increases sharply. Thus, the statistical information provides evidence for a transition between one- and two-layer models at ~3.5 min. σ for the two-layer model remains as low as can be expected (<0.004 throughout the growth process) given the constraints of experimental precision and accuracy. Just as the presence of a second layer is not supported by the data for t<3.5 min, a third-layer is also not supported by the data throughout the subsequent deposition. Under certain conditions of film growth (e.g. sputtering at high pressure), σ(t) for the two layer model increases significantly during the growth of a 250 Å film, indicating the presence of a graded void structure buried in the growing film. In fact, such conditions are those for which columnar microstructure is expected (Thornton, 1977).

For quantitative microstructural information, we will concentrate on the evolution of d_s and d_b in the two-layer model since these parameters exhibit the most interesting behavior with regards to the deposition process dependence. In Figure 14(a) (points), results are presented for the remote PECVD a-Si:H of Figs. 9, 11(b), and 13. In the earliest stages of film growth (t<3.5 min), d_b is less than 2.5 Å, indicating that the first bulk monolayer has not yet formed. Strictly, according to Fig. 13, the two-layer model is an overinterpretation of the data in this regime, and the one-layer model should be applied. The open triangles in Fig. 14(a), denote results for d_s obtained with the one-layer model ($d_b\equiv0$). Thus, the difference between the one- and two-layer interpretations is negligible, and the surface layer in the two-layer model simulates the nucleating clusters in the one-layer model. At t~3.5 min, d_b~2-3 Å and d_s reaches a maximum thickness of 20 Å. Near this time, the first bulk monolayer has formed and subsequent flux is incorporated predominantly as part of the bulk layer. This behavior supports the conclusions based on an inspection of σ(t) in Fig. 13. The aspect of greatest interest in Fig. 14(a) is the fact that once the first bulk monolayer forms, subsequent flux smoothens the nucleation related surface morphology. We conclude that the initial nuclei reach a thickness of 20 Å just prior to contact, and after contact the surface smoothens by 8 Å on the time scale of Fig. 14(a). Further gradual smoothening by another 2 Å occurs beyond the time scale depicted in Fig. 14(a) until the deposition is terminated after 380 Å. The remaining 10 Å surface layer after 380 Å most likely includes contributions from sub-microscopic roughness, but also surface SiH_n bonding which gives rise to the blue shift in both the ε_2 peak and optical gap for the 21 Å clustered film [see Figs. 10 and 11(a)].

To underscore the importance of these results, Fig. 14(b) shows corresponding data for pure a-Si deposited by low pressure magnetron sputtering onto oxidized c-Si at 300°C (An et al., 1990). In this case also, σ(t) exhibits behavior indicating a transition from a one-layer to two-layer structure at t=5.5 min. After this transition, the two-layer σ remains low throughout the deposition, whereas the one-layer σ increases significantly. Figure 14(b) shows that the initial a-Si clusters reach a thickness of 13 Å just prior to bulk film formation. The first bulk monolayer forms abruptly thereafter (at t=5.5 min where σ also shows a transition). In contrast to the results for PECVD a-Si:H, the nucleation-related surface morphology on the PVD a-Si is enhanced as a function of time. It should also be noted that the thickness at which the initial nuclei make contact, 20 Å for PECVD a-Si:H and 13 Å for PVD a-Si films on oxidized c-Si substrates, provides a measure of the nucleation density at this time. The value obtained is somewhat model-dependent because an analysis of f_v in the cluster growth stage is required to estimate the shape of nuclei upon contact. For the PECVD a-Si:H, about 60% of the substrate surface is covered in the first monolayer, and the void volume fraction increases gradually from 0.4 to 0.5 as the clusters grow and make contact at 20 Å. For the PVD a-Si, about 40% of the surface is covered in the first monolayer, and the void volume fraction decreases weakly from 0.6 to 0.45 as the clusters grow and make contact at 13 Å. Both sets of results for f_v are consistent with disc-like structures on the substrate. But because $f_v=f_{vs}$~0.45-0.50 when the first bulk monolayer is formed, it is reasonable to suggest that the disc-like clusters have evolved into hemispheres at this point (for which f_v would be 0.476 for a square grid of nucleation sites). With this geometry, a spacing between clusters and a nucleation density of 40 Å (26 Å) and 6×10^{12} cm^{-2} (1.5×10^{13} cm^{-2}) for PECVD a-Si:H (PVD a-Si) are obtained.

The differences in the growth behavior for PECVD and PVD amorphous semiconductors arise from the differences between surface processes in chemical and physical vapor deposition. It is instructive to consider theoretical models developed to investigate the stability of one-dimensional thin film surface profiles in response to imposed sinusoidal perturbations of wavelength λ_r (Mazor et al., 1988; Karunasiri et al., 1989). Effects of finite atomic size and flux shadowing by asperities have been proposed to enhance the surface perturbations whereas adatom surface diffusion damps them. The surface relaxes to a smooth profile for an adatom diffusion length λ_o such that $\lambda_o > \lambda_r$. In the experimental situation the perturbation is the nucleation-generated surface morphology. In light of the above considerations, the smoothening behavior for the PECVD a-Si:H may arise from an adatom surface diffusion length much greater than ~40 Å (the dominant roughness wavelength at cluster contact). The roughening for the PVD a-Si then arises from a surface diffusion length much less than ~25 Å. This difference is reasonable given the proposals of PECVD as occurring from SiH_3 precursors which are mobile on an H-terminated surface (Robertson and Gallagher, 1986). In contrast, for the PVD a-Si, precursors have a tendency to stick where they land, and any surface diffusion originates from the initial energy of the incoming Si, which is maximized at low sputtering gas pressure. Under these conditions, surface diffusion is sufficient to prevent trapped voids from forming at the substrate interface and hence to prevent subsequent columnar morphology, but it is insufficient to smoothen the surface after the onset of bulk film growth. This general interpretation of the results of Fig. 14 in terms of differences in surface diffusion length is supported by additional recent studies of PECVD a-Si:H at different temperatures. The extent of surface smoothening is found to decrease with decreasing substrate temperature as shown by the typical results in Fig. 14(a) (lines) obtained at 120°C. This effect is understood in terms of thermally-assisted diffusion, and the observations are consistent with the single-photon energy studies described in Sec. 3.1.

Further remarks are appropriate in comparing the above analysis of remote PECVD a-Si:H to corresponding single-photon energy studies discussed in Sec. 3.1. Similarities are noted, for example, the deductions of disc-shaped initial nuclei and surface smoothening phenomena used to describe the lobe feature. The exact magnitude of the surface smoothening effect is impossible to establish without a full spectroscopic analysis that simultaneously determines the three parameters of the two-layer film growth model (Fig. 8). Of great importance is the ability first to establish the minimum number of physical layers required to explain the data (through Fig. 13); second to extract detailed microstructural information with this model, allowing all free parameters to vary (Fig. 14); and third to propose a geometrical model *a posteriori*. This three-step procedure places the ellipsometric analysis on a rigorously quantitative footing for unprecedented microstructural chracterization. The procedure also dispels questions concerning the uniqueness of the final model.

4.   APPLICATIONS IN THE CHARACTERIZATION OF III-V SEMICONDUCTOR HETEROSTRUCTURES

4.1   Overview

Ellipsometry at a single photon energy has also been applied to characterize the growth of GaAs/AlGaAs heterostructures in real time (Theeten et al., 1979; Laurence et al., 1981, Aspnes et al., 1990a). In comparison to the case of amorphous semiconductors in which the evolution of microstructure and thickness has been sought from such measurements, the evolution of Al composition and thickness has been the goal in corresponding studies of III-V semiconductor heterostructure fabrication. The analysis of single-photon energy ellipsometry data for GaAs/AlGaAs is more straightforward owing to the epitaxial growth achieved for this system. Thus, one can neglect surface roughness, which would require a more complex two-layer model for the growing film. Any roughness that is present presumably arises from steps on the surface, and a recent theoretical study suggests that the contribution of biatomic steps to the amplitude reflection coefficients at 1.96 eV and normal incidence is <0.1% when the step spacing is greater than 20 Å (Aspnes, 1990).

The first ellipsometry measurements of GaAs/AlGaAs were undertaken on heterostructures fabricated by organometallic vapor phase epitaxy (VPE) (Theeten et al., 1979). In these studies, a conventional rotating polarizer ellipsometer with a HeNe laser source was used to probe the surface at 1.96 eV. For the growth of AlGaAs on GaAs (or vice versa), spiral trajectories in (ψ,Δ) were obtained having similar overall features to that of Fig. 5. The stable points in (ψ,Δ) space correspond to the opaque materials, and in theory, the bulk dielectric function, $(\varepsilon_1,\varepsilon_2)$, can be calculated from these points through Eq. (9) with the left side replaced by $\varepsilon_1+i\varepsilon_2$. From $(\varepsilon_1,\varepsilon_2)$, the Al composition can also be obtained as long as a prior calibration has been performed. This calibration generally entails preparing a series of opaque layers of various Al compositions, determining their endpoints by real time ellipsometry, and characterizing the Al compositions by another technique (e.g. x-ray diffractometry or Auger electron spectroscopy) after removing the samples from the reactor. Such a calibration in terms of the optical properties is valid only at the temperature of growth and at the single photon energy of the probe. In practice, however, the absolute accuracy of the first experimental results was poor owing to the constraints of the VPE reactor including birefringence in the windows, deposition on the windows, alignment problems, and angle of incidence determination. Thus, ψ values measured relative to the GaAs endpoint were used for compositional calibration. The Δ values were not considered as they tend to be less accurate in general, and the accuracy depends strongly on Δ itself. These techniques provided the Al composition in opaque layers of $Al_xGa_{1-x}As$ with an estimated accuracy of ±0.05 in x.

In these first studies, calculated trajectories were obtained assuming that the $Al_xGa_{1-x}As$ grows epitaxially (within a one-layer model for the alloy) with thickness-independent optical properties and, thus, a thickness-independent Al composition, x (Theeten et al., 1979). This calculation uses optical functions for both the substrate and the film from the opaque starting and ending points of the trajectory. As long as reasonable agreement between the experimental and calculated trajectories is obtained (such as in Fig. 5 for d>125 Å), then the thickness versus time can be deduced in the regime of semitransparency by associating each experimental data point to the nearest value on the calculated trajectory (along which the thickness is known). In the VPE process, experimental trajectories showed pronounced deviations from the corresponding calculated trajectories in the early stages of growth. These deviations were attributed to a profile in the Al composition near the interfaces. For example, in the early stages of the growth of AlGaAs on GaAs, some time is required for the Al composition to stabilize. In the growth of GaAs on AlGaAs, cross-contamination of the initial GaAs by Al may occur. An analysis procedure was developed to determine the profile in the optical properties (Laurence et al., 1981). In this procedure, the initial two (ψ,Δ) points along the experimental trajectory are fit to a trajectory in which the free parameters are the real and imaginary parts of the dielectric function of the layer. Once the fit is obtained, the final thickness at the second point can be established by correspondence with the best-fit trajectory. This layer is left behind to serve as a substrate in an analysis of the next two points. The process is continued, and the deposited material is described as a succession of homogeneous layers of known optical functions and thicknesses. The practical thickness resolution for this approach was found to be 50 Å. In this way, transition widths for interfaces of ~100 Å for AlGaAs on GaAs and vice versa, were detected in the early VPE studies.

Over the last decade significant progress has been made in the instrumentation, measurement, and analysis techniques for single-photon energy ellipsometry. As a result, the problem of monitoring composition during AlGaAs growth on GaAs by single-photon energy ellipsometry has been reconsidered, this time concentrating on the more easily controllable organometallic molecular beam epitaxy technique (MBE) (Aspnes et al., 1990a). For example, in the early stages of growth a compositional sensitivity of ±0.03 for x>0.2 has been achieved in 10 Å thickness increments by analyzing the slope of the pseudo-dielectric function trajectory at 2.6 eV. As before, the thickness is also determined by the distance traversed along the trajectory. Over extended periods of time after the AlGaAs has become opaque, an equivalent compositional sensitivity at the level of ±0.001 has been achieved by monitoring the drift in the "stable" endpoint. An optimum choice of photon energy made possible with a broad-band Xe source, improved optical access to the growth environment made possible with novel strain-free windows (Studna et al., 1989), and finally, improved reference dielectric functions collected versus x, have all contributed to the recent advances.

Furthermore, to interpret the single-photon energy ellipsometry trajectory at any point during the growth process, a computational shortcut has also been developed recently for estimating the optical functions in the near surface region of the growing film, and hence the Al composition there (Aspnes et al., 1990b). It involves analyzing the pseudo-dielectric function as arising from a three-medium sample (bulk/film/ambient) in which the film is now defined by a virtual boundary at a constant depth, d, below the growing surface. Because d is chosen to be small, and the amplitude reflection coefficients for the virtual film-substrate boundary are likewise small, a first order expansion of the full three-medium complex amplitude reflection coefficients in these two parameters is appropriate. It is found that the reflection coefficients for the film-substrate boundary, which contain all previous growth history of the sample, do not appear in the first order expansion. As a result, the film optical functions can be estimated solely from d and the pseudo-dielectric functions obtained before and after the deposition of the surface material of thickness, d. In practice d is chosen to be ~50 Å, and this represents the thickness resolution. The ease with which the computation can be made, relative to the earlier analysis, permits near-surface composition to be estimated in real time from ε_2. This in turn has led to the first closed loop control system based on an ellipsometric probe, which in this case regulates the flow of triethylaluminum. In this recent work it has been noted that a real time spectroscopic capability is required to optimize loop stability with respect to other possible changes during growth that are unrelated to composition as noted in Section 1.

In the following part, the most recent applications of real time spectroscopic ellipsometry to III-V semiconductor processing will be reviewed (Heyd et al., 1991). Studies performed thus far have concentrated on information which can be deduced during etching. This is a problem which is more demanding than epitaxial growth because of the surface damage, interface broadening, and surface temperature changes that occur, and thus is impossible to characterize completely at a single photon energy. In these initial studies, we have concentrated on sputter-etching with an Ar ion beam.

4.2   Real Time Spectroscopic Ellipsometry of III-V Semiconductor Etching

Figure 15 shows the evolution of the pseudodielectric function obtained in real time during Ar ion beam etching of an MBE-prepared GaAs/AlGaAs/GaAs structure. A Kaufman source generated 200 eV Ar ions at a dose rate of 6.5×10^{15} cm^{-2}s^{-1}, and the structure was not intentionally heated during etching. Pairs of spectra in ($<\varepsilon_1>,<\varepsilon_2>$), each consisting of 80 spectral positions, were collected at 16 s intervals with an angle of incidence of 70°. Figure 15 is constructed from approximately 75 such pairs of spectra. As in the studies of Section 3, raw data were obtained as an average over a 3.2 s duration, allowing a resolution on the order of a few monolayers in the etching process.

In order to interpret the experimental results in Fig. 15 by LRA, dielectric functions for the anticipated components of the structure are required. Selected results from the references are included in Fig. 16. These include pure GaAs (Aspnes and Studna, 1983), Al$_{0.20}$Ga$_{0.80}$As (Aspnes et al., 1986), amorphous GaAs (a-GaAs) (Gheorghiu and Theye, 1981), and the native oxide on GaAs (Aspnes et al., 1981). The a-GaAs is expected to be a component of the damage layer on the GaAs, and the oxide is present on the starting structure. In order to use the Al composition as a free parameter in the spectroscopic analysis, the AlGaAs reference dielectric functions of Aspnes et al. (1986) were expressed in terms of wavelength-independent parameters using a harmonic oscillator model for the dielectric function (Erman et al., 1983). The oscillator parameters (i.e. the strength, energy, and width) were all smooth functions of Al composition x, and this provided an interpolation scheme in the linear regression analysis (LRA), allowing the dielectric function to be determined for any value of x. Attempts were also made to include the effects of ion beam-induced increases in surface temperature in the analysis. Thus, the dielectric function of bulk GaAs and Al$_{0.20}$Ga$_{0.80}$As materials were measured versus temperature from 25°C to 150°C. The room temperature

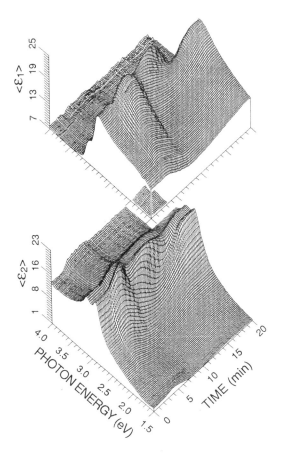

Fig. 15 A three-dimensional plot of the evolution of the pseudo-dielectric function obtained during ion beam etching of a GaAs/AlGaAs/GaAs structure. About 75 pairs of spectra are shown, each consisting of ~80 points between 1.5 and 4.0 eV (after Heyd et al., 1991).

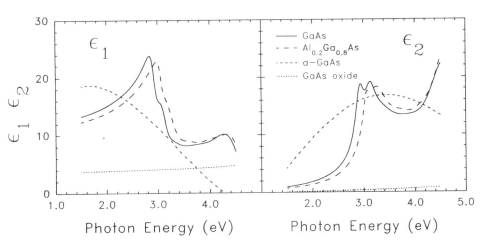

Fig. 16 Room temperature optical functions for GaAs, $Al_{0.2}Ga_{0.8}As$, a-GaAs, and native GaAs oxide used in the analysis of the experimental data of Fig. 15.

Ambient

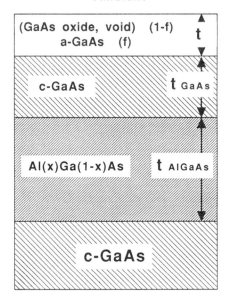

Fig. 17 Sample structure used in the analysis of the data of Fig. 15. In the initial stages of etching the surface layer is assumed to be an a-GaAs/oxide mixture whereas in the later stages, after the oxide has been sputter-etched away, an a-GaAs/void mixture is assumed. In addition to the three thicknesses and two compositions, the temperature of the structure is also a free parameter.

(ψ, Δ) spectra in each case were subjected to LRA in order to deduce the oxide layer thickness, assumed to be GaAs oxide in each case. The oxide thickness was then numerically extracted from all spectra to obtain the true dielectric function of the bulk material, under the assumption that the oxide optical properties and thickness did not change upon heating. This was verified by remeasurement upon cooling back to room temperature. The bulk dielectric functions vs. temperature were also parameterized with the harmonic oscillator model for purposes of interpolation. It was safe to assume that the temperature dependence of AlGaAs oscillator parameters were independent of x over the relatively narrow range of x required in the LRA modeling ($0.18 \leq x \leq 0.23$). It was also safe to neglect the effect of temperature on the surface damage layer, because it is thin and its optical properties are expected to be much less strongly dependent on temperature than the crystalline components of the structure.

Figure 17 shows the sample structure used in the LRA modeling. There are six free parameters which were varied in the analysis (not necessarily all at the same time), and some were not needed once the layers of the structure were etched away. The parameters included three thicknesses: those of the surface damage layer, the topmost GaAs, and the underlying AlGaAs. Two compositions were required: those of the AlGaAs and the damage layer, which was modeled in the first two minutes of etching as an a-GaAs/oxide mixture, and later as an a-GaAs/void mixture. Finally the sample near-surface temperature was the sixth parameter.

The time regime beyond the three-dimensional plot in Fig. 15 was analyzed first since both layers were removed and only three parameters were required in the fitting. Figures 18 and 19 show these parameters: near-surface temperature (T) and surface damage thickness and composition. The damage layer thickness and void volume fraction (the latter relative to the a-GaAs reference) were found to be stable at 34 Å and 0.03-0.04, respectively, whereas T increased in a roughly linear manner. The damage layer thickness tends to correlate with T such that a 1 Å increase in the former resembles a ~10°C increase in the latter. The fact that the damage layer thickness obtained here matches literature results for Ar ion bombardment of bulk GaAs (Aspnes, 1988) and Ge (Aspnes and Studna, 1980), as well as the fact that it is stable for t>20 min, suggests that the correlations have been successfully decoupled. Even so, such correlations give rise to the relatively large absolute confidence limits of ±8-10°C on the

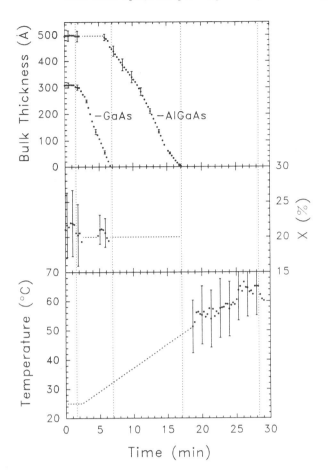

Fig. 18 Bulk thicknesses of GaAs and AlGaAs (top), $Al_xGa_{1-x}As$ composition (center), and temperature (bottom, estimated to be within ~400 Å of the surface) deduced from the experiment of Fig. 15. The error bars represent the 90% confidence limits in the absolute values, and the bold dashed lines indicate the regime over which the parameters were fixed in the linear regression analysis. The vertical dotted lines, from left to right, denote the times of beam ignition, the two interface crossings, and beam termination.

temperature. The precision in the measurement of T appears to be much better, however, ~±2°C. Figure 19 includes the unbiased estimator of the mean square deviation, defined by Eq. (10), which is a measure of the quality of the fit, and Fig. 20 shows a typical fit in to $<\varepsilon_2>$ in this later time regime. Including T as a free parameter improves the quality of the fit by 30%, and allows one to match the positions of the E_1 and $E_1+\Delta_1$ features in the experimental spectra. Because the spectral region from 2.6 to 3.4 eV is most sensitive to changes in T, we anticipate that the temperature value represents an average over the top 400 Å of the structure, this being 2x the optical penetration depth of GaAs at 3.0 eV. Finally it is important to note that the overall features of the structural model should be correct in order to obtain an accurate value of T. In other words, a good match to the overall magnitude of $(<\varepsilon_1>,<\varepsilon_2>)$ must be obtained before the LRA can match the positions of the features in the calculated spectra to those in the data. For reasons that will become clear shortly, it is not possible to use T as a free parameter throughout the full etching process. As a result, the roughly linear increase in T is extrapolated linearly back to room temperature, and this profile is assumed to be fixed throughout the remainder of the analysis.

Prior to igniting the Ar ion beam at t=1.6 min, the starting substrate was analyzed with the temperature fixed at 25°C and the remaining five parameters free. The following values were obtained: oxide thickness, 10±1 Å; topmost GaAs thickness 310±10 Å; underlying AlGaAs thickness 498±20 Å; and Al composition 0.215±0.045. The latter three results overlap the

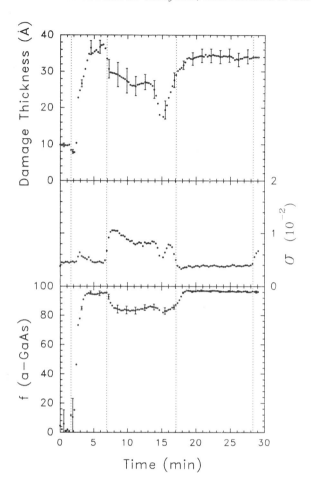

Fig. 19 Damage layer thickness (or oxide thickness for t < 1.6 min, top), damage layer composition (bottom), and the unbiased estimator of the mean square deviation deduced from an analysis of the data of Fig. 15. In the early stages of etching, the damage is assumed to be an a-GaAs/oxide mixture and in the later stages, an a-GaAs/void mixture is used. The error bars and vertical lines are as in Fig. 18.

nominal values of 300 Å, 500 Å, and 0.23 estimated from flux calibration in the MBE facility. Ex situ spectroscopic ellipsometry studies of similar structures suggest that improved confidence limits would be obtained at an angle of incidence closer to 75°C (Snyder et al., 1986). In view of the results obtained prior to igniting the beam voltage, the value of x was fixed at 0.20 throughout much of the analysis of the spectra collected during etching, based on the assumption that the Al composition is uniform within the absolute confidence limits.

Again referring to Figures 18 and 19, analysis of data collected after Ar^+ beam ignition at 1.6 min reveals the following behavior. First, ~2 min of ion exposure are required to eliminate the damage layer oxide component which originates from the 10 Å oxide on the initial structure. Because it is impossible to distinguish between oxide and void in this thin layer, a possible alternative explanation of the results at the bottom of Fig. 19 in terms of damage layer densification can be ruled out on the basis of additional experiments which show that such an effect occurs very rapidly (≤0.5 min). In the first 2 min, the damage layer evolves to 35 Å a-GaAs, and 80-90 Å of underlying GaAs are removed. The latter value provides a measure of the ability of Ar^+ impact to bury O atoms below the surface in a process of near-surface atomic mixing. Over the time intervals from 2.5 to 7 min and from 6 to 17 min, the GaAs and $Al_{0.2}Ga_{0.8}As$ are removed at average rates of 65 Å/min and 45 Å/min. The lower etch rate for the latter was found to be reproducible, observed in a number of experiments over an Ar^+

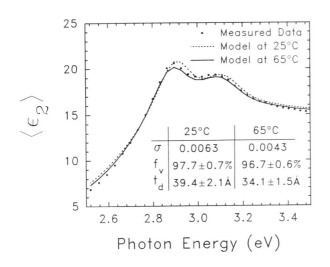

	25°C	65°C
σ	0.0063	0.0043
f_v	97.7±0.7%	96.7±0.6%
t_d	39.4±2.1Å	34.1±1.5Å

Photon Energy (eV)

Fig. 20 The imaginary part of the experimental pseudo-dielectric function in the vicinity of the E_1 and $E_1+\Delta_1$ transitions collected at 28 min (see Figs. 18 and 19), after both GaAs and AlGaAs layers have been etched away (points). The lines represent best-fit models to these data including T as a free parameter in one case (solid line) and fixing it at 25°C in the other (broken line). In both cases, the damage layer is included in the model, and the best fit free parameters and the quality of the fit are provided in the inset.

energy range of 100 to 300 eV. It is clear from the result at the top of Fig. 18 that ~30 Å or more of AlGaAs have been removed after the overlying GaAs thickness has been reduced to zero. It is this material that either contributes to the damage layer on the AlGaAs or is sputtered from the surface.

It is clear from Fig. 19 that the damage layer on AlGaAs is different in character than that on GaAs; it appears to be thinner by 5-10 Å and is best characterized within the model as a 0.85/0.15 mixture of a-GaAs and void. The origin of the apparent decrease in thickness of this layer in approaching the lower GaAs/AlGaAs interface is unclear at present. Yet further information on atomic scale mixing at the surface can be deduced by observing the evolution of the void content (which simulates the effect of Al in the damage layer) upon crossing the lower interface. Figure 19 (bottom) shows that about 1 min is required before the damage composition returns to a value typical of that on pure GaAs, and in this time, an estimated 65 Å of the lower GaAs material is removed. Similar to the earlier observations of the ion impact-induced mixing of O, this provides an accurate measure of the mixing of Al in the near surface region. In spite of the useful information provided by the simple model of damage on AlGaAs, an assessment of the statistical data available indicates that the simple a-GaAs/void model does not successfully simulate the optical properties of the AlGaAs damage layer. Foremost, the value of σ (Fig. 19 at center) is about 2x higher throughout the etching of AlGaAs in comparison to the fits for earlier and later times. Secondly the confidence limits on the AlGaAs damage layer are relatively larger. These poorer fits prevent a simultaneous determination of x and T during the etching of AlGaAs. Thus, the most accurate value of x is obtained after much of the topmost GaAs is etched away but before the AlGaAs is damaged (see Fig. 18, center). Here we verify the previous result of 0.21 but within narrower limits of ±0.02, which is a measure of our absolute confidence in the composition. Future improvements in characterizing such etching processes require a better model for the AlGaAs damage layer, and an optimized angle of incidence which could be employed in a deposition system dedicated to III-V semiconductor materials.

5.   SUMMARY

A unique approach to ellipsometry has been developed recently which has led to the first real time spectroscopic studies of surfaces and interfaces during semiconductor growth and etching. Previous real time ellipsometry studies of evolving semiconductor surfaces have been restricted to single or, at most, a few selected photon energies owing to the time required for sequential acquisition of complete spectra. The novel real time spectroscopic ellipsometer is designed in the rotating polarizer configuration and utilizes a multichannel analysis system, consisting of a fixed polarization analyzer, spectrograph, Si photodiode array, and detector controller for parallel acquisition of spectra. With a state-of-the-art detector controller, complete ellipsometric spectra, $\{\psi(h\nu),\Delta(h\nu)\}$, consisting of 128 pairs of points over the near-infrared to the near-ultraviolet spectral range, may eventually be collected in a time as short as 5 ms. The spectra in $\{\psi(h\nu),\Delta(h\nu)\}$ can be computed from four successive read-outs of the array, each one representing the integrated irradiance striking the detector surface over a 90° quadrant of the rotating polarizer optical cycle.

In the semiconductor growth and etching studies discussed here, sensitivity to changes in film thickness typically below the monolayer level was achieved using an average of 80 optical cycles, requiring a 3.2 s data acquisition time. Improvements in the precision of the system have been made more recently, and as a result similar performance on the 160 ms time scale has been achieved. In addition, improvements in accuracy are being made continually, and procedures have been designed to correct for detection system non-linearity, effective image persistence, and stray light; source polarization; polarizer optical activity; stress birefringence in the vacuum chamber windows; and deposition-related background light. Such corrections are particularly important when the beam reflected from the surface is near-linearly polarized, e.g. for semiconductors below the onset of direct absorption. Improvements in accuracy are paramount if the data are to be analyzed by linear regression analysis techniques. Two applications of real time spectroscopic ellipsometry have been reviewed in detail here, amorphous semiconductor growth and crystalline III-V semiconductor etching, and the advantages of the spectroscopic approach in each of these applications have been stressed.

In studies of amorphous semiconductor growth, real time spectroscopic ellipsometry is particularly advantageous owing to its ability to characterize both the optical properties and microstructure. As long as the growth process can be represented in terms of a two-layer model [substrate/(bulk layer)/(surface layer)/ambient], then it becomes straightforward to extract the bulk layer dielectric function and microstructural evolution from the ellipsometric spectra. As a result, the optical band gap has been deduced for amorphous silicon [a-Si(:H)] films with bulk layer thicknesses as low as 30 Å. This represents the first determination of the optical gap of an amorphous semiconductor from real time observations, and the final result is free of artifacts due to the surface condition of the film, i.e. uncorrected roughness and oxide layers, that generally influence ex situ measurements. The ability to determine optical gap in ultrathin layers is extremely important for modeling and understanding the performance of photovoltaic devices fabricated from such layers. Furthermore the demonstrated ability to determine the evolution of the optical gap of a thin film in real time during hydrogenation may find application in tailoring band profiles at interfaces in electronic devices.

Once the a-Si(:H) optical functions are known, then linear regression analysis (LRA) can be performed on the ellipsometric spectra to deduce the wavelength-independent structural parameters (e.g. thicknesses and void volume fractions) that characterize the nucleation and growth behavior. An inspection of the quality of the LRA best fit as a function of time of during growth establishes the minimum number of layers needed to characterize the structural evolution under different deposition conditions and over different growth regimes. For example, for optimum plasma-enhanced CVD a-Si:H deposited on native oxide-covered c-Si at 250°C, a one-layer model is appropriate in the cluster stage, but after the first bulk density monolayer forms, a two-layer model (bulk/surface) is required. It is found that the initial nuclei form as disc-shaped structures with a nucleation density of 6×10^{12} cm^{-1}. The nuclei make full contact and the first bulk monolayer is formed after a nuclei height of 20 Å. The resulting 20 Å of surface modulation, that remains after bulk monolayer formation, relaxes (or smoothens) by 8 Å in thickness in the first 50 Å of bulk film growth, then relaxes by another monolayer after deposition to a thickness of 400 Å. Such a surface relaxation process indicates microstructural coalescence and a relatively high adatom surface diffusion length. Coalescence is weaker for PECVD a-Si:H at lower substrate temperatures and is absent for sputtered materials. Thus, the ellipsometric studies provide insights into the monolayer scale phenomena responsible for the observed process/property relationships.

For hydrogenated amorphous semiconductors, there are a wide range of possible atomic structures depending on the deposition conditions. As a result, the optical functions at one photon energy are insufficient to establish the parameter of greatest interest, the optical gap. Thus, the spectroscopic probe in this case has major advantages over a single-photon energy probe. This situation is in contrast to that for III-V semiconductor heteroepitaxy, AlGaAs deposition on GaAs, for which the optical functions at one photon energy uniquely define the alloy composition. For such problems, the real time spectroscopic capability is expected to improve accuracy, provide surface temperature in real time, and study proximity effects that lead to a modified band structure in the initial stages of growth. It is expected to play a more important role in studying pseudomorphic growth processes and strained layer superlattices. Initial real time spectroscopic ellipsometry studies of III-V semiconductors have focused on the more complex problem of ion beam etching of heterostructures. In these studies, the instantaneous thicknesses, etch rates, and compositions of the components of the structure can be determined as well as the thickness and composition of the surface damage layer. Once the structure has been etched away then a measurement of the temperature of the top ~400 Å of the surface is possible. The damage layer is of great interest as it provides information on the degree of penetration, the uniformity of etching, and the extent of atomic intermixing as interfaces are crossed. Future advances are required in characterizing the spectroscopic signatures of damage layers so that surface temperature and other more detailed information can be extracted throughout the full etching process.

6. ACKNOWLEDGMENTS

The authors express their gratitude to Professors K. Vedam, R. Messier, and D.L. Miller and Dr. S.S. Bose for their encouragement and participation in this research. We gratefully acknowledge financial support from the National Science Foundation under Grant Nos. DMR-8901031 and DMR-8957159. Assistance was also provided by the Electric Power Research Institute, the Solar Energy Research Institute (subcontract No. XG-1-10063-10), and the State of Pennsylvania under the Ben Franklin Centers of Excellence Program.

7. REFERENCES

An I and Collins R W 1991 *Rev. Sci. Instrum.* **62** 1904.
An I, Nguyen H V, Nguyen N V, and Collins R W 1990 *Phys. Rev. Lett.* **65** 2274.
An I, Nguyen H V, Nguyen N V, and Collins R W 1991 *J. Vac. Sci. Technol. A* **9** 622.
Aspnes D E 1974 *J. Opt. Soc. Am.* **64** 812.
Aspnes D E 1976 *Optical Properties of Solids: New Developments* ed B O Seraphin (Amsterdam: North-Holland) p 801.
Aspnes D E 1981 *Proc. Soc. Photo-Opt. Instrum. Eng.* **276** 188.
Aspnes D E 1988 *Proc. Soc. Photo-Opt. Instrum. Eng.* **946** 84.
Aspnes D E 1990 *Phys. Rev. B* **41** 10334.
Aspnes D E and Studna A A 1975 *Appl. Opt.* **14** 220.
Aspnes D E and Studna A A 1980 *Surf. Sci.* **96** 294.
Aspnes D E and Studna A A 1983 *Phys. Rev. B* **27** 985.
Aspnes D E, Theeten J B, and Hottier F 1979 *Phys. Rev. B* **20** 3292.
Aspnes D E, Schwartz G P, Gualtieri G J, Studna A A, and Schwartz B 1981 *J. Electrochem. Soc.* **128** 590.
Aspnes D E, Studna A A, and Kinsbron E 1984 *Phys. Rev. B* **29** 768.
Aspnes D E, Kelso S M, Logan R A, and Bhat R 1986 *J. Appl. Phys.* **60** 754.
Aspnes D E, Quinn W E, and Gregory S 1990a *Appl. Phys. Lett.* **56** 2569.
Aspnes D E, Quinn W E, and Gregory S 1990b *Appl. Phys. Lett.* **57** 2707.
Azzam R M A and Bashara N M 1977 *Ellipsometry and Polarized Light* (Amsterdam: North-

Holland).
Bruggeman D A G 1935 *Ann. Phys. (Leipzig)* **24** 636.
Cody G D 1984 *Semiconductors and Semimetals, Vol. 21 B* ed J I Pankove (New York: Academic) p. 11.
Cody G D, Brooks B G, and Abeles B 1982 *Solar Energy Mater.* **4** 231.
Collins R W 1988 *Amorphous Silicon and Related Materials* ed H Fritzsche (Singapore: World Scientific) p 1003.
Collins R W 1990 *Rev. Sci. Instrum.* **61** 2029.
Collins R W and Yang B-Y 1989 *J. Vac. Sci. Technol. B* **7** 1155.
Collins R W, An I, Nguyen H V, and Gu T 1991 *Thin Solid Films* in press.
Cong Y, An I, Vedam K, and Collins R W 1991 *Appl. Opt.* **30** 2692.
Drevillon B 1989 *J. Non-Cryst. Solids* **114** 139.
Drude P 1889 *Ann. Phys. Chem.* **36** 865.
Erman M, Theeten J B, Vodjdani N, and Demay Y 1983 *J. Vac. Sci. Technol. B* **1** 328.
Forouhi A R and Bloomer I 1986 *Phys. Rev. B* **34** 7018.
Gheorghiu A and Theye M-L 1981 *Philos. Mag. B* **44** 285.
Heyd A R, An I, Collins R W, Cong Y, Vedam K, Bose S S, and Miller, D L 1991 *J. Vac. Sci. Technol. A* **9** 810.
Hottier F and Theeten J B 1980 *J. Cryst. Growth* **48** 644.
Karunasiri R, Bruinsma R, and Rudnick J 1989 *Phys. Rev. Lett.* **62** 788.
Kim Y-T, Collins R W, and Vedam K 1990 *Surf. Sci.* **233** 341.
Knights J C 1980 *J. Non-Cryst. Solids* **35&36** 159.
Laurence G, Hottier F, and Hallais J 1981 *Rev. Phys. Appl. (Paris)* **16** 579.
Lucovsky G, Tsu D V, Rudder R A, and Markunas R J 1991 *Thin Film Processes II* ed J L Vossen and W Kern (New York: Academic) p 565.
Malitson I H 1965 *J. Opt. Soc. Am.* **55** 1205.
Mazor A, Srolovitz D J, Hagan P S, and Bukiet B 1988 *Phys. Rev. Lett.* **60** 424.
Messier R and Yehoda J E 1985 *J. Appl. Phys.* **58** 3739.
Mui K and Smith F W 1988 *Phys. Rev. B* **38** 10623.
Muller R H 1976 *Surf. Sci.* **56** 19.
Muller R H and Farmer 1984 *Rev. Sci. Instrum.* **55** 371.
Nguyen N V, Pudliner B S, An I, and Collins R W 1991 *J. Opt. Soc. Am. A* **6** 919.
Paul W and Anderson D A 1980 *Solar Energy Mater.* **5** 229.
Robertson R and Gallagher A 1986 *J. Appl. Phys.* **59** 3402.
Ross R C, Johncock A G, and Chan A R 1984 *J. Non-Cryst. Solids* **66** 81.
Snyder P G, Rost M C, Bu-Abbud G H, Woollam J A, and Alterovitz S A 1986 *J. Appl. Phys.* **60** 3293.
Studna A A, Aspnes D E, Florez L T, Wilkens B J, Harbison J P, and Ryan R E 1989 *J. Vac. Sci. Technol. A* **7** 3291.
Tauc J, Grigorovici R, and Vancu A 1966 *Phys. Status Solidi* **15** 627.
Theeten J B, Hottier F, and Hallais J 1979 *J. Cryst. Growth* **46** 245.
Thornton J A 1977 *Ann. Rev. Mater. Sci.* **7** 239

Chapter 4

X-ray reflectivity from heteroepitaxial layers

Paul F. Miceli

Bellcore, Red Bank, NJ 07701

ABSTRACT: The ability of x-rays to probe both the surface and interior of solids makes x-ray scattering a critical tool for the study of semiconductor heterostructures and their buried interfaces. Recent work has shown that the extended range reflectivity, which is measured from grazing angles out through the Bragg reflections, can achieve monolayer sensitivity for buried interfaces. The GaAs/ErAs/GaAs(001) epitaxial system displays rich morphological behavior and demonstrates the power of such measurements which reveal pin-holes for 2 atomic layers (AL) of ErAs, atomically uniform films at 5 AL, Gaussian thickness fluctuations for intermediate thickness ErAs and ultimately exponential thickness fluctuations resulting from lattice relaxation for the thickest films. I address the concepts underlying interface sensitive scattering with a focus on specular reflectivity and the models used to interpret the data. Of particular importance is the inclusion of atomically discrete interface fluctuations as well as a proper treatment of the diffuse scattering which is a necessary consequence of disorder.

1. Introduction

Artificially layered materials that are grown epitaxially have provided new challenges to x-ray scattering, which has always played a vital role in understanding the structure of solids and their attendant defects. The structural characteristics of the resulting buried interfaces are important for controlling their growth, physical and electrical properties, as well as determining their performance in technological applications. As a non-destructive probe, x-rays can penetrate deep into the sample, however, the surface/interface sensitivity of x-rays is well documented. Indeed, ever since the first experiments by Marra, Eisenberger and Cho (1979) which demonstrated depth sensitive scattering on the Al/GaAs buried interface, there has been an explosion of x-ray work investigating surfaces and buried interfaces (eg. recent conference proceedings: Bienfait and Gay 1989 and Zabel and Robinson 1992). This work has benefited greatly from synchrotron x-ray radiation sources which provide the necessary brightness to study the relatively weak scattering coming only from the surface or interface atoms.

Today, grazing incidence x-ray diffraction (GIXD) to investigate surface and interface crystallography is virtually routine and there are a number of reviews on GIXD:

Fuoss, Liang and Eisenberger (1992), Robinson (1991), Fuoss and Brennan (1990), Feidenhans'l (1989), Als-Nielsen (1987) and Nielsen (1985). Although it is a microscopic probe, high resolution x-ray scattering can also provide morphological information. Examples are faceting (Ocko 1988), islanding (Williams 1991, Miceli 1992), atomic surface steps (Renaud 1992) and surface roughening (Held 1987, 1989, Mochrie 1987, Liang 1987). Thus, x-ray scattering experiments cover a broad range of length scales: from less than 10^{-3}Å, the accuracy to which interatomic distances can be determined, to over 10^4Å where long range correlations can be investigated.

X-ray scattering offers advantages over traditional electron diffraction methods because it has superior resolution and the measurements can be interpreted using a simpler kinematic theory due to the weaker interaction with matter. Buried interfaces, however, can only be probed by x-rays because electrons would be absorbed. Transmission electron microscopy can give local information but does not give a sample averaged result and one must make a sample cross section which is not only destructive to the sample but the cross sectioning itself may alter the results. Certainly, electron microscopy, ion scattering and various spectroscopies (Feldman 1986) can provide valuable structural information. However, x-ray scattering gives the Fourier transform of the pair correlation function, which is the quantity of interest for structure determination (Lipson 1966, Stout 1968).

In this review, I will discuss some of the underlying concepts involved in surface sensitive x-ray scattering and then focus on specular reflectivity as it pertains to heteroepitaxial layers of interest to the semiconductor field. Specular reflectivity that is conventionally performed at low angles gives information on density variation and interfaces (Als-Nielsen 1987) but with length scales typically no shorter than ~5Å because of the low angles. Shorter length scale information can only be obtained by measuring the extended range specular reflectivity -- ie. out through higher angles, between and perhaps including Bragg reflections. The measurements are independent of the in-plane structure and all layers in the system contribute. Interpretation of these data must include interface fluctuations on an atomic scale and at every interface in the laminar structure. Specular reflectivity provides a very simple but exceedingly effective way to obtain a large amount of information. For example, one can obtain layer thicknesses, thickness fluctuations, strain, degree of crystallinity and detect the presence of reacted layers. Extremely quantitative information is also available if careful modeling is performed. I will present applications which demonstrate the methods as well as discuss models used to interpret the data.

The organization of the paper is as follows. As an introduction, the dielectric nature of scattering at low angles will be discussed followed by a look at crystal truncation rods which are a natural extension of this when crystal symmetry is included. The remainder of the paper will focus on specular reflectivity which includes interference effects near Bragg reflections as well as the extended range reflectivity and examples are given for determining growth morphology starting at atomic scale thicknesses. Finally, a discussion is presented on the lateral correlations of disorder which affect the line shape transverse to the rod of reflectivity and, consequently, an extended range reflectivity measurement.

2. Reflectivity

The geometry for specular reflectivity is shown in the inset of fig. 1. At low angles, the scattering can be understood "optically" as electromagnetic radiation reflecting from the surface of a homogeneous dielectric medium. Because of the small momentum transfer, $Q_z=(4\pi/\lambda)\sin(\theta)$, long length scales ($d=2\pi/Q_z$) are probed and the scattering is insensitive to crystallinity.

Since x-ray energies are above all or most of the absorption resonances, the refractive index for x-rays is slightly less than unity, $n = 1 - \delta$, where (Agarwal 1979, James 1965),

$$\delta = \lambda^2 \frac{r_e \rho}{2\pi}[f + \Delta f' + i\Delta f''] \quad , \tag{1}$$

and is typically of the order $\sim 10^{-6}$. Here, λ is the x-ray wavelength, r_e is the classical electron radius, ρ is the atomic number density, and f is the atomic form factor. $\Delta f'$ and $\Delta f''$ are the energy dependent real and imaginary anomalous dispersion corrections to f due to the presence of absorption edges, away from which these corrections are usually small. Substantial changes in the form factors can be achieved, however, using the energy tunability at synchrotron radiation facilities and this has allowed the determination of complex structures through anomalous dispersion (Karle 1989). It was recently demonstrated that anomalous x-ray reflectivity can be used to obtain model-independent density profiles in thin films (Sanyal 1992).

Although δ is very small, refraction effects become significant at low angles (Miceli 1986), since,

$$\sin^2(\theta) = \sin^2(\theta') + 2\delta \quad , \tag{2}$$

where θ' is the internal angle. Furthermore, total external reflection will occur below the critical angle, $\theta_c=\sqrt{2\delta}$. From Maxwell's equations, one can obtain the Fresnel coefficients (Jackson 1975a) which describe the transmission and reflection at a single dielectric interface. The Fresnel reflectivity is given as,

$$R_F = \left|\frac{Q_z - Q_z'}{Q_z + Q_z'}\right|^2 \quad , \tag{3}$$

where Q_z' is the momentum transfer inside the dielectric medium determined according to eq. (2). As shown by the solid curve in fig. 1, when $Q<Q_c=\frac{4\pi}{\lambda}\theta_c$ total reflection, $R_F=1$, occurs and far from Q_c, $R_F \sim (\frac{Q_c}{2Q})^4$.

The reflectivity calculated according to Maxwell's equations (often called "dynamical scattering") is rigorously correct, however, when more complex density profiles are desired this method may be more tedious. In the limit of weak scattering a kinematical calculation can be used which is derived from the Born approximation[1]

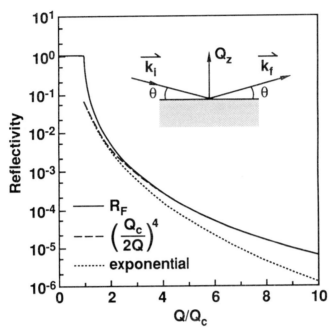

Figure 1. Reflectivity from a homogeneous dielectric. The solid curve is the Fresnel reflectivity for an abrupt surface and the dashed is calculated from the kinematic approximation, showing that it approaches the Fresnel result when $Q \gg Q_c$. The dotted curve is for an exponentially terminated surface and deviates from an abrupt surface at large Q.

and gives the specular ($Q_{x,y}=0$) reflectivity (Als-Nielsen 1987),

$$R(Q_z) = \frac{P(\theta)(Q_c/2)^4}{Q_z^2 \rho_0^2} \, |\int dz \, \rho(z) \, e^{iQ_z z}|^2 \quad , \tag{4}$$

where $P(\theta)$ is the polarization factor (Warren 1969) and ρ_0 is the average electron density. This is a significant simplification since any density profile, $\rho(z)$, can be calculated. For a sharp interface, the integral gives ρ_0/iQ_z and the kinematic calculation, shown by the dashed curve in fig. 1, approaches R_F at large Q ($P(\theta) = 1$ in eq. (3)). As can be seen from fig. 1, the validity of this approximation is good as long as Q is sufficiently far from Q_c, or equivalently, the reflectivity is not too large.

Density variations at the sample surface will affect the shape of the reflectivity. The dotted curve in fig. 1 shows the reflectivity calculated using eq. (4) for an exponentially decaying surface density with a decay length of $0.2/Q_c$ which would be

---

1. The reflectivity is related to the differential scattering cross section (Jackson 1975b), $\frac{d\sigma}{d\Omega}$, through

$$R = \frac{\sigma}{A_{inc}} = \frac{1}{A_{inc}} \int \frac{d\sigma}{d\Omega} \frac{d^2\vec{Q}_\parallel}{k^2 \sin(\theta_2)}$$

where A_{inc} is the incident beam area, $k=2\pi/\lambda$, θ_2 is the angle of the scattered beam, and $d^2\vec{Q}_\parallel$ is an integration over in-plane momentum transfer.

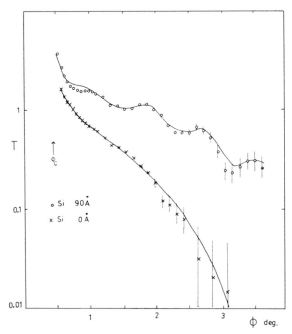

Figure 2: Reflectivity multiplied by the fourth power of the scattering angle, $T=R\phi^4$, for Si wafers having ~90Å oxide and native oxide (from Cowley and Lucas 1989).

~5Å for Ge. Significant deviation from the reflectivity expected for a flat surface is observed at large Q, demonstrating that reflectivity measurements are quite surface sensitive.

For measurements confined to $Q\ll2\pi/d$ (d=interatomic spacing) the scattering will not depend on the detailed interatomic arrangements and exact dynamical calculations can be done assuming a laminar sample is a stratified collection of dielectric slabs. Parratt (1954) developed a recursion algorithm where boundary conditions are matched at each interface. In principle, any density profile can be modeled in this way since an arbitrary number of thin dielectric slabs can be used in the calculation (Ankner 1992).

As an example, fig. 2 shows the reflectivity multiplied by the fourth power of the scattering angle for SiO_2/Si (Cowley 1987, 1989). The upper data is for a Si wafer heated in oxygen and distinct intensity oscillations appear as the result of a 98Å layer of SiO_2 with sharp interfaces. The solid line is a least squares fit to a two layer model (overlayer and semi-infinite substrate) based on Parratt's method. No detectable roughness was found at the inner interface and 5Å rms roughness was found at the top surface. More recent grazing incidence diffraction studies (Renaud 1991) have shown that an interfacial phase of ~5Å thickness occurs at the SiO_2/Si interface. Since the density difference between the interfacial phase and Si or SiO_2 is not large and the reflectivity data does not go out to high enough Q, one would not expect to detect such a phase in this reflectivity data. The lower data in fig. 2 shows a wafer with native oxide. A three layer model was required to fit the data where, in addition to 16Å of SiO_2, a top "contaminating" water layer was needed.

Similar measurements can be done using thermal neutrons (Hayter 1981, Majkrzak 1990). Since neutrons scatter from nuclei, light elements are more easily observed than by x-rays and isotopic substitutions can provide contrast, as for example H and D in polymers (Russell 1990) and other hydrogenous materials.   Also, neutrons carry a magnetic moment and scatter from magnetically polarized electronic shells. The utility of such measurements for magnetic films (Felcher 1981, Majkrzak 1991, 1986) as well as superconductors (diamagnetic) (Felcher 1984, Mansour 1989) has been demonstrated.

When Q becomes large, atomic arrangements must be considered. Dynamical calculations near Bragg reflections from ideal crystals (Zachariasen 1945) and strained crystals (Takagi 1962, Taupin 1964, Bartels 1986) are well known, however, these cannot be solved for general types of disorder such as crystalline interface fluctuations. For a thin film system two factors will work in favor of kinematic theory: the scattering power of a sufficiently thin film will be very low and the intensity oscillations will be more slowly varying with Q, allowing measurements to be made farther away from regions of dynamical scattering. Thus, as long as θ is several times θ_c (or a Darwin width in the case of a substrate Bragg peak) eq. (4) is adequate and this will be a crucial approximation for describing disorder phenomena in buried layers and interfaces.

3. Crystal Truncation Rods

Since a surface is an entity which is localized in real space (surface normal along $\hat{z}$), it will produce a rod of scattering in reciprocal space along Q_z. For example, the crystallography of a reconstructed surface is revealed by the positions and integrated intensities of these rods which occur for fractional order in-plane reflections. There have been numerous GIXD studies of reconstructed clean bulk crystal surfaces of semiconductors such as Ge(001) (Eisenberger 1981), Ge(111) (Feidenhans'l 1988), Si(111) (Robinson 1988b), InSb(111) (Bohr 1985), GaSb(111) (Feidenhans'l 1987), and GaAs(001) (Sauvage-Simkin 1989), as well as metals such as Au(001) (Mochrie 1990), Au(110) (Robinson 1983) and Pt(110) (Vlieg 1990). Scans along the rods will show little intensity variation unless there are vertical displacements within the surface structure.

Using translational symmetry as a natural extension of the above specular reflectivity discussion, each bulk Bragg reflection (ie. integer order) will also exhibit scattering that is extended along Q_z. These are known as crystal truncation rods (Andrews 1985, Robinson 1986a) and differ from fractional order rods in that they arise from the termination of a bulk crystal lattice and, therefore, exhibit singularities along Q_z at the bulk Bragg positions. Scans along crystal truncation rods reveal vertical displacements or non-ideal surface termination. Many of these surface/interface crystallography problems have been reviewed by Feidenhans'l (1989).

The situation for a thin overlayer incommensurate with a substrate is schematically represented in fig. 3 where crystal truncation rod scattering is observed for both the thin film and substrate. Because of the incommensurate lattices, the film and

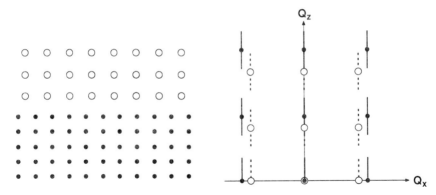

Figure 3: Schematic showing crystal truncation rod scattering (right) for a thin film and substrate which have incommensurate lattices (left). Solid circles refer to the substrate and the open circles refer to the film. Q_z is along the surface normal.

substrate truncation rods occur at different positions of Q_x. Therefore, a non-specular reflectivity measurement along a crystal truncation rod with the in-plane momentum transfer fixed at an in-plane Bragg position, $\vec{G_{\parallel}}$, (where $\vec{Q}=\vec{G_{\parallel}}+Q_z\hat{z}$) can yield information on the structure and termination properties of a *specific* buried layer. Other layers in the heterostructure which have a different in-plane lattice parameter or are amorphous will not contribute to that particular truncation rod.

The reflectivity along a crystal truncation rod for a semi-infinite crystal is given by the Fourier transform of the density[2] in eq. (4) which is now a sum over lattice planes (spacing d) plus an additional term that accounts for whatever non-ideal termination may exist,

$$R(Q_z) = \frac{P(\theta)(Q_c/2)^4}{Q_z^2} \, |F(\vec{Q}) \, d \sum_{n=0}^{-\infty} e^{i\,Q_z\,d\,n} + d \sum_{j=1}^{\infty} \tau_j \, e^{i\,Q_z\,z_j} |^2 \quad , \tag{5}$$

where $F(\vec{Q})$ is the structure factor, z_j and τ_j are the coordinates and occupancies, respectively, of any additional overlayers. For a perfectly terminated surface, the second term is zero, yielding,

$$R(Q_z) = \frac{P(\theta)(Q_c/2)^4}{Q_z^2} \, |F(\vec{Q})|^2 \, \frac{(d/2)^2}{\sin^2(Q_z d/2)} \tag{6a}$$

$$= \frac{P(\theta)(Q_c/2)^4}{Q_z^2} \, |F(\vec{Q})|^2 \, | \sum_{l=-\infty}^{\infty} \frac{(-1)^l}{Q_z - \frac{2\pi}{d}l} |^2 \quad . \tag{6b}$$

This gives Bragg reflections at $Q_z d=2\pi$ with $1/q_z^2$ singularities where $\vec{q}=\vec{Q}-\vec{G}$ is a

---

2. We may continue to use eq. (4) but with the understanding that the scattering density specifically has an in-plane periodicity corresponding to $\vec{G_{\parallel}}$. Also, the factor of $1/Q_z^2$ in eq. (4) arises from geometrical considerations and is valid for non-specular scattering (ie. $Q_x \neq 0$) as long as $\vec{k_i}$ and $\vec{k_f}$ make the same angle to the sample surface. The general form of this geometric factor is given by Gibbs et. al. (1988).

reduced wavevector measured from a Bragg position. When Q_z approaches zero, the Q_z^{-4} dependence discussed earlier in connection with the reflectivity from a smooth homogeneous dielectric is recovered.

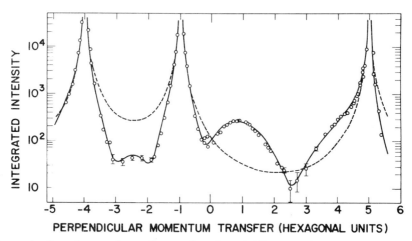

Figure 4: Crystal truncation rod scattering along $(10l)_h$ as a function of l from the α–Si/Si(111) buried interface. The dashed curve is calculated for a perfectly terminated surface and the solid curve is a best fit according to the stacking fault model, as discussed in the text (from Robinson et. al. 1986b).

The application of crystal truncation rods to buried semiconductor interfaces has been demonstrated by Robinson et. al. (1986b) where the SiO_2/Si and α-Si/Si (111) interfaces were investigated. Fig. 4 shows their data taken along the $(10l)_h$ truncation rod for α-Si/Si and the dashed curve indicates the expected behavior for an ideally terminated interface, according to eq.(6).[3] Hexagonal crystallographic units are used where, $\frac{1}{3}(111)_c=(00l)_h$, $\frac{1}{3}(42\overline{2})_c=(100)_h$ and $\frac{1}{3}(22\overline{4})_c=(010)_h$. Negative l is obtained in fig. 4 from the $(01l)_h$ rod since symmetry considerations require $(10\overline{l})_h\equiv(01l)_h$. The logarithmic intensity scale makes it quite apparent that the interface is not ideally terminated. Their proposed interface structure originates from the observation that broad features occur for $l = +1, +4$ and that Bragg reflections occur for $l = -1, -4$, suggesting a symmetry reversal along l. The data could be described (solid curve) by a model which has two coexisting regions (three monolayers in thickness), one with normal stacking and one with a stacking fault atop the dimer layer. This stacking fault finds a natural explanation from the clean Si(111) 7x7 reconstructed surface where such a reversed layer occurs and, evidently, the low temperature preparation of the α-Si does not completely destroy the symmetry of the reconstructed surface. Similar measurements (Robinson 1986b) on the SiO_2/Si interface show no such stacking fault behavior, consistent

---

3. Note that $F(\vec{Q})\to 0$, thereby canceling singularities in eq.(6) for all l except $l = 5, -1, -4$ where Si Bragg reflections occur.

with the oxidation process. Ag deposited on Si(111) at room temperature is found to preserve the 7x7 surface (Hong 1992) but Cu on Si(111) forms Cu_3Si (Walker 1991). GaAs/Si(001), which has an interface consisting of a grid of misfit dislocations, has also been investigated (Specht 1991).

In these examples, it is clear that the scattering far from the Bragg reflections is sensitive to the detailed crystal termination. This property has long been exploited in reflection high energy electron diffraction (RHEED) where intensity oscillations are observed during layer by layer epitaxial growth. Similar oscillations have recently been demonstrated (Vlieg 1988, Van Silfhout 1989) using x-rays which has the advantage of allowing quantitative analysis. Fig. 5 shows x-ray specular reflectivity data (Van Silfhout 1989) measured during the homoepitaxial growth of Ge(111). The intensity oscillates with the period of a monolayer and the "peak to valley" intensity ratio is largest near L=0.5. The solid curves are a fit to a model which allows for the fractional coverage of three atomic layers. X-rays can also be used in non-vacuum situations such as in organometallic vapor phase epitaxy, as demonstrated by Fuoss et. al. (1989).

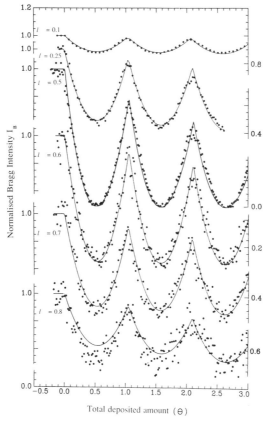

Figure 5: Specular reflectivity measured during the homoepitaxial growth of Ge(111) for various perpendicular momentum transfers (in hexagonal reciprocal lattice units). Solid curves are calculated from a model which allows for three incomplete atomic layers (from Van Silfhout et. al. 1989).

4. Specular and Non-Specular Reflectivity

The simple, idealized situation presented in fig. 3 suggests that non-specular reflectivity can independently probe each layer. By comparison, the specular reflectivity which is measured with $Q_x=0$ has all of the layers contributing to the total scattering amplitude *regardless of their in-plane crystal structure*. In this case, interface fluctuations and any other disorder present in each layer must be accounted for.

The non-specular reflectivity from a heteroepitaxial structure, however, is not as simple as implied by fig. 3. For example, interface fluctuations from preceding layers will propagate atomic displacements, thereby affecting the non-specular reflectivity from the layer of interest. If some of the layers are pseudomorphic (commensurate in-plane) a non-specular reflectivity measurement (ie. along a now co-linear set of crystal truncation rods) will exhibit interference effects from neighboring truncation rods. Layers that are incommensurate may have an interfacial region which is disordered or has a different (or perhaps as of yet unknown) in-plane structure. This may contribute to diffuse scattering or scattering somewhere else in reciprocal space so that rod scans will reflect a crystalline thickness rather than the compositional thickness of that layer.

These points illustrate the richness of specular and non-specular scattering experiments. In general, several different x-ray scattering measurements may be needed depending on the specific information desired for that system. In the following, we focus on the specular reflectivity which includes interference effects, interface fluctuations and a discussion of diffuse scattering. These considerations are generic in that they are fundamental to the non-specular reflectivity as well. It is worth pointing out that specular reflectivity is the starting point for most investigations as it can provide a significant amount of useful information.

5. Bragg Reflectivity

Crystal truncation rods from a semi-infinite crystal were discussed above. It is also useful to consider the scattering from a thin crystal which is obtained by summing the contribution from each of N atomic planes with spacing d (Warren 1969):

$$R \propto | \sum_{n=1}^{N} e^{i Q_z d n} |^2 = \frac{\sin^2(NQ_z d/2)}{\sin^2(Q_z d/2)} \quad . \tag{7}$$

The finite thickness gives scattering which is extended in reciprocal space and the result is essentially that obtained for the truncation rod of a single perfect interface (eq. 6a) multiplied by an oscillatory factor due to the interference resulting from two interfaces. The period of oscillation is $\Delta Q \sim 2\pi/Nd$ where Nd is the thickness of the crystal.

As an example, fig. 6 shows (002) Bragg reflectivity data for 140Å of ErAs grown on GaAs(001). The sample has a 40Å cap layer of highly textured Al(111) which does not contribute significantly in this data range. In addition, the zinc-blende structure factor for the (002) GaAs reflection ($2\Theta = 14.415°$) is very weak. Therefore, the ErAs scattering dominates and the data in fig. 6 are essentially that of a free standing thin

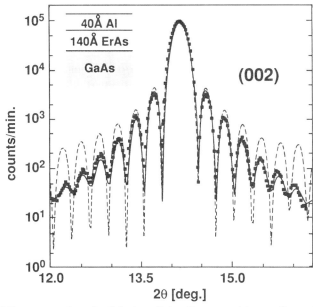

Figure 6: (002) Bragg specular reflectivity from ErAs. Because of the small contribution from the GaAs and Al scattering, this data is essentially that of a free standing ErAs film. Dashed curve is calculated from eq. (7) and intensity oscillations occur due to the finite thickness. The solid curve is a fit based on a binomial film thickness fluctuation model discussed in the text, giving 2.5 atomic layers for these fluctuations (after Miceli et. al. 1991).

crystal. The dashed curve is calculated from eq. (7) and, although it reproduces the period of oscillation, the rate of intensity decrease is poorly represented. Fluctuations in film thickness can entirely account for this discrepancy as shown by the solid curve which is calculated using binomial (essentially Gaussian) fluctuations, as discussed in the next section. The magnitude of the fluctuations is quite small, ~2.5 monolayers, demonstrating that data such as these are exceedingly sensitive to fluctuations.

In heteroepitaxial systems, interference effects between the various layers are important, particularly when Bragg positions are close or when there is significant overlap of the truncation rod scattering. Fig. 7 shows results from Tapfer and Ploog (1989) for a quantum well structure (see inset) consisting of two $Al_{0.4}Ga_{0.6}As$ cladding layers each having the same thickness of 1.06μm separated by 50Å of GaAs. Because of the small difference in scattering contrast between the layers, intensity oscillations with a period corresponding to the total thickness appear about the Bragg position of the cladding layers (E). Insertion of the 50Å GaAs layer, however, causes a phase shift which then modulates the reflectivity and, in this case, every other interference fringe alternates in intensity due to the equal thickness of the cladding layers. Since the layers are rather thick, the authors used a dynamical theory. The differences between data and the calculation are probably due to resolution effects which were not convoluted with the calculation.

When the total thickness of the layers is comparable to the extinction length,

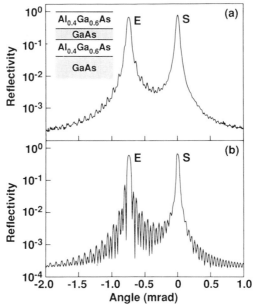

Figure 7: (a) Experimental and (b) simulated (004) Bragg reflectivity for two 1.06μm $Al_{0.4}Ga_{0.6}As$ cladding layers separated by 50Å of GaAs and grown on a GaAs substrate. (E) indicates the epilayer and (S) is the GaAs (from Tapfer and Ploog 1989).

dynamical scattering effects become important. Dynamical models for Bragg reflections in heteroepitaxial systems have been considered by a number of authors (Bartels 1986, Berreman 1976, Chu 1986, Fewster 1987, Hill 1985, Macrander 1988, Tapfer 1989, Takagi 1962, Taupin 1964), although, interface fluctuations have not been considered. Dynamical theory for a thin crystal predicts intensity oscillations, known as Pendellosung fringes, which have been studied for both the x-ray Laue (transmission) (Kato 1959) and Bragg (reflection) (Batterman 1968, Renninger 1968) cases as well as for the neutron Laue case (Shull 1968). These oscillations have essentially the same origin as those in the kinematical theory except that in the dynamical regime the reflectivity depends on both the thickness and the structure factor (Zachariasen 1945). Shull (1968) used this effect to accurately determine the neutron cross section of Si. Macrander (1988) investigated strain-graded semiconductor heterostructures and observed interference fringes resulting from the strain gradient. Although a dynamical model was used, the essential interference features can be understood from kinematical considerations (Macrander 1988). Crystalline garnet films strained by ion-implantation have also been studied by Speriosu (1981) using a kinematical model which was shown to be valid for reflectivities below 6%.

Another example of interference in Bragg reflection is shown in fig. 8 for 33Å of $NiSi_2$ on Si(111) (Robinson 1988). The scattering amplitudes of the substrate and film (shown independently in fig. 8a) will have maximum interference when their amplitudes are comparable. The interference depends on the interfacial spacing, d, so that the total scattering amplitude is, $F_{tot}=F_{Si}+e^{iQ_z d} F_{NiSi_2}$. As shown in fig. 8b, the scattering is symmetrical when d=a, where a is the substrate lattice spacing which is approximately the same for the NiSi$_2$. Changing the d/a ratio from unity introduces an asymmetry and allows an accurate determination of the interfacial spacing. A best fit to the data in fig. 8c, which also includes the first pair of interference fringes, gives d/a = 1.1.

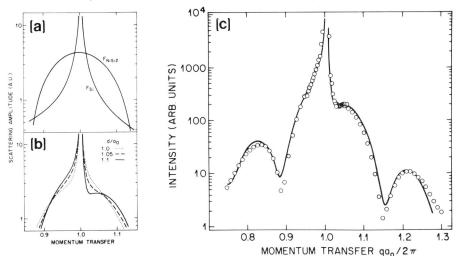

Figure 8: Reflectivity near the (111) position for NiSi$_2$/Si(111). (a) Scattering amplitudes shown separately. (b) Total amplitude for different interfacial separation, d, between NiSi$_2$ and Si. (c) Bragg reflectivity data (circles) and model (solid curve) for d/a = 1.1 (from Robinson et. al. 1988).

6. Extended Range Specular Reflectivity

In this section a model for the specular reflectivity is presented. These ideas are demonstrated through the study of the growth morphology of the epitaxial system, ErAs/GaAs(001) (Palmstrøm 1988, 1989, 1990). Measurements performed over the extended range of specular reflectivity, that is, measurements from grazing angles through the regions of Bragg scattering at higher angles, allow one to obtain information on short length scales which is particularly useful for ultra-thin heteroepitaxial layers (Miceli 1992). The interpretation of these measurements require careful consideration of two points: interface fluctuations and the scattering transverse to the rod of reflectivity. Regarding the latter, a two component line shape, delta and diffuse scattering, is generally observed and the relative contributions are expected to change with Q_z, as discussed in the next section. The model presented below assumes that only the delta component is measured everywhere along the extended reflectivity. Furthermore, it is essential that a crystalline interface is modeled by discrete fluctuations (Chrzan 1986, Clemens 1987, Fullerton 1992, Lucas 1988, Miceli 1992) since these account for the translational symmetry intrinsic to the disorder and provide a correct description of the intensity decay away from the Bragg positions.

The specular reflectivity can be calculated, in the kinematical approximation, by summing the contributions to the scattering from each layer in the system,

$$R_{Spec} = P(\theta) \frac{16\pi^2 r_e^2}{Q_z^2} |<A>|^2 = P(\theta) \frac{16\pi^2 r_e^2}{Q_z^2} |\sum_{l=0}^{L} <A_l e^{i\phi_l}>|^2, \tag{8}$$

where A_l is the scattering amplitude of the lth layer ($l = 0$ is the substrate), $\phi_l = Q_z \sum_{j=1}^{l-1} t_j$ is the phase shift to that layer and t_j is the thickness of a layer. $<...>$ indicates an average over the in-plane dimensions of the film, which can be accomplished by considering the vertical configuration of the heterostructure at a given (x,y) position and then averaging over all possible configurations (which are discrete in the case of a crystalline layer). Notice that the reflectivity is given as the square of the average amplitude rather than an average of the amplitude squared. As discussed in the next section, this is a consequence of measuring only the delta component (diffuse scattering is subtracted) which arises from in-plane long range order.

If the fluctuations are statistically independent between different layers then,

$$<e^{i\phi_l}> = \prod_{j=0}^{l-1} V_j, \tag{9}$$

where,

$$V_0 \equiv <e^{i Q_z h_0}>, \quad V_l \equiv <e^{iQ_z t_l}>.$$

$l=0$ denotes the substrate with h_0 describing the substrate surface height fluctuations.

The average scattering amplitude, $<A_l>$, must be evaluated for each layer in the system and for a crystalline layer it is given by,

$$<A_l> = \rho_l d_l f_l \frac{V_l - 1}{1 - e^{-i Q_z d_l}}, \tag{10a}$$

where f_l is an atomic or molecular form factor, ρ_l is the average number density, and d_l is the lattice constant (along $\hat{z}$) of the lth layer. The thickness, t_l, is $d_l N_l$ where N_l is the number of atomic planes. For an amorphous layer the average scattering amplitude is given by

$$<A_l> = \rho_l f_l \frac{V_l - 1}{i Q_z}. \tag{10b}$$

For a crystalline layer containing a random distribution of point defects (where the scattering by defects themselves is neglected; eg. H in metal), the average scattering amplitude is given as,

$$<A_l> = \rho_l d_l f_l \psi \frac{V_l - 1}{\psi - 1}, \tag{10c}$$

where $\psi \equiv <e^{iQ_z d}>_d$ is an average over the possible d spacings and $V_l = <\psi^N>$. In principle, the average scattering amplitude for any type of layer can be calculated numerically, as for example, a linear strain gradient.

The evaluation of eq. (8) is now very simple since the effect of fluctuations reduces to one average phase factor, V_l, for each layer in the heterostructure. For a crystalline layer, this average must be performed over all possible discrete layer thicknesses,

$$V_l^{cryst} = <e^{i\,Q_z\,d_l\,N_l}> = \sum_{n=0}^{\infty} P_n^l e^{i\,Q_z\,d_l\,n} \quad , \tag{11}$$

where P_n^l is the probability to find n atomic planes in the lth layer. Because eq. (11) is a discrete Fourier transform, the translational symmetry of a buried crystalline layer is preserved. Furthermore, any laminar system consisting of amorphous and crystalline layers can be properly handled in this formalism and the reflectivity can be calculated for an arbitrary range of Q_z. It only remains to specify the form of the interfaces which are given by P_n^l. For a few atomic layers, the P_n can be independently specified. When there are many atomic layers, both binomial and exponential fluctuations will be considered.

Binomial fluctuations imply that an atomic layer will be completed with some probability, p, or it will not with probability, 1-p, and that the atomic layers behave independently. This is akin to layer by layer growth and, in the limit of many atomic layers, a Gaussian interface will be obtained. Evaluating eq. (11) for the binomial coefficients gives,

$$V_l = [(1-\sigma_l^2/N_l)e^{i\,Q_z\,d_l} + \sigma_l^2/N_l]^{\frac{N_l^2}{N_l - \sigma_l^2}} \quad , \tag{12}$$

where N_l is the average number of layers and σ_l^2 is the variance. The solid curve in fig. 6 was obtained from a least squares fit to this model and provides a good description of the data.

Exponential fluctuations have been observed at the bulk crystal surfaces of Pt, W, and InSb (Robinson 1986a), during homoepitaxial growth of Ge (Vlieg 1988), and in thicker epitaxial layers of ErAs (Miceli 1992). For the bulk crystal surfaces the roughness was relatively small, typically 1 or 2 monolayers, where a new layer begins before the lower layers are completed. In the case of ErAs, thin layers which are pseudomorphic exhibit Gaussian like fluctuations and a transition to exponential fluctuations occurs when there is a high degree of lattice relaxation for thicker layers (~300Å). This indicates that lattice relaxation causes a change in growth morphology to a more extended interface perhaps with a tendency for islanding or faceting. The situation for ErAs was well represented by some number of perfect layers followed by an exponential interface. Exponential fluctuations at a surface can be expressed as $P_n=(1-f)f^n$ giving,

$$V_0 = \frac{1-f}{1-fe^{iQ_z d}} \quad , \tag{13}$$

where f is a parameter between 0 and 1.

An important test of the above ideas lies not only in the ability to give the reflectivity near a Bragg position as in fig. 6, but also to describe the extended range of reflectivity. Fig. 9 shows a scan along [00L] for 35Å of ErAs buried beneath 200Å of α-Si. Intensity oscillations are present both at low and high angles due to the 35Å ErAs layer. The low angle data are sensitive to all of the layers in the system whereas only the crystalline layers contribute to the Bragg reflectivity at high angles.

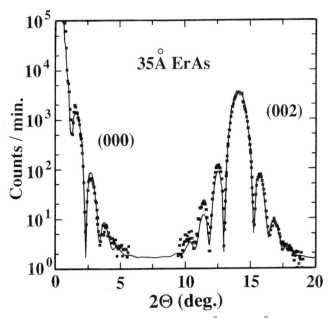

Figure 9: Extended range specular reflectivity for 200Å α–Si / 35Å ErAs / GaAs(001). The model (solid curve) assumes binomial thickness fluctuations and provides a good fit over the extended data range. The interface roughness of the substrate is 1 AL. The film thickness fluctuations are 1 AL for ErAs and 10 Å for the α–Si (from Miceli et. al. 1992).

The model calculation is shown by the solid curve which is the result of a least squares fit using binomial fluctuations for the ErAs layer and Gaussian fluctuations for the amorphous Si cap. A very sensitive measure of the layer roughness is obtained since the damping of the intensity oscillations increases rapidly as one moves away from the Bragg positions. As is evident from the figure, a single curve provides a good fit to the data for *both* low and high angles. In particular, there is no rescaling of the data at low and high angles.

The extended reflectivity has been used to study the growth morphology of ErAs on GaAs (Miceli 1992). Fig. 10 shows reflectivity data collected out through the (004) position for nominally 2 atomic layers (AL) of ErAs buried under 500Å of GaAs. Because of the small quantity of buried ErAs the low angle reflectivity does not directly measure the thin layer; in fact, this portion of the reflectivity could be equally well modeled using two GaAs slabs separated by a gap. In contrast, the GaAs scattering is much reduced at the (002) so that the ErAs layer directly produces a broad feature there. The above model was fitted to the data to obtain the thickness coefficients, P_n: $P_1 = 0.0 \pm 0.02$, $P_2 = 0.04 \pm 0.08$, $P_3 = 0.41 \pm 0.07$, $P_4 = 0.19 \pm 0.07$, $P_5 = 0.07 \pm 0.04$, and $P_n \equiv 0$ for $n \geq 6$. The result is shown by the solid curve which is in excellent agreement with the data over the entire range of reflectivity. The average coverage is given by $\sum n P_n = 2.4$ AL (± 0.3) which is consistent with growth conditions. These results show, however, that the film is predominantly consisting of 3 AL or more, suggesting that there are regions of no coverage ("pin-holes") in the ErAs film. Since the P_n are normalized the uncovered fraction is directly obtained, $P_0 = 0.3 \pm 0.2$. All three orders of measured reflectivity were required to obtain these

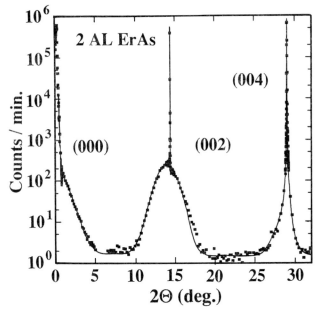

Figure 10: Extended range specular reflectivity for nominally 2 AL of ErAs buried beneath 500Å of GaAs. The solid curve is a fit to the atomic layer model, finding that the ErAs film is discontinuous with predominantly 3 AL islands (from Miceli et. al. 1992).

quantitative results, particularly for P_0.

Fig. 11 shows data for nominally 5 AL of ErAs buried by 500Å of GaAs where there are three subsidiary maxima clearly visible, indicating that the ErAs layer is very uniform. A least squares fit to the data yields (to within ±0.03) $P_3 = 0.01$, $P_4 = 0.41$, $P_5 = 0.55$, $P_6 = 0.03$, and all other $P_n \equiv 0$. The average coverage is 4.6 AL with nearly equal amounts of 4 and 5 AL regions only, indicating that the layer is indeed uniform with single monolayer steps arising from the half monolayer terminated growth.

Summarizing the ErAs/GaAs growth morphology, it is found that films grow with pinholes for 1 or 2 atomic layers of coverage and evolve into continuous, atomically uniform films by ~ 5 AL coverage. This explains the behavior of the reflection high energy electron diffraction (RHEED) intensity oscillations measured during film growth which are weak at 2 AL, but are strongest around 4 or 5 AL. Film growth evolves with Gaussian-like thickness fluctuations (σ typically on the order of 6% of the thickness) until ~300Å where exponential thickness fluctuations occur, signaling a characteristic change in the growth mode morphology which is due to the concomitant lattice relaxation.

It is seen that the growth morphology of epitaxial layers can be determined by extended range reflectivity methods which give structural detail at the sub-monolayer length scale. The ErAs/GaAs system was studied, *ex-situ*, beneath a capping layer using a conventional rotating anode laboratory x-ray source and, certainly, the strong scattering cross section of Er makes this a favorable case. However, *in-situ* and *ex-*

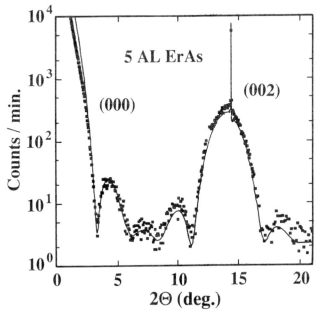

Figure 11: Extended range specular reflectivity for nominally 5 AL of ErAs buried beneath 500Å of GaAs. There are three subsidiary maxima which indicate an atomically uniform ErAs layer. The solid curve is a fit to the atomic layer model discussed in the text (from Miceli et. al. 1992).

situ studies of this type using synchrotron radiation should be a valuable tool for exploring the properties of epitaxial systems.

The final topic for this section will be a special case of interdiffusion between layers. Although the high heat of formation for ErAs prevents interdiffusion in the above examples, some degree of interdiffusion is expected in most heteroepitaxial systems. The crux of the problem is as follows: suppose two adjacent layers have a perfectly uniform total thickness, $t_l + t_{l+1}$, but their adjacent interface is not uniform, then the thickness fluctuations of the individual layers are correlated and the derivation of the above model must be reconsidered. For simplicity, some of the topological problems associated with interdiffusion will be neglected such as cross substitution and interstitials. Without loss of generality, crystalline layers will be considered so that,

$$<A_l\,e^{i\phi_l}> = \rho_l\,d_l\,f_l\ \frac{<e^{i\phi_{l+1}}> - <e^{i\phi_l}>}{1 - e^{-i\,Q_z\,d_l}} \ . \tag{14}$$

Let the number of atomic planes (locally) in the layer be given as $N_l + n_{l,l-1} + n_{l,l+1}$ where the $n_{l,l-1}$ represent interface fluctuations due to interdiffusion and are statistically independent of N_l. Since material conservation requires that $n_{l,l+1} + n_{l+1,l} = 0$, the phase shift is,

$$\phi_l = Q_z\left\{h_0 + \sum_{j=1}^{j=l-1} d_j N_j\right\} + Q_z\left\{\sum_{j=1}^{j=l-1} (d_j - d_{j-1})n_{j,j-1}\right\} + Q_z d_{l-1} n_{l-1,l} \ , \tag{15a}$$

which gives,

$$<e^{i\phi_I}> = W_I \left[\prod_{j=1}^{j=I-1} U_j \right] \left[\prod_{j=0}^{j=I-1} V_j \right] . \qquad (15b)$$

Notice that the factor, $W_I \equiv <e^{iQ_z d_{I-1} n_{I-1,I}}>$, which is due to interdiffusion occurs *non-cumulatively*. This means that it will have no effect on the average phase factor when $k \neq I$ just as expected, since, the total amount of material is conserved for interdiffusion. The other new factor, $U_I \equiv <e^{iQ_z(d_I - d_{I-1})n_{I,I-1}}>$, which does occur cumulatively results from misfit strain induced by interdiffusion. However, this is not formally new since it could have been included in an additional translation, δ_I, between layers I and $I-1$ to accommodate the interface chemistry (eg. fig. 8) with $U_I = <e^{iQ_z \delta_I}>$. Also, any fluctuation in δ_I at the interface will tend to be continuous yielding a Debye-Waller factor attenuation which is monotonic in Q_z.

7. Transverse Scattering

Up to now, the effects of disorder have been considered only for scattering along Q_z, however, the *lateral correlations* of disorder will be revealed along $\vec{Q}_\perp$, transverse to the rod of reflectivity (this would be along Q_x in fig. 3). X-ray scattering from a rough surface has been considered by Sinha et. al. (1988, 1991) and by Andrews and Cowley (1985). Fig. 12 shows a rather typical behavior of this type of scan (inset) for an ErAs sample of 5 atomic layers. There are two components, one which is resolution limited that we will refer to as the "delta" component, and another broad diffuse component. The delta component is often referred to as the "specular" line, although, we will avoid this terminology since a similar two component line shape can occur for non-specular crystal truncation rods as well. Essentially infinite range correlations in the plane of the film are responsible for the delta component whereas the diffuse scattering arises from disorder which is correlated over a finite range. Another feature in fig. 12 is that the diffuse scattering exhibits side peaks called "angel's wings" or "Yoneda scattering" (Yoneda 1963, Sinha 1988, Kortright 1991, Russell 1990). Because the data are taken at rather low Q_z ($2\theta \sim 4.6°$), the incident or outgoing beam can approach θ_c, where the increased electric field at the surface enhances the diffuse scattering. Thus, a dynamical scattering model must be used even for the diffuse scattering when it is measured at low angles and a quantitative fit to a line shape such as in fig. 12 is desired. This has been done quite successfully for liquid surfaces (Sanyal 1991, Schwartz 1990, Tidswell 1991). The general ideas behind the delta and diffuse components can, however, be understood in the context of a kinematic theory which also becomes more rigorous as one goes to larger 2θ where the Yoneda scattering is less important.

To investigate the origin of the two component line shape, consider the differential scattering cross section:

$$\frac{d\sigma}{d\Omega} = P(\theta) \, f^2 \, r_e^2 \, |\int d^3\vec{r} \, e^{i\vec{Q}\cdot\vec{r}} \rho(\vec{r})|^2 \qquad (16)$$

$$= P(\theta) \, f^2 \, r_e^2 \int d^3\vec{r} \int d^3\vec{r}' \, e^{i\vec{Q}\cdot(\vec{r}-\vec{r}')} \rho(\vec{r}) \, \rho(\vec{r}') .$$

Because of the infinite extent of the sample in the plane of the film, $\vec{\xi}_\perp = \vec{r}_\perp - \vec{r}_\perp'$ is translationally invariant[4]. However, the same is not true of $z - z'$ since there are

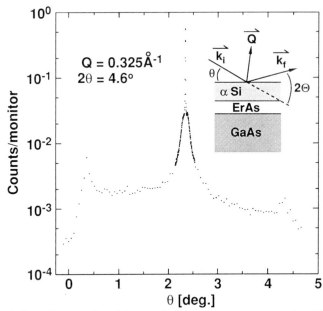

Figure 12: Inset shows the experimental geometry of a transverse scan measured across the rod of specular reflectivity by "rocking" the sample with the detector at fixed 2Θ. There is a resolution limited "delta" component due to long range flatness followed by a broader diffuse component arising from surface/interface roughness. The increased scattering near θ equal to 0 and 2Θ is due to the diffuse scattering which is enhanced by dynamical effects as $\vec{k}_i$ and $\vec{k}_f$ approach θ_c, respectively.

surface and/or interface boundary conditions and eq. (16) becomes

$$\frac{d\sigma}{d\Omega} = P(\theta)f^2 r_e^2 \, A_{ir} \int dz \int dz' e^{iQ_z(z-z')} \int d^2\vec{\xi}_\perp \, e^{i\vec{Q}_\perp \cdot \vec{\xi}_\perp} < \rho(\vec{\xi}_\perp, z)\rho(0, z') > \ , \qquad (17)$$

where <> indicates an in-plane average and A_{ir} is the irradiated area.

As an example, consider a film exhibiting conformal roughness -- that is, it has uniform thickness but "rides" on top of a rough substrate surface. Neglecting the scattering from the substrate eq. (17) becomes,

$$\frac{d\sigma}{d\Omega} = P(\theta)f^2 r_e^2 \, A_{ir} \, | \int dz \, e^{iQ_z z}\rho(0,z)|^2 \int d^2\vec{\xi}_\perp \, e^{i\vec{Q}_\perp \cdot \vec{\xi}_\perp} \, e^{-\frac{1}{2}Q_z^2 <h(\vec{\xi}_\perp) - h(0)>^2} \ , \qquad (18)$$

where h, assumed to be a Gaussian random variable, is the displacement caused by the substrate. At some very long distance the height-height correlation function, $\sigma(\vec{\xi}_\perp) \equiv \sqrt{<h(\vec{\xi}_\perp) - h(0)>^2}$, will saturate to a constant value, σ_∞, and eq. (18) becomes,

---

4. For simplicity it is assumed that $\vec{Q}_\parallel$ is near zero. A similar result can be obtained when translational symmetry is included for a crystal lattice.

$$\frac{d\sigma}{d\Omega} = (2\pi)^2 P(\theta) f^2 r_e^2 \, A_{ir} \, | \int dz \, e^{iQ_z z} \rho(0,z) |^2 \tag{19}$$

$$\times \left\{ e^{-\frac{1}{2}Q_z^2 \sigma_\infty^2} \delta(\vec{Q_\perp}) + \frac{1}{(2\pi)^2} \int d^2\vec{\xi_\perp} \, e^{i\vec{Q_\perp} \cdot \vec{\xi_\perp}} [e^{-\frac{1}{2}Q_z^2 \sigma^2(\vec{\xi_\perp})} - e^{-\frac{1}{2}Q_z^2 \sigma_\infty^2}] \right\} .$$

The first term will be observed as a resolution limited (delta) component with its contribution depending on the degree of long range flatness. The second term is explicitly nonsingular and gives the diffuse scattering arising from the lateral short-range correlation of fluctuations. As a direct consequence of conformal roughness, both components are modulated in the same way by the reflectivity along Q_z (eg. intensity oscillations). This fact was exploited in a recent study of liquid layers on a substrate where it was found that thin layers conformed to the substrate roughness but as the layers became thicker the layers fluctuated independently of the substrate (Tidswell 1991).

A liquid surface is described by capillary wave fluctuations which give no true long range order and a delta component arises only from gravitational effects. Quantitative agreement between theory and experiment has been achieved for liquids (Braslau 1988, Sanyal 1991, Schwartz 1990, Tidswell 1991) and liquid crystals (Pershan 1987, Tweet 1990, Holyst 1990). The situation for solid surfaces is not always as clear, since, these can be inherently more complex and may exhibit a variety of phenomena. For example, there can be surface roughening (Held 1987, 1989, Liang 1987), periodically spaced surface steps (Renaud 1992, Alerhand 1990) which will give diffuse scattering from spacing fluctuations, and there can be disorder in surface reconstructions as in the missing row reconstruction of Pt (110) (Fenter 1985, Robinson 1989). Fractal surfaces, which might arise from non-equilibrium situations, will also produce diffuse scattering (Chiarello 1991, Sinha 1988, Pfeifer 1990). In general, eq. (17) is useful as a starting point for any of these situations.

In addition to diffuse scattering arising from surface or interface disorder, it may be seen from eq. (17) that fluctuations in $\rho(\vec{\xi_\perp}, z)$ arising from defects within the bulk can also produce diffuse scattering, particularly as one goes out to higher Q_z near Bragg reflections (Miceli 1991, 1992b). Fig. 13 shows a transverse scan taken across the (002) Bragg reflection from a 140Å film of ErAs (same sample as in fig. 6) where a two component line shape is observed. There is a sharp, resolution limited (~0.002°) delta component which suggests some character of flatness over a correlation range of at least several microns and a broad diffuse component which is due to mosaic-like rotational fluctuations. The latter is inferred from the angular width of the diffuse component that is found to be constant with Q_z (ie. at (004) and (006) etc.). As shown in the inset, the intensity of the delta component decreases with increasing film thickness. Concomitantly, the diffuse intensity increases evolving into a conventional mosaic line shape (Zachariasen 1945) for the thickest films. The thickness dependence of this line shape is a consequence of the misfit dislocation density which increases with film thickness and relieves the 1.6% lattice mismatch between the film and substrate.

In order to obtain detailed information on both interface and bulk disorder it is of interest to study the evolution of diffuse scattering (intensity and line shape) as a function of Q_z. This has not been done to date and it is clear that studies of diffuse

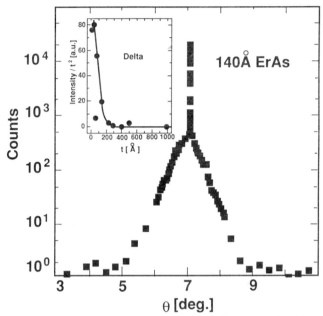

Figure 13: Transverse scan across the center of the (002) Bragg reflectivity (fig. 6) for an ErAs film. A delta component occurs due to long range flatness but the diffuse scattering arises from bulk disorder connected to the mosaic-like rotational fluctuations associated with lattice relaxation. As shown in the inset, the delta component decreases with film thickness due to the increasing dislocation density (from Miceli et. al. 1991, 1992b).

scattering from epitaxial layers will be an important research activity in the future.

Independent of whether the two component line shape arises from bulk or interface disorder it is fair to ask: What impact does such a line shape have on a reflectivity measurement? Typically, when one measures "specular" reflectivity the delta component is obtained by subtracting the diffuse scattering that is derived from a second scan with the sample slightly misoriented. Such data is modeled according to eq. (8). However, it is clear from the above discussion that a delta component is not necessarily present at larger Q_z, although, there could be a measurable (and possibly useful) amount of diffuse scattering -- certainly, this will complicate a measurement of the extended range specular reflectivity. The best way to address this situation is to perform transverse scans at every Q_z to obtain the transversely integrated reflectivity. From eq. (17) it is seen that the average over disorder is done differently for the specular and transversely integrated cases,

$$R_{spec} \propto |< \int dz \; e^{iQ_z z} \; \rho(0,z) >|^2 \qquad (20a)$$

$$R_{int} \propto <| \int dz \; e^{iQ_z z} \; \rho(0,z) |^2 > . \qquad (20b)$$

For a crystalline film these are,

$$R_{spec} \propto |V_0|^2 \frac{|1-V_1|^2}{|1-e^{-i\,Q_z\,d}|^2} \qquad (21a)$$

$$R_{int} \propto \frac{2-V_1-V_1^*}{|1-e^{-i\,Q_z\,d}|^2} \qquad (21b)$$

where the effects of conformal roughness induced by the substrate and the film thickness fluctuations are included in V_0 and V_1, respectively. The substrate scattering has been neglected for simplicity.

Conformal roughness affects R_{spec} only and V_1 enters differently into the two reflectivities. Although the discussion is specific to interface roughness these results also hold for a mosaic film. It can be shown that rotational fluctuations in the bulk of the film are essentially "conformal" in that they affect V_0 only. The film thickness dependence in fig. 13 arises from the diminution of $|V_0|^2$ in eq. (21a).

The difference between R_{spec} and R_{int} depends on the magnitude of the film thickness fluctuations. Neglecting conformal roughness ($|V_0|=1$), the difference is $R_{int}-R_{spec}\propto(1-|V_1|^2)/|1-e^{-i\,Q_z\,d}|^2 \geq 0$, which tends to be larger far from the Bragg position where $|V_1|\ll1$. Since fluctuations generally increase with film thickness, these considerations are particularly important for the extended reflectivity of heteroepitaxial films and perhaps less so for a crystal truncation rod measurement of a buried substrate interface which typically involves only a few atomic layers.

A clear distinction between R_{spec} and R_{int} can be experimentally demonstrated (Miceli 1992). Fig. 14 shows the (002) Bragg reflectivity for a 400Å film of ErAs where the lattice relaxation is 75% complete and the transverse scattering is a mosaic line shape (no delta component observed). Fig. 14 is essentially transversely integrated reflectivity which is trivially obtained from a mosaic line shape. The data are well described by the solid curve which is a fit using R_{int}, exponential thickness fluctuations and a linear strain gradient. The exponential thickness fluctuations arise from a change in growth morphology, as discussed above, and the linear strain gradient results from the incomplete lattice relaxation. By contrast, the dashed curve is calculated from the same model parameters but now using R_{spec}. There is a pronounced asymmetry resulting from the strain gradient which is clearly absent from the data (trying to fit R_{spec} to the data does not improve the situation nor is a satisfactory fit obtained if the strain gradient is omitted). For thinner films where a delta component is present R_{spec} provides a good description of the scattering, such as in fig. 6 which was measured along the delta component of fig. 13. Therefore, it has been explicitly demonstrated that (1) a reflectivity measurement along the delta component is described by R_{spec}, (2) reflectivity which is transversely integrated must be described by R_{int} and (3) a mosaic line shape is intrinsically diffuse scattering.

The stage is now set for more complicated heteroepitaxial systems that are of common practical interest. For example, one may have a substrate and buffer layer which are of high crystalline quality followed by a layer(s) which is somewhat disordered. At large Q_z the disordered layer may not have a delta component, although, this situation could be masked by the delta components coming from the substrate and buffer layer. Therefore, R_{int} should be modeled and measured using

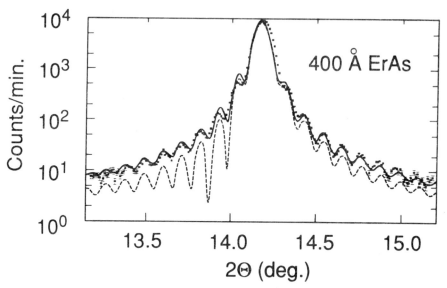

Figure 14: The (002) Bragg reflectivity is shown for a mosaic sample of 400Å ErAs where there is no delta component in the transverse line shape. The solid curve is a fit using the transversely integrated reflectivity, R_{int}, a strain gradient and exponential thickness fluctuations. Using R_{spec}, the reflectivity is calculated for the same structure as shown by the dashed curve (from Miceli et. al. 1992a).

synchrotron radiation (due to the large amount of data and weak diffuse scattering for some regions of Q_z). Moreover, the fact that the extended range reflectivity has been demonstrated for a case which was clearly specular reflectivity gives us some confidence that the extended range reflectivity of more complex systems should be amenable to this type of analysis.

8. Concluding Remarks

It has been shown that x-ray scattering can provide a variety of useful structural information on heteroepitaxial systems ranging from the crystallography of buried interfaces to the overall film morphology. As a step beyond the electron diffraction techniques that are traditionally used in UHV film growth chambers, the marriage of synchrotron radiation with film deposition systems can have a profound impact on our understanding and applications of film growth in the future. It is possible, for example, to determine whether a buried interface reconstructs during deposition, if there is interdiffusion between layers, or if the layers are growing to the correct thickness and uniformly -- without ever having to remove the sample from the growth chamber. More importantly, such an *in-situ* arrangement would allow one to investigate the conditions (temperature, pressure etc.) under which certain interface structures, reactions, islanding (wetting) and faceting occur. The virtue of x-rays is their unique range of surface and bulk sensitivity as well as the relatively weak interaction which allows a quantitative interpretation of experiments using kinematic scattering theory. Therefore, x-ray scattering will continue to play a vital role in the understanding of surface and interface physics in the future.

I would like to thank C. J. Palmstrøm, M. Sanyal, S. K. Sinha and J. F. Ankner for numerous enlightening discussions and their collaborations.

9. References

Agarwal B. K. 1979 *X-ray Spectroscopy* (Springer-Verlag, Berlin)

Alerhand O. L., Berker A. N., Joannopoulos J. D., Vanderbilt D., Hamers R. J. and Demuth J. E. 1990 Phys. Rev. Lett. **64** 2406

Als-Nielsen J. 1987 *Structure and Dynamics of Surfaces II: Phenomena, Models and Methods*, ed. W. Schommers and P. von Blanckenhagen, Topics in Current Physics, Vol. 43, (Springer-Verlag, Berlin) p. 181

Andrews S. R. and Cowley R. A. 1985 J. Phys. C **18** 6427

Ankner J. F. 1992 in *Surface X-ray and Neutron Scattering*, ed. H. Zabel and I. K. Robinson, (Springer-Verlag, Berlin) 105

Bartels W. J., Hornstra J. and Lobeek D. J. W. 1986 Acta. Cryst. A**42** 539

Batterman B. W. and Hildebrandt G. 1968 Acta. Cryst. A **24** 150

Berreman D. W. 1976 Phys. Rev. B **14** 4313

Bienfait M. and Gay J. M., eds., 1989 *X-ray and Neutron Scattering from Surfaces and Thin Films*, Colloque de Physique C7, Tome 50, Suppl 10, Marseille

Bohr J., Feidenhans'l R., Nielsen M., Toney M., Johnson R. L. and Robinson I. K. 1985 Phys. Rev. Lett. **54** 1275

Braslau A., Pershan P. S., Swislow G., Ocko B. M. and Als-Nielsen J. 1988 Phys. Rev. A **38** 2457

Chiarello R., Panella V., Krim J. and Thompson C. 1991 Phys. Rev. Lett. **67** 3408

Chrzan D. and Dutta P. 1986 J. Appl. Phys. **59** 1504

Chu X. and Tanner B. K. 1986 Appl. Phys. Lett. **49** 1773

Clemens R. M. and Gay J. G. 1987 Phys. Rev. B **35** 9337

Cowley R. A. and Ryan T. W. 1987 J. Phys. D, Appl. Phys. **20** 61

Cowley R. A. and Lucas C. 1989 *X-ray and Neutron Scattering from Surfaces and Thin Films*, ed. M. Bienfait and J. M. Gay, Colloque de Physique C7, Tome 50, Suppl 10, Marseille, p. 145

Eisenberger P. and Marra W. C. 1981 Phys. Rev. Lett. **46** 1081

Feidenhans'l R., Nielsen M., Grey F., Johnson R. L. and Robinson I. K. 1987 Surf. Sci. **186** 499

Feidenhans'l R., Pedersen J. S., Bohr J., Nielsen M., Grey F. and Johnson R. L. 1988 Phys. Rev. B **38** 9715

Feidenhans'l R. 1989 Surf. Sci. Rep. **10** 105

Felcher G. P. 1981 Phys. Rev. B **24** 1595

Felcher G. P., Kampwirth R. T., Gray K. E. and Felici R. 1984 Phys. Rev. Lett. **52** 1539

Feldman L. C. and Mayer J. W. 1986 *Fundamentals of Surface and Thin Film Analysis* (North-Holland)

Fenter P. and Lu T. M. 1985 Surf. Sci. **154** 15

Fewster P. F. and Curling C. J. 1987 J. Appl. Phys. **62** 4154

Fullerton E. E., Schuller I. K., Vanderstraeten H. and Bruynseraede Y. 1992 preprint.

Fuoss P. H., Kisker D. W., Brennan S., Kahn J. L., Renaud G. and Tokuda K. L. 1989 *X-ray and Neutron Scattering from Surfaces and Thin Films*, ed. M. Bienfait and J. M. Gay, Colloque de Physique C7, Tome 50, Suppl 10, Marseille, p. 159

Fuoss P. H. and Brennan S. 1990 Annu. Rev. Mater. Sci. **20** 365

Fuoss P. H., Liang K. S. and Eisenberger P. 1992 *Synchrotron Radiation Research: Advances in Surface Science*, ed. R. Z. Bachrach (Plenum), in press

Gibbs D., Ocko B. M., Zehner D. M. and Mochrie S. G. J. 1988 Phys. Rev. B38 7303

Hayter J. B., Highfield R. R., Pullman B. J., Thomas R. K., McMullen A. I. and Penfold J. 1981 J. Chem. Soc, Faraday Trans. 1, **77** 1437

Held G. A., Jordan-Sweet J. L., Horn P. M., Mak A. and Birgeneau R. J. 1987 Phys. Rev. Lett. **59** 2075

Held G. A., Jordan-Sweet J. L., Horn P. M., Mak A. and Birgeneau R. J. 1989 *X-ray and Neutron Scattering from Surfaces and Thin Films*, ed. M. Bienfait and J. M. Gay, Colloque de Physique C7, Tome 50, Suppl 10, Marseille, p. 245

Hill M. J., Tanner B. K., Halliwell M. A. G. and Lyons M. H. 1985 J. Appl. Cryst. **18** 446

Holyst R., Tweet D. J. and Sorensen L. B. 1990 Phys. Rev. Lett. **65** 2153

Hong H., Aburano R. D., Lin D. S., Chen H., Chiang T. C., Zschack P. and Specht E. D. 1992 Phys. Rev. Lett. **68** 507

Jackson J. D. 1975a *Classical Electrodynamics* (Wiley, New York) p. 281-2

Jackson J. D. 1975b *ibid.* p. 420-2

James R. W. 1965 *The Optical Principles of the Diffraction of X-rays* (Cornell University Press, Ithaca, New York)

Karle J. 1989 Physics Today **42** 22

Kato N. and Lang A. R. 1959 Acta. Cryst. **12** 787

Kortright J. B. 1991 J. Appl. Phys. **70** 3620

Liang K. S., Sirota E. B., D'Amico K. L., Hughes G. J. and Sinha S. K. 1987 Phys. Rev. Lett. **59** 2447

Lipson H. and Cochran W. 1966 *The Determination of Crystal Structures* (Bell, London)

Lucas C. A., Hatton P. D., Bates S. and Ryan T. W. 1986 J. Appl. Phys. **63** 1936

Macrander A. T. Ann. Rev. Mater. Sci. 1988 **18** 283

Majkrzak C. F. 1986 Physica B **136** 69

Majkrzak C. F. and Felcher G. P. 1990 MRS Bulletin **15**(11) 65

Majkrzak C. F., Kwo J., Hong M., Yafet Y., Gibbs D., Chien C. L. and Bohr J. 1991 Adv. Phys. **40** 99

Mansour A., Hilleke R. O., Felcher G. P., Laibowitz R. B., Chaudhari P. and Parkin S. S. P. 1989 Physica B **156-157** 867

Marra W. C., Eisenberger P. and Cho A. Y. 1979 J. Appl. Phys. **50** 6927

Miceli P. F., Neumann D. A. and Zabel H. 1986 Appl. Phys. Lett. **48** 24

Miceli P. F., Palmstrøm C. J. and Moyers K. W. 1991a Appl. Phys. Lett. **58** 1602

Miceli P. F., Moyers K. W. and Palmstrøm C. J. 1991b Mat. Res. Soc. Symp. Proc. **202** 579

Miceli P. F., Palmstrøm C. J. and Moyers K. W. 1992a, preprint

Miceli P. F., Palmstrøm C. J. and Moyers K. W. 1992b in *Surface X-ray and Neutron Scattering*, ed. H. Zabel and I. K. Robinson, (Springer-Verlag, Berlin) 203

Mochrie S. G. J. 1987 Phys. Rev. Lett. **59** 304

Mochrie S. G. J., Zehner D. M., Ocko B. M. and Gibbs D. 1990 Phys. Rev. Lett. **64** 2925

Nielsen M. 1985 Z. Phys. B **61** 415

Ocko B. M. and Mochrie S. G. J. 1988 Phys. Rev. B **38** 7378

Palmstrøm C. J., Tabatabaie N. and Allen S. J. 1988 Appl. Phys. Lett. **53** 2608

Palmstrøm C. J., Garrison K. C., Mounier S., Sands T., Schwartz C. L., Tabatabaie

N., Allen S. J. Jr., Gilchrist H. L. and Miceli P. F. 1989 J. Vac. Sci. Technol. B **7** 747

Palmstrøm C. J., Mounier S., Finstad T. G. and Miceli P. F. 1990 Appl. Phys. Lett. **56** 382

Parratt L. G. 1954 Phys. Rev. **95** 359

Pershan P. S., Braslau A., Weiss A. H. and Als-Nielsen J. 1987 Phys. Rev. A **35** 4800

Pfeifer P. 1990 New J. Chem. **14** 221. Also: Pfeifer P. 1988 in *Chemistry and Physics of Solid Surfaces VII* ed. Vanselow R. and Howe R. F. (Springer-Verlag, Berlin) 283

Renaud G., Fuoss P. H., Ourmazd A., Bevk J., Freer B. S. and Hahn P. O. 1991 Appl. Phys. Lett. **58** 1044

Renaud G., Fuoss P. H., Bevk J. and Freer B. S. 1992, preprint.

Renninger Von M. 1968 Acta. Cryst. A **24** 143

Robinson I. K. 1983 Phys. Rev. Lett. **50** 1145

Robinson I. K. 1986a Phys. Rev. B **33** 3830

Robinson I. K., Waskiewicz W. K., Tung R. T. and Bohr J. 1986b Phys. Rev. Lett. **57** 2714

Robinson I. K., Tung R. T. and Feidenhans'l R. 1988a Phys. Rev. B **38** 3632

Robinson I. K., Waskiewicz W. K., Fuoss P. H. and Norton L. J. 1988b Phys. Rev. B **37** 4325

Robinson I. K., Vlieg E. and Kern K. 1989 Phys. Rev. Lett. **63** 2578

Robinson I. K. 1991 *Handbook on Synchrotron Radiation* Vol. 3 ed. G. Brown and D. E. Moncton (Elsevier Science Publishers)

Russell T. P. 1990 Mat. Sci. Rep. **5** 171

Sanyal M. K., Sinha S. K., Gibaud A., Huang K. G., Carvalho B. L., Rafailovich M., Sokolov J., Zhao X. and Zhao W. 1992, preprint

Sanyal M. K., Sinha S. K., Huang K. G. and Ocko B. M. 1991 Phys. Rev. Lett. **66** 628

Sauvage-Simkin M., Pinchaux R., Massies J., Calverie P., Jedrecy N., Bonnet J. and Robinson I. K. 1989 Phys. Rev. Lett. **62** 563

Schwartz D. K., Schlossman M. L., Kawamoto E. H., Kellogg G. J. and Pershan P. S. 1990 Phys. Rev. A **41** 5687

Shull C. G. 1968 Phys. Rev. Lett. **21** 1585

Sinha S. K. Sirota E. B., Garoff S. and Stanley H. B. 1988 Phys. Rev. B**38** 2297

Sinha S. K. 1991 Physica B **173** 25

Specht E. D., Ice G. E., Peters C. J., Sparks C. J., Lucas N., Zhu X. M., Moret R. and Morkoc H. 1991 Phys. Rev. B **43** 12425

Speriosu V. S. 1981 J. Appl. Phys. **52** 6094.

Stout G. H. and Jensen L. H. 1968 *X-ray Structure Determination* (Macmillan, New York)

Takagi S. 1962 Acta Crystallogr. **15** 1311; 1969 J. Phys. Soc. Jpn. **26** 1239

Tapfer L. and Ploog K. 1989 Phys. Rev. B **40** 9802

Taupin D. 1964 Bull. Soc. Fran. Miner. Cryst. **87** 469

Tidswell I. M., Rabedeau T. A., Pershan P. S. and Kosowsky S. D. 1991 Phys. Rev. Lett. **66** 2108

Tweet D. J., Holyst R., Swanson B. D., Stragier H. and Sorensen L. B. 1990 Phys. Rev. Lett. **65** 2157

Van Silfhout R. G., Frenken F. W. M., Van Der Veen J. F., Ferrer S., Johnson A.,

Derbyshire H., Norris C. and Macdonald J. E. 1989 *X-ray and Neutron Scattering from Surfaces and Thin Films*, ed. M. Bienfait and J. M. Gay, Colloque de Physique C7, Tome 50, Suppl 10, Marseille, p. 159

Vlieg E., Van Der Gon A. W. D., Van Der Veen J. F., Macdonald J. E. and Norris C. 1988 Phys. Rev. Lett. **61** 2241

Vlieg E., Robinson I. K. and Kern K. 1990 Surf. Sci. **233** 248

Walker F. J., E. D. Specht and McKee R. A. 1991 Phys. Rev. Lett. **67** 2818

Warren B. E. 1969 *X-ray Diffraction* (Addison-Wesley, Reading)

Williams A. A., Thornton J. M. C., Macdonald J. E., van Silfhout R. G., van der Veen J. F., Finney M. S., Johnson A. D. and Norris C. 1991 Phys. Rev. B **43** 5001

Yoneda Y. 1963 Phys. Rev. **131** 2010

Zabel H. and Robinson I. K., eds., 1992 *Surface X-ray and Neutron Scattering*, Springer Proceedings in Physics, Vol. 61 (Springer-Verlag, Berlin)

Zachariasen W. H. 1945 *Theory of X-ray Diffraction in Crystals* (Wiley, New York). Also: (Dover, 1967).

Chapter 5

Spontaneous and stimulated emissions from optical microcavity structures

H. Yokoyama[a], S. D. Brorson[b]*, E. P. Ippen[b], K. Nishi[a], T. Anan[a], M. Suzuki[c]**, and Y. Nambu[a]

a) Opto-Electronics Research Laboratories, NEC Corporation, 34 Miyukigaoka, Tsukuba 305, Japan
b) Department of Electrical Engineering and Computer Science and Research Laboratory of Electronics, Massachusetts Institute of Technology, Cambridge, MA 02139, USA
c) Functional Devices Research Laboratories, NEC Corporation, 4-1-1 Miyazaki, Miyamae-ku, Kawasaki 213, Japan

* Present address: Max–Planck–Institut für Festkörperforschung, Heisenbergstrasse 1, D-7000 Stuttgart 80, Germany
** Present address: Max–Planck–Institut für Polymerforschung, Postfach 3148, D-6500 Mainz, Germany

ABSTRACT: We describe the alteration of spontaneous emission of materials in optical microcavities having dimensions on the order of emitted wavelength. Particular attention is paid to one-dimensional optical confinement structures with pairs of planar reflectors (planar microcavities). The presence of the cavity causes great modifications in the emission spectrum and spatial emission intensity distribution accompanied by changes in the spontaneous emission lifetime. Experimental results are shown for planar microcavities containing GaAs quantum wells. Also discussed are the laser oscillation properties of microcavities. A remarkable increase in the spontaneous emission coupling into the laser oscillation mode is expected in microcavity lasers. A rate equation analysis shows that increasing the coupling of spontaneous emission into the cavity mode causes the disappearance of the lasing threshold in the input-output curve. This is experimentally verified using planar optical microcavities confining an organic dye solution instead of a semiconductor. The coupling ratio of spontaneous emission into a laser mode increases to be as large as 0.2 for a cavity having the half a wavelength distance between a pair of mirrors. At this point, the threshold becomes quite fuzzy. Differences between the spontaneous emission dominant regime and the stimulated emission dominant regime are examined with emission spectra and emission lifetime analyses.

1. INTRODUCTION

Lately, much interest has been focused on the spontaneous emission properties of materials in microcavities. These structures are resonators having at least one dimension's size to the order of a wavelength. In the framework of the Fermi golden rule, the effect of optical confinement in one or more dimensions is understood as a rearrangement of the usual free space density of photon states (mode density) [Purcell 1946, Kleppner 1981]. The mode density at some frequencies will be increased, whereas at others, it will be decreased. Furthermore, this increase or decrease will be accompanied by a spatial redistribution of mode density. Thus, if a photon emitting medium is introduced into such a cavity, its spontaneous emission rate and spatial emission intensity distribution will be altered, depending on the cavity-mode density at the emission frequency. In the last decade, much work has been done in this research field, which is called "cavity quantum electro-dynamics (cavity QED)", particularly as a means of studying the interaction of matter with vacuum field fluctuations. To date, many experiments have demonstrated such effects, using Rydberg atoms, organic dyes [Haroche et al 1988], and semiconductors [Yablonovitch et al 1988, Yokoyama et al 1989, Yamamoto et al 1989].

Altering the spontaneous emission, however, is also interesting from the device application point of view. Of particular interest is the concept of a threshold-less laser proposed by Kobayashi et al [1982, 1985]. Recent successful demonstration of controlling spontaneous emission and nearly threshold-less laser operation [Yokoyama et al 1991] using condensed materials hold technological promise for constructing ultralow power consumption semiconductor lasers. It should be noted that after the first success in the current injection vertical cavity surface emitting laser (VCSEL) [Soda et al 1979], marked progress has been seen in constructing high performance VCSELs. For example, submilliampere VCSELs with a very short cavity structure have been fabricated [Schere et al 1989, Geels et al 1990]. Further technological progress in these VCSEL will be naturally combined with the cavity QED approach. Changes in spontaneous emission properties could play an important role in these devices.

In this article, we describe the spontaneous emission and laser oscillation properties of optical microcavities. In Section 2, the basic principles of the spontaneous emission alteration in microcavities are discussed within the framework of the Fermi golden rule. It is shown that two- or three-dimensionally confined microcavity structures have to be employed to induce large changes in the spontaneous emission rate. However, strong modifications of spontaneous emission pattern occur even for one-dimensional confinement structures with pairs of planar reflectors (planar microcavities). Thereafter, particular attention is paid to planar microcavities since all the experiments described were carried out using planar microcavities. Experimental studies are described in Section 3, on spontaneous emission from planar microcavities containing GaAs quantum wells. Section 4 describes the operation principle of microcavity lasers, in which controlled spontaneous emission plays an important role. A very large coupling of spontaneous emission into a cavity mode results in a laser which has no apparent threshold in the input–output curve, and this can be called the threshold-less laser. Nearly threshold-less laser operation is shown in Section 5, employing planar microcavities containing a dye solution. Section 6 is the summary. Prospects for microcavity devices are described.

2 . ALTERATION OF SPONTANEOUS EMISSION IN MICROCAVITIES

2.1 Fermi golden rule

First, we discuss the enhancement of the spontaneous emission rate in a closed microcavity like a wavelength sized sphere or cube surrounded by a highly reflective material. Here, we assume that only one resonant cavity mode overlaps the emission bandwidth of a light emitting medium because of the very small (wavelength sized) cavity. Consider the situation in which the gain bandwidth is much smaller than the cavity mode band width. In this case, according to the Fermi golden rule , the spontaneous emission rate A in a resonant cavity is represented in this case by

$$A = \frac{2\pi}{\hbar^2 c} |<f|H|i>|^2 \rho(k) = FA_f, \qquad (1)$$

with

$$F = \frac{\rho(k)}{\rho_f(k)} = \frac{2Q\pi^2}{Vk^3}, \qquad (2)$$

where A_f is the spontaneous emission rate in free space (hereafter, we use the word "free space" as the meaning of "without a cavity"), $\rho(k)$ $(\rho_f(k))$ is the mode density for a final photon state in a cavity (in free space) at transition frequency v $(k = 2\pi v/c)$, Q is the cavity quality factor, c is the velocity of light, V is the mode volume (in this case, cavity volume), H is an interaction hamiltonian, $|i>$ is the initial state without photons, and $|f>$ is the final state with one photon. Then, F represents the enhancement of the spontaneous emission rate caused by the cavity [Kleppner 1981]. If the cavity is off-resonant, the mode density is remarkably reduced, and spontaneous emission will be suppressed. Classically, this mode density increase (or decrease) can be understood as the resonant enhancement (or anti-resonant destruction) of the emitted electro-magnetic field in the cavity. Thus, the enhanced (or suppressed) spontaneous emission is the self-reaction process of an oscillating dipole.

Alteration of spontaneous emission is induced not only in a closed (three dimensionally confined) microcavity, but also in a two-dimensionally confined waveguide structure (optical wire) or a one-dimensionally confined planar cavity. In the following two sub-sections, we describe this issue assuming a cavity formed by perfectly conducting mirrors. This simplification gives us analytical solution forms for the spontaneous emission rate alteration [Brorson et al 1990].

2.2 Spontaneous emission rate in a planar cavity

The spontaneous emission rate given by (1) can be re-expressed in an another form as

$$A = \frac{2\pi}{\hbar^2 c} |M|^2 g(k), \qquad (3)$$

where $g(k)$ is the effective mode density factored as

$$g(k) = \frac{1}{V} \sum_k P(k) \rho(k). \qquad (4)$$

Assuming a field–electric dipole interaction, $P(k)$ describes the angle dependence of the dipole matrix element $|ed \cdot E(k)|^{-2}$, and $\rho(k)$ is the mode density of the field in the k direction at wavenumber k, where ed is the electric dipole moment and E is the electric field. In a microcavity, $\rho(k)$ will in general depend on the direction of the k vector. The sum proceeds over all states of wavenumber k and is normalized by the volume of the box in which the field is presumed to exist. Thus, since all the direction-dependent information is contained in $g(k)$, the matrix element M can be factored into two scalar parts; one is the atomic dipole matrix element for the transition from the upper to the lower level $<\Psi_1 \mid ed \mid \Psi_2>$ and the other is the electric field matrix element corresponding to the creation of a photon in a previously empty cavity $< 1 \mid (ck/\hbar\varepsilon)^{1/2} \, a^\dagger \mid 0 >$. In this form, the quantization volume V is not included in the matrix element M because it has been moved inside $g(k)$ as shown in (5). The effective mode density for given k is found by counting the increases in the number of the allowed k vectors with unit increase in the radius of a spherical shell having the radius k. Using this formula, the effective mode density seen by a dipole radiating into free space is calculated to be

$$g_f(k) = \frac{k^2}{3\pi^2} .$$

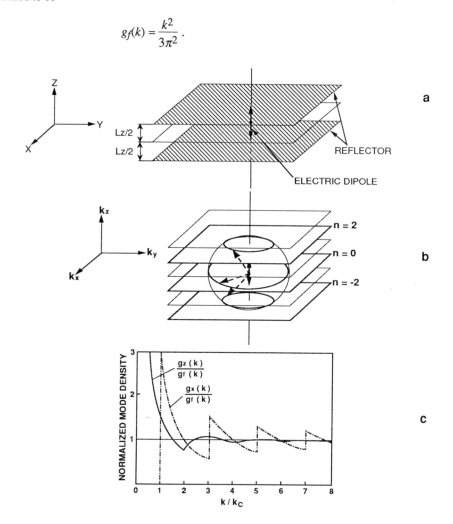

Fig. 1   (a) Physical setup of the planar mirror microcavity. The dipole sits exactly halfway between two planar conducting mirrors. The z axis oriented dipole is shown. (b) The **k** space seen by a z axis oriented dipole in the planar microcavity. The allowed modes are planes perpendicular to the z axis, which are labeled by even *n* ; the modes having the wavenumber *k* exist on the thick circles. (c) Effective mode densities for the *x* and *z* axes oriented dipoles normalized by the free space mode density.

We now adopt the above prescription to calculate the effective mode density seen by dipoles in a cavity consisting of two parallel plane mirrors. The physical system is depicted in Fig. 1 (a). The dipole is situated at the origin of a *xyz* coordinate system, whereas the mirrors are defined by the $z = \pm L_z/2$ planes. We must consider two possible orientations for the dipole: the vertical one (parallel to z axis) and the horizontal one (perpendicular to z axis). Any other orientation for the dipole can be represented as a superposition of these two cases.

Consider the structure of k space seen by the dipole between two planar mirrors. The mirrors at $z = \pm L_z/2$ impose periodicity in the z direction on the real-space fields. Thus, in k space, allowed values of k_z exist only at a series of points centered at $k_z = 0$ and separated by a distance of π/L_z. On the other hand, the cavity is unbounded in the *x-y* plane; therefore, under Fourier transform, k_x and k_y can take continuous values. Note that an allowed k at a certain value of k can exist on a circle (or circles) in an allowed *x-y* plane (or *x-y* planes) intersected by the sphere of radius k. This situation is shown in Fig. 1 (b). Therefore, the entire set of allowed k values is a set of planes intersecting the k_z axis at $k_z = n\pi/L_z$ with integer n.

Carrying out the mode counting, the effective mode density for a z axis oriented dipole is given by

$$g_z(k) = \frac{1}{V} \sum_k P(k)\, \rho(k)$$

$$= \frac{k_c k}{2\pi^2} \sum_{n \text{ even}} \left\{ 1 - (n\frac{k_c}{k})^2 \right\}. \tag{5}$$

Here, the quantization volume is taken to be $V = L^2 L_z$, and k_c is the cut off wavenumber defined as π/L_z. The last term in the second equation in brackets represents the function corresponding to $P(k)$. It should be noted that the z axis oriented dipole positioned midway between the mirrors couples only to even-order modes. This occurs because the z component of the electric field E_z varies as $\cos[n\pi(z/L_z - 1/2)]$ to match boundary conditions. The summation is taken over all integer $|n| < k/k_c$.

The effective mode density for a horizontal electric dipole is calculated in the same manner. Considering an x axis oriented dipole, the effective mode density is obtained as

$$g_x(k) = \frac{k_c k}{4\pi^2} \sum_{n \text{ odd}} \left\{ 1 + (n\frac{k_c}{k})^2 \right\}. \tag{6}$$

Note that this dipole couples only to modes that have a field maximum at $z = 0$ and have a form of $\sin[n\pi(z/L_z - 1/2)]$. Thus, only odd-order modes are included in the sum.

The effective mode densities $g_x(k)$ and $g_z(k)$ normalized by $g_f(k)$ are shown in Fig. 1 (c). The curves in this figure give the increase or decrease in the atomic transition rate when all the electric dipoles are parallel to only one axis. If the orientation of the dipoles is random, the total effective mode density is given by

$$g(k) = \frac{2}{3} g_x(k) + \frac{1}{3} g_z(k) ,$$

since $g_y(k) = g_x(k)$.

Figure 1 (c) indicates that, with a wavelength sized planar cavity, the increase and the decrease in the spontaneous emission rate are respectively at most three and two. It should be noted that the result shown here is equal to the standard electrodynamic field solution [Brorson et al 1990], and also to the previously reported QED calculations [Stehle 1970, Milloni et al 1973, Philopott 1973]. However, the simplicity of the present calculation recommends it as the preferable method for analyzing microcavities of any geometry.

2.3 Spontaneous emission rate in a waveguide

Here, we apply the method described in **2.2** to calculate the mode density of a two dimensional confinement waveguide (optical wire) structure. In Fig. 2 (a), an optical wire and the corresponding allowed modes in k space are schematically shown. It is assumed that the active material is located in the center of the y-z cross section. If the wave guide is completely surrounded by a perfect conductor, the y and z components of all the allowed k vectors become discrete. When the length of the cavity L is much larger than the wavelength (i.e. $L >> L_y, L_z$), the x component of k can be regarded as a continuous quantity. The allowed k space modes then form a series of parallel lines in the k_x direction separated by π /L_y in the k_y direction, and by π /L_z in the k_z direction. Depending on the orientation of the radiating dipole, only certain modes will be excited, while others will be disallowed. The mode density is obtained by counting the number of allowed modes contained on the surface of a sphere of radius $|k|$ centered on the origin in k space. For an electric dipole parallel to z axis, the volume normalized effective mode density $g_z(k)$ is

$$g_z(k) = \frac{1}{V} \sum_k P(k)\, \rho(k)$$

$$= \frac{1}{2\pi\, L_y\, L_z} \sum_{m\text{ odd}} \sum_{n\text{ even}} \frac{2k}{\sqrt{k^2-((mk_{cy})^2 + (nk_{cz})^2)}} \cdot \left\{ 1-(\frac{nk_{cz}}{k})^2 \right\}. \qquad (7)$$

In the first equation, this time, the quantization volume is taken to be $V= LL_yL_z$. In the second equation, $k_{cy} = \pi /L_y$ and $k_{cz} = \pi /L_z$ are respectively the cut off wavenumbers for y and z directions, and the summation is taken over all integer m and n satisfying $k^2 \geq ((mk_{cy})^2 + (nk_{cz})^2)$. The allowed modes are odd m and even n because of the boundary conditions introduced by the perfect conductor, and the dipole is situated exactly in the middle of the waveguide cross section. If the dipole is parallel to the y axis, m and n, as well as k_{cy} and k_{cz} are interchanged.

When the dipole is parallel to the x axis (parallel to the open axis of the wire), the effective mode density becomes

$$g_x(k) = \frac{1}{2\pi\, L_y\, L_z} \sum_{m\text{ odd}} \sum_{n\text{ odd}} \frac{2k}{\sqrt{k^2-((mk_{cy})^2 + (nk_{cz})^2)}}$$

$$\cdot \left\{ \frac{(mk_{cy})^2 + (nk_{cz})^2}{k^2} \right\}. \qquad (8)$$

This is the situation depicted in Fig. 2(b).

In Fig. 2 (c), the normalized effective mode densities $g_x(k)/g_f(k)$ and $g_z(k)/g_f(k)$ are plotted for the case $L_y = L_z$. When the orientation of the light emitting dipoles is random, the total effective mode density is given by

$$g(k) = \frac{1}{3}(\, g_x(k) + g_y(k) + g_z(k)\,).$$

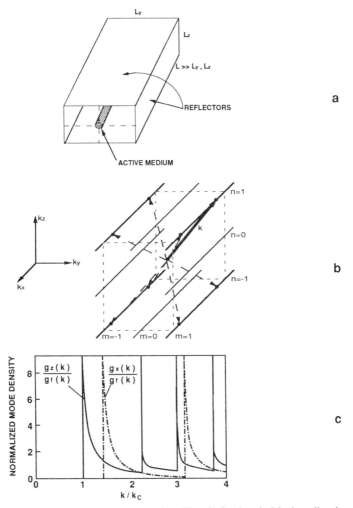

Fig. 2   (a) Schematic drawing of the optical wire waveguide. The wire has length L in the x direction and is bounded by reflecting mirrors in the y and z planes. The dipole (optically active medium) is assumed to sit exactly in the center of the waveguide. (b) The k space seen by an x directed dipole in the optical wire waveguide. The allowed modes form a series of lines in the x direction. Since k_x orientation is assumed, the dipole can only couple to lines indexed by m odd and n odd. For a given k, each line is intersected twice, as indicated by the dots. (c) Effective mode densities for the x and z axes oriented dipoles normalized by the free space value.

As depicted in Fig. 2 (c), the mode density ratio becomes infinite when $k = k_c$. For a real waveguide structure, however, the finite loss of the reflector should be taken into account. This results in a broadening of the mode peaks. Furthermore, for broad emission width materials like semiconductors, the spontaneous emission rate A is represented by

$$A = \int_{0}^{\infty} g(k)R(k)dk \ , \tag{9}$$

where $R(k)$ is essentially a factor corresponding to the optical transition matrix element, and it approximately corresponds to the spectral shape of the free space spontaneous emission. Thus, media having broad emission spectra will display smaller emission enhancement than those having narrow emission spectra, since they will sample a larger spectral region than just the peak of the effective mode density.

An estimate of the effect of an emitting medium with finite bandwidth can be obtained by considering the spectral width of GaAs, where $\Delta\lambda/\lambda_0 \sim 0.02$. Assuming $L_y = L_z = \lambda/2$ for the waveguide, the enhancement factor η is ~ 12. On the other hand, η is only 2.5 for the planar cavity of $L = \lambda/2$. This simple calculation illustrates the importance of restricted dimensionality in increasing the spontaneous emission rate. In other words, the confinement of spontaneously emitted photons into a smaller volume can induce a larger emission rate change.

A word should be said about the adaptivity of the present calculations to experimental realizations of microcavities. These will be quite adequate in microwave region because it is rather easy to make an extremely low loss cavity with a superconductor. On the other hand, in optical regions, it may be necessary to use multilayer dielectric stacks for highly reflective mirrors which have a large penetration depth and a very limited reflection bandwidth, as well as rather strong incident angle dependence of the reflectivity. Thus, a quantitative comparison between the present calculations and the experiments will be difficult. However, the qualitative results we find will still be applicable to optical experiments.

Recently, a few authors have tried to calculate the spontaneous emission rate change by more realistic optical microcavity structures made by dielectric multi-layer reflectors [Björk et al 1991, Baba et al 1991]. Progress in these theoretical works will enable quantitative analyses of experimental results.

2.4 Radiation pattern alteration in a planar waveguide

In the previous two subsections, attention has been focused on alteration of the spontaneous emission rate. Here, we discuss the change in spatial distribution of spontaneous emission intensity caused by a microcavity. Again, we focus the discussion on planar cavity structures because all the experiments performed were carried out with planar cavity structures.

For simplicity, an imaginary optical planar cavity similar to Fig. 1 (a) is again assumed, but a finite transmission loss is introduced in the present analysis. Then, the spatial radiation pattern of the spontaneous emission is calculated using the reversible principle [Drexhage 1974], i.e. seeing how the atomic absorption of radiation is modified by the cavity. Equivalently, this can be considered from the quantum mechanical point of view, as due to the change in zero-point fluctuation amplitude inside the cavity [Yamamoto et al 1989]. For example, assuming an x axis oriented electric dipole located at the midway point inside the present planar cavity and incident field polarized in x-z plane, change in the absorptivity (or the zero-point fluctuation intensity) seen from y axis is represented by the following equation, which describes the enhancement (or suppression) of incident electromagnetic field in a FP étalon.

$$\frac{I}{I_0} = \frac{(1 - R)\ (1 + R + 2\sqrt{R}\ \cos(kL_z\ \cos\theta + \pi))\ \cos^2\theta}{1 + R^2 - 2R\ \cos(2kL_z\ \cos\theta + 2\pi)}. \tag{10}$$

Where, I_0 is the incident field intensity and I is the field intensity seen by the dipole, R is the power reflectivity of the reflectors, and θ is the incident angle in x-z plane. Here, it is assumed that the absorbance inside the cavity is extremely small, then the absorptivity of an atom or the zero-point fluctuation intensity is factored by (10).

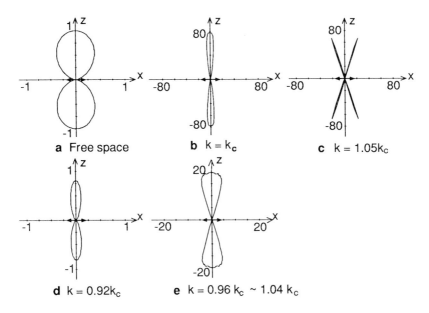

a Free space   **b** $k = k_c$ **c** $k = 1.05k_c$

d $k = 0.92k_c$ **e** $k = 0.96\ k_c \sim 1.04\ k_c$

Fig. 3   Emission intensity distribution patterns of dipoles directed to x axis; the views from y axis. (a) free space emission, (b) - (e) emission from the dipoles located at the midway inside the $\lambda/2$ planar cavity and having different transition frequencies. It is assumed that the field is polarized in x-z plane.

The curves depicted in Fig. 3 show this function at different dipole frequencies for a $\sim\lambda/2$ cavity assuming the reflectivity $R = 0.95$. The distance between a point on the curve and the origin corresponds to the emission intensity. The emission intensity distribution in free space is also shown in Fig. 3 (a) for comparison; free space emission intensity into z axis direction is normalized to be unity. Note that the same emission intensity distribution pattern is expected in a large cavity having a separation much larger than the emission wavelength. For a monochromatic dipole having a wavenumber equal to the cut-off wavenumber $k_c = \pi/L_z$, the emission intensity around the cavity axis direction is enhanced by a factor of $2\ (1 + R)\ /\ (1 - R) \approx 80$ as shown in Fig. 3 (b), and the emission is greatly suppressed in the other directions. When the dipole's wavenumber is slightly larger than k_c, as is shown in Fig. 3

(c), the emission cone spreads off the cavity axis. On the other hand, if the dipole's wavenumber is smaller than k_c (Fig. 3 (d)), the emission intensity is decreased in all the directions; this is consistent with "inhibited spontaneous emission [Kleppner 1981]" below k_c as shown in Fig. 1 (c). Considering again such condensed materials as semiconductors or dyes, we must remember that the emission will be broadband rather than monochromatic. Figure 3 (e) corresponds to this situation. A ±4 % emission width is assumed. This result indicates that, even with a ~λ/2 cavity, the directionality of the spontaneous emission in the cavity axis deteriorates if the laser medium has a very broad emission bandwidth. However, some directionality is still expected in comparison with a conventional "large" cavity case corresponding to Fig. 1 (a). It should be noted that, even if randomly oriented dipoles are assumed instead of single axis oriented dipoles, the characteristics described do not change significantly.

3. SPONTANEOUS EMISSION EXPERIMENT WITH PLANAR MICROCAVITIES

Although many successful cavity QED experiments have been carried out in the microwave region, experiments in the optical region have proven to be more difficult. The reason for this is that the cavity's dimension should be of the micron or sub-micron scale in order to produce a strong cavity effects. On this size scale it is then appropriate to use condensed materials as light emitters instead of atomic gasses that can be employed with larger microwave cavities. This in turn sacrifices the spectral purity of the atomic optical transitions and makes quantitative comparison with theory difficult. In addition, only planar cavity structures, i. e. one-dimensionally confined optical microcavity structures, are available at present for controlling spontaneous emission. Dielectric microspheres work as laser cavities of high quality factor (Q) for certain modes; but in these structures a large portion of spontaneous emission can be radiated into free space without seeing strong cavity effects.

The first report of the altering spontaneous emission with a fabricated microcavity structure was made by Drexhage [1974]. His microcavities consisted of Langmuir-Blodgett (LB) films deposited on metal mirrors. Using an LB film offers the considerable advantage of allowing the thickness to be controlled with monomolecular layer precision, while simultaneously giving reproducible, high quality films. Dye molecules were embedded in single monolayers of LB films and were optically excited. Drexhage observed dramatic modifications of the radiation distribution patterns induced by the presence of the microcavities. He did not observe, however, a clear change in spontaneous emission lifetime, because he did not fashion a cavity of high enough quality factor (Q factor). He also encountered the problem of charge transfer to a metal mirror. More recently, an LB film microcavity experiment has been carried out employing pairs of highly reflective dielectric mirrors [Suzuki et al 1991]. In this experiment, decreases and increases of the spontaneous emission rate of a factor of two were observed depending on the mirror separation.

Progress in ultrathin semiconductor crystal growth technology has also made it possible to fabricate monolithic semiconductor microcavity structures. This is quite important from a device application point of view since semiconductors can be excited electrically rather than optically with an external laser. Using molecular beam epitaxy (MBE) technology, several groups have fabricated microcavities containing GaAs quantum wells (QWs) combined with monolithic layered AlGaAs/AlAs reflectors [Yokoyama et al 1989, 1990, Yamamoto et al 1989, 1991]. It is noteworthy that quite similar structures are utilized in vertical cavity surface emitting semiconductor lasers, independently of the purpose of spontaneous emission control. When an experiment is performed at low temperature, a narrow band excitonic transition can dominate the spontaneous emission process, instead of the broad band-to-band electron–hole recombination. In this situation, a Fabry–Perot resonance curve can cover the entire spectral width of an excitonic emission. In this case, the excitonic emission can be regarded as quasi-monochromatic, and the emission power along the cavity axis direction is considerably increased under the on resonance condition. This intensity increase in the spectral domain corresponds in the spatial domain to a concentration of the spontaneous emission power into the cavity axis as shown in Fig. 3 (b). On the other hand, when $d < \lambda/2$, emission is significantly suppressed overall. This corresponds to Fig. 3 (d). These features of an excitonic emission in a monolithic microcavity have been observed experimentally by Yamamoto et al [1989, 1991]. Broadband spontaneous emission at room temperature emission is also modified by a microcavity as shown in Fig. 3 (e). In this situation, the bandwidth narrowing is observable in the cavity axis direction due to the cavity resonance restriction accompanying a marked intensity enhancement within the cavity resonance curve. Furthermore, even the spectrally integrated emission intensity can be increased by the microcavity. This occurs when only one Fabry–Perot cavity resonance curve exists within the material's emission bandwidth [Yokoyama and Brorson 1989]. This is because of the absence of allowed emission modes within the material's bandwidth. These features of a QW microcavity have been experimentally observed [Yokoyama et al 1989, 1990].

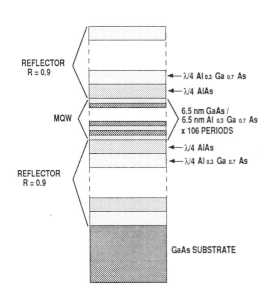

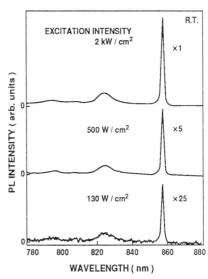

Fig. 4   Schematic of an all MBE grown mono-lithic MQW FP étalon structure.

Fig. 5   Room temperature PL spectra of an MQW étalon observed from the cavity axis direction under different excitation intensities.

Initially, we had observed a spontaneous emission alteration of GaAs QWs in a monolithic nonlinear Fabry–Perot (FP) étalon structure made by molecular beam epitaxy (MBE). The structure and the photoexcitation luminescence (PL) spectra observed from the cavity axis are shown respectively in Fig. 4 and Fig. 5.   Since the spectral shape does not change by decreasing the excitation intensity to a very low level, this strong modulation of the QWs' PL spectra is due to spontaneous emission modified by the cavity structure rather than laser oscillation.   In the initial stage of VCSEL research, there were several reports misinterpreting this kind of modulated spontaneous emission as laser oscillation.

In order to see more clearly the influence of the cavity on the spontaneous emission spectrum, the FP cavity employing an external reflector shown in Fig. 6 was also constructed. In this structure, one reflector was epitaxially grown under the multiple quantum well (MQW) layer, and the surface of the MQW layer was anti-reflection coated. In Fig. 7, emission spectra are shown for different cavity lengths. By decreasing the cavity length, it is seen that the emission intensity is concentrated in a few cavity resonance modes. (The heights and widths of the resonance peaks at multiple peaks situation are limited by the system spectral resolution of ~0.5 nm.) At the shortest cavity length available in this structure, spontaneous emission intensity is gathered into a single resonance peak reminiscent of laser oscillation. Note that the spectrally integrated PL intensity is slightly increased in this single mode situation compared to longer cavity cases. This seems to be the enhanced spontaneous emission described by (9) within a small solid angle around the cavity axis. In an FP cavity structure, averaging over the standing wave effect, the mode density of a resonance peak is enhanced by a factor of $(1 + R) / (1 - R)$ if absorption loss is negligible. Instead of this, the resonance width is approximately given by the ratio of free spectral range and the above factor.   In another words, the "mode" condenses into a narrow resonance peak. Thus, in the situation of single resonance peak within the material's emission width, the integral of (9) in the cavity axis direction can be larger than the integral without the cavity modulation of mode density; i.e. an increase in the spectrally integrated emission intensity <u>in the cavity axis direction</u> will be observed.

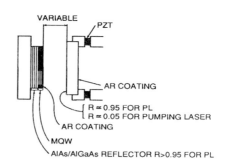

Fig. 6   Schematic illustration of a GaAs MQW FP cavity employing an external reflector.

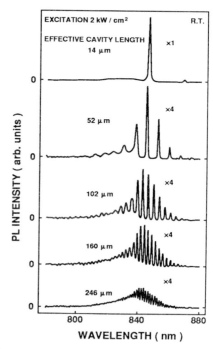

Fig. 7   Room temperature PL spectra of an external mirror GaAs MQW FP cavity detected from the cavity axis direction for different cavity lengths.

Then (9) is approximately represented as $P(E_0') \Delta E$ for a cavity and $P(E_0) \Delta P$ without the cavity, where $P(E)$ is the energy-dependent transition rate at photon energy E, E_0' and E_0 are respectively the photon energies at the cavity resonance peak and free-space emission peak, and ΔE and ΔP are respectively the FP mode separation, and the full width at half maximum (FWHM) of $P(E)$. Therefore, the intensity ratio is approximately given by

$$\eta = P(E_0') \Delta E / P(E_0) \Delta P . \tag{11}$$

If $E_0' = E_0$, the emission intensity is enhanced by a ratio of $\sim\Delta E/\Delta P$. Note that if $E_0' \neq E_0$, and $P(E_0)/P(E_0') > \Delta E/\Delta P$, the microcavity causes the spontaneous emission into the cavity axis direction to be suppressed instead of enhanced.

In an experiment using an atomic beam, the enhancement and suppression in the cavity axis direction have been clearly shown [Heinzen et al 1987]. Note, however, that if there are multiple cavity resonance peaks within the emission width, enhancement of the spontaneous emission does not occur because the mode density increase caused by the resonance peaks is canceled out by the mode density decrease between the resonance peaks. Thus, a quite short cavity is necessary to observe an enhancement for such broadband materials as semiconductors or organic dyes because of their very broad emission widths.

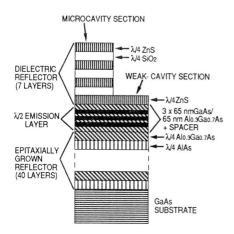

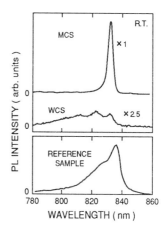

Fig. 8   Schematic view of a GaAs MQW microcavity structure.

Fig. 9   Room temperature PL spectra for a GaAs MQW microcavity structure. As a reference, a PL spectrum for a MQW without reflectors is also shown. Excitation He-Ne laser light intensity is 1 kW cm^{-2}, with ~24 μm focal spot diameter. PL detection solid angle is ~10^{-2} π.

It is possible to increase the enhancement of spontaneous emission described by (11) by further decreasing the cavity length because of an increase in the cavity resonance width. Therefore, we fabricated another microcavity structure shown in Fig. 8 [Yokoyama et al 1989, 1990]. The optical thickness of the light emitting layer is $\lambda/2$. Only the bottom reflector was made by the epitaxial growth, and this was designed to yield a reflectivity of 0.98 flat over approximately 20 nm. Half of the sample wafer, hereafter called the "microcavity section" (MCS), was covered by a seven layer ZnS/SiO$_2$ upper reflector having a reflectivity of ~0.9 for the 740-900 nm wavelength region, while showing reflectivity of less than 0.1 for wavelengths shorter than 700 nm. The other half of the wafer, hereafter called the "weak-cavity section" (WCS), had only one ZnS upper layer as an antireflection (AR) coating. Note that there is a weak cavity effect in this section because of the epitaxially grown reflector and the incomplete AR coating. For optical measurements, a small sample including both the MCS and WCS was extracted from a wafer of 40 mm in diameter. To avoid cracking the coated dielectric layers, all the measurements were carried out at room temperature.

Figure 9 shows the static PL spectra for the sample under the excitation with a He-Ne laser (the excitation spot diameter is ~20 μm). The reference spectrum was obtained from a MQW sample without any reflectors. Since this comes from another wafer, its amplitude should not be compared to those from the microcavity sample. In the measurement, PL is detected along the cavity axis perpendicular to the sample surface within a solid angle of ~ $10^{-2} \pi$. The PL spectral width for the MCS is ~4 nm FWHM around the lowest quantized electron-heavy hole transition. The PL spectrum for the WCS is modified by a residual cavity effect. The spectral shapes for both MCS and WCS do not change when the excitation intensity is varied from 1 kW cm^{-2} to 100 W cm^{-2} (the detection limit for spectra). It should be noted that the MCS and WCS, from which PL data are obtained, are less than 1 mm distant from each other on the wafer in order to assure similarity of the layer thickness and the quality of the MQW.

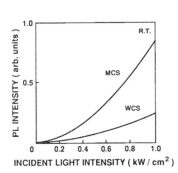

Fig. 10   Spectrally integrated room temperature PL intensities for an MCS and a WCS as a function of excitation intensity.

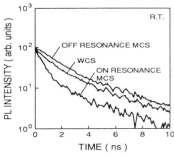

Fig. 11   Room temperature PL decay traces for a GaAs MQW microcavity measured by the combination of visible diode laser's ~100 ps pulse excitation and single photon counting technique. The excited initial carrier density is ~10^{18} cm^{-3}. The decay times are estimated from the slope of the decay in the initial 2 ns portion of the curves.

Figure 10 shows the spectrally integrated PL intensities of the MCS and the WCS emitted into the cavity axis direction. The PL intensity of the MCS is increased by a factor of 3.6 compared with that of the WCS. From (12) we expect an enhancement factor of ~7 taking into account the effective cavity separation (~3λ). However, considering the absorption of the MQW, the observed factor 3.6 seems to be reasonable. Note that with increasing the excitation intensity, the PL intensities for both sections show a quadratic increase. This is the feature of bimolecular radiative carrier recombination when nonradiative processes dominate the overall recombination [Fouquet and Burnham 1986]. In fact, the excited carrier density is estimated to be ~10^{17} cm^{-3} at 1 kW cm^{-2} excitation intensity, taking into account the measured nonradiative carrier lifetime of ~2ns. Net stimulated emission does not occur at this carrier density. Furthermore, even if we assume a gain of 10^3 cm^{-1} under intense excitation ($>> 10^{18}$ cm^{-3} carrier density), the single pass gain of ~4 x 10^{-3} can not compensate the single pass cavity loss of 0.06 for the present rather low quality factor cavity. That means stimulated emission is negligible in the present excitation condition. Therefore, the observed emission intensity is clearly due to the enhanced spontaneous emission in the cavity axis direction.

Thus far, we have discussed changes of the PL intensity in the cavity axis direction. However, the presence of the cavity can also alter the total spontaneous emission rate as discussed in **2**. We have examined these effects with time-resolved PL measurements. An AlGaInP visible diode laser driven by nanosecond pulse current was used as the excitation source, which generate one hundred picosecond optical pulses of 660 nm wavelength; the decay of the emitted PL was monitored in time using the single photon-counting technique [Yokoyama et al 1988]. From the time-resolved PL measurement, the nonradiative carrier recombination lifetime of the MQW structure has been determined to be ~2 ns at room temperature. This indicates that the well/barrier interfacial recombination velocity is rather large. Although a radiative recombination lifetime change cannot be observed at an initial carrier density lower than 10^{17} cm^{-3} because of the very short nonradiative lifetime, the lifetime difference between the MCS and WCS becomes measurable at a higher excitation condition. As explained previously, the contribution of stimulated emission is quite unimportant in the present structure. In Fig. 11, the decay of the spectrally integrated PL intensity is shown for ~10^{18} cm^{-3} initial carrier density. The slope of the initial PL decay trace gives a decay time constant of ~0.6 ns for the on-resonant MCS, and ~1 ns for the WCS. In order to see the influence of cavity resonance condition, the off-resonant MCS, which comes from a different portion of the wafer and having a thicker light emitting layer, has been also examined and is shown in Fig. 11; the decay time constant is almost the same as that of the WCS. The observed PL lifetime τ_T is related to the nonradiative lifetime τ_{nr}, and the radiative lifetime τ_{rad} by $1/\tau_T = 1/\tau_{nr} + 1/\tau_{rad}$. Although the estimated value of radiative lifetime τ_{rad} sensitively depends on the value of τ_{nr}, using the above relation and the measured nonradiative lifetime τ_{nr} ~2 ns, we find radiative lifetimes around 2 ns for the WCS and off-resonant MCS, and 1 ns for the on-resonant MCS. The reduction of the radiative lifetime from 2 ns to 1 ns directly reflects the spontaneous emission rate enhancement by the microcavity. However, a clear increase in the radiative lifetime, i.e. the suppressed spontaneous emission has not been detected with any samples extracted from the present wafer.

In order to see more clearly the effect of a planar optical microcavity on the spontaneous emission lifetime change, we have also studied the microcavities having rhodamine dye embedded Langmuir-Blodgett films (LB films) [Suzuki et al 1991]. Clear increase and decrease, but within a factor of two, in the spontaneous emission lifetime were observed depending on the cavity length. Therefore, we conclude that a dielectric multilayer microcavity can also induce the spontaneous emission rate change, although a quantitative theoretical analysis is difficult at present.

It should be noted that we did not observe a measurable lifetime change when we carried out a similar experiment using several dye solutions instead of LB films. This is inconsistent with the results of DeMartini et al [1987] who reported large changes in the spontaneous emission lifetime of solutions contained between dielectric mirrors. In the dye LB films, electric dipoles are located midway between the mirror pair, and could be oriented parallel to the layers, in which case the cavity effect should be stronger than in the dye solution experiment.

4. RATE EQUATION ANALYSIS OF MICROCAVITY LASERS

In the above two sections, we have concentrated on controlling spontaneous emission. However, controlling spontaneous emission will also induce remarkable changes in the laser oscillation properties. The most drastic effect of controlling spontaneous emission in a laser may be the threshold-less laser oscillation [Kobayashi et al 1982, 1985].

Suppose that a light emitting material has a single emission band with an extremely high quantum efficiency. In the mode point of view, the excited atoms are mostly coupled with free space modes in a conventional large sized cavity, even though there is only one cavity mode within the emission bandwidth. That is to say that most of the spontaneous emission radiates out the side of a conventional laser cavity. In that situation, the cavity mode photon number can only increase rapidly above "threshold" due to stimulated emission. Thus, the phase transition (threshold) appears in the cavity mode output. On the other hand, in the ideal microcavity, all the photons emitted couple into the single cavity resonance mode. Therefore, increasing pumping, the emission process gradually changes from spontaneous to stimulated emission without a phase transition (threshold) in the input-output curve.

In order to completely confine spontaneous emission into a single cavity mode, a closed microcavity structure must be ideal. However, note that a very large coupling of spontaneous emission into the laser mode can be also expected with a planar microcavity because spontaneous emission concentrates around the cavity axis as already discussed.

In the following, we will pay attention to changes of laser oscillation properties with the change in the fraction of spontaneous emission coupling into the laser mode.

Although our interest has been focused on spontaneous emission in Section 2, emission rate alteration is also expected in stimulated emission. This becomes obvious with quantization of the electromagnetic field. In this procedure, the overall photon emission rate R_e for an atom in a closed cavity is expressed as

$$R_e = A(s+1), \tag{12}$$

where s represents the number of photons in the cavity mode in the initial state.

As discussed in **2.3,** the spontaneous emission rate change based on the golden rule is expected also in such broad transition linewidth systems as organic dyes, certain solid state laser materials, and semiconductors, as long as the cavity resonance width is broader than the inverse of the radiative lifetime. However, a breakdown of the "golden rule" often occurs in atomic systems. In this situation, coherent effects, such as Rabi oscillations [Meschede et al 1985], or "one atom maser" operation [Rempe and Walther 1987] occur.

To use rate equations based on the golden rule, we must insure that the adiabatic approximation is valid. That is, no transient coherent effects occur. The phase coherence time of organic dyes and semiconductors are in the femtosecond range, while the inverse of the Rabi frequency in a cavity will be on the order of $1 \sim 10$ ps for usual optical pumping rates (< 1 MWcm^{-2}). Thus, transient coherent phenomena will not easily occur for these materials, and a rate equation approach is valid.

To begin, we may study the rate equations of a single mode microcavity laser, which is completely enclosed by the reflector. For such a device, the spontaneous emission rate is given by (1). Assuming an ideal four-level laser material (the decay rates of the highest state to the upper laser state, and of the lower laser state to the lowest state are extremely fast), with no nonradiative processes and no inversion saturation, the rate equations can be written as

$$\frac{dn}{dt} = p\text{-}A(s+1)n , \tag{13}$$

$$\frac{ds}{dt} = A(s+1)n - \gamma s , \tag{14}$$

where n is the number of excited atoms (molecules) in the cavity of volume V, p represents the pumping rate, and γ is the damping rate for photons from the passive cavity [Yokoyama and Brorson 1989]. The static solution of these equations is simple but noteworthy:

$$s = \frac{p}{\gamma} , \quad \text{and} \quad n = \frac{\gamma p}{A_c(p+\gamma)} .$$

We see that the light output increases linearly with increasing pumping for all pumping rates. In other words, this device works as a "threshold-less laser", as long as we focus our attention on the output versus input characteristics. As we will show, this occurs because all photons are emitted into the one single cavity mode. Note that n does not proportionally increase with an increase in pumping, and this behavior is different from that of ordinary spontaneous emission, in which the excited state population n linearly increases with pumping increase. This threshold-less laser operation is different from that of the "one atom maser (laser)", in which a gain by single atom population inversion overcomes an extremely low cavity loss.

Although enhanced spontaneous emission $(A > A_f)$, is not the necessary condition for the absence of a threshold, the consequent increase in the spontaneous emission rate has some great advantages from the device point of view. For one thing, the response speed of the device to dynamic modulation will be improved, as a result of the increased spontaneous emission rate. An another interesting feature of the threshold-less laser is that relaxation oscillations will not occur under low pumping levels. This happens because the pumping energy efficiently couples into the laser mode. Thus, the mechanism for storing energy in the laser medium, which is necessary for relaxation oscillations, is weakened. This suppressed relaxation oscillation feature is confirmed by a standard small signal analysis.

So far, we have considered the case of a completely closed cavity resonator. Now we would like to generalize to the case of an open resonator. We assume there is still one cavity mode, but now other modes exist which correspond to photons leaving the open cavity. We assume that the spontaneous emission into the cavity mode can still be enhanced, but the free space modes have the free space spontaneous emission rate. This corresponds to the case discussed by Heinzen et al [1987]. We take the fraction of the solid angle subtended by the cavity mode to the free space modes to be β. Thus, β is proportional to the inverse of the mode volume V; from another view point, it is the light-material interaction strength because larger β results in smaller beam cross-section. If a concentric cavity [Heizen and Feld 1987] is assumed, the value of β simply corresponds to the solid angle which an atom sees the cavity mirrors at the cavity center. Also taking into account nonradiative depopulation processes, the rate equations can be represented as,

$$\frac{dn}{dt} = p - (1-\beta)A_f n - \beta A(1+s)n - \Gamma n, \tag{15}$$

$$\frac{ds}{dt} = \beta A(1+s)n - \gamma s. \tag{16}$$

Here, s is now the number of photons coupled to the cavity mode, and Γ is the nonradiative depopulation rate. Note that in a broad bandwidth material, $F = A_f/A$ depends on the cavity mode separation width as discussed in **3.1**. Thus, F depends on the cavity size, as does β. Therefore, to get a large βA value, the cavity should be quite small, and to avoid the photon lifetime $(1/\gamma)$ decrease, the reflectivity of cavity mirrors should be quite high. A planar microcavity with wavelength dimensions could easily provide a rather large value for β (> 0.1). Full confinement of spontaneous emission into the single cavity mode might be realized with microsphere or microcube cavity structures.

We have carried out numerical analysis using (15) and (16). Steady-state solutions for an ideal four-level laser ($\Gamma = 0$), are shown in Fig. 12 with logarithmic scales. β is the parameter in this calculation.

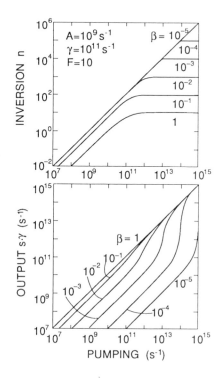

When the spontaneous emission coupling β is quite small, clear thresholds are observed in the input-output curves. (Note that in conventional semiconductor laser devices, β is ranging from 10^{-6} to 10^{-5} per cavity mode.) It is seen, however, that the threshold becomes unclear as β increases, and it disappears in the input-output curve at $\beta = 1$. Although it may not be meaningful to distinguish spontaneous emission and stimulated emission if there is no threshold and all the photon emission processes are controlled by the cavity, for convenience, we distinguish the emission rate proportional to s in the equation as stimulated emission. If we pay attention to the behavior of the population inversion, the difference between the spontaneous emission dominant region and the laser oscillation region is separated in the figure even for $\beta = 1$. With increasing pumping, n linearly increases within the spontaneous emission dominant region. On the other hand, in the laser oscillation region, n is clamped at the lasing threshold level. Thus, in that sense, a fuzzy threshold still exists although it does not appear in the input-output curves.

Fig. 12   Calculated light output S_{out} and population inversion n versus pumping p of microcavity four level lasers in logarithmic scales. $A = 10^9$ s^{-1}, $F = 10$, $\gamma = 10^{12}$ s^{-1} and $\Gamma = 0$.

The threshold in the population inversion n ($= \gamma/\beta A$) and also in the pumping rate (i.e. input power) decreases with increasing β. This is because of decrease in the mode volume. For a concentric cavity, in which the laser medium is located at the focal point, increasing β corresponds to an increase in the mirror size keeping the cavity length constant. This can be considered as an equivalent microcavity effect. Ujihara has recently given a formula for the mode volume for a planar microcavity [1991]. Based on his formula, it is shown that the mode volume is proportional to the square of the cavity length for constant mirror reflectivity. Thus, only shortening the planar cavity's length can also decrease the threshold population inversion n because the photon dumping rate γ is only proportional to the inverse of cavity length.

It should be noted that if $\beta = 1$, the threshold pumping rate remains at $p_{th} = \gamma$ independent of A although $n_{th} = \gamma/A$ decreases with increasing A. This is reasonable since more input power is necessary to give the same n_{th} value under an increased spontaneous decay rate A.

In the operation of a microcavity laser, a very large nonradiative depopulation rate ($\Gamma >> A$) will disturb the threshold-less laser action even at $\beta = 1$ since only stimulated emission decay can provide efficient light emission. Therefore, it is also very important to use a high quantum efficiency light emitting material to achieve a threshold-less laser operation.

Although (9) and (10) are valid for a four-level laser system, they are also approximately applicable to intrinsic semiconductors (with bimolecular radiative recombination). There, the spontaneous emission rate is represented by $A = B_r n$ (under the Boltzman carrier distribution approximation), where B_r is the bimolecular carrier recombination coefficient. Furthermore, even though the inversion parameter is less than 1 (i.e. the material is absorptive at a low excitation), the condition of $\beta = 1$ can still provide threshold-less laser operation if the nonradiative decay process is negligible. When the calculation is performed using this expression for A, the features are qualitatively the same as for the case of four-level lasers.

5. LASER OSCILLATION OF PLANAR MICROCAVITIES CONTANING A DYE SOLUTION

In order to achieve threshold-less laser operation, the preferable cavity geometry is a three-dimensionally confined closed one. However, a planar cavity may provide a rather large confinement of spontaneous emission in a single cavity resonance mode.

A "zero–threshold laser" action was previously reported, in which a dielectric planar FP microcavity containing an organic dye solution was utilized [DeMartini et al 1988]. However, the combination of a planar cavity and material having a very broad emission width such as a dye solution cannot insure a nearly full confinement of spontaneous emission into a single cavity mode. Furthermore, a rather high population inversion density in a very thin layer is necessary to achieve laser action compensating the ~1 % single pass cavity loss. Thus, understanding the details of emission properties for the planar microcavity is a very relevant issue. We here show the results for an experimental study on the transition from spontaneous emission to laser oscillation using carefully designed planar FP microcavities containing a Rhodamine 6G solution [Yokoyama et al 1991]. The reasons for using Rh6G dye solution are its extremely high light emission quantum efficiency and stability. Semiconductor materials like MQWs in a very thin microcavity structure presently have the problem of a rather large nonradiative recombination rate, thus high quantum efficiency is not available in the spontaneous emission regime. Although dye embedded LB films are excellent media for examining the spontaneous emission properties, they degrade rapidly under the rather intensive excitations necessary for laser oscillation.

The microcavity structure is schematically shown in Fig.13. In order to precisely control the thickness of dye solution, titanium films having 200 nm thicknesses were partially deposited on one of the multi-layer dielectric mirrors of a F-P cavity as a spacer to introduce a dye solution (refractive index n ~1.5). Furthermore, an SiO_2 film having a thickness of an integer multiple of the half wavelength was also deposited on the remaining mirror of the F-P cavity. By changing the SiO_2 film thickness, the total distance between the mirrors was varied while keeping the dye solution thickness constant. The dielectric mirrors used consisted of 19 layers of ZnS (refractive index n = 2.3) /SiO_2 (n = 1.49) λ/4 stacks for highly reflective sides on 5 mm thick BK7 glass substrates of λ/10 flatness precision. This coating shows 0.996 reflectivity for light having a wavelength within 530 – 630 nm, and less than 0.2 reflectivity for wavelengths of less than 500 nm for normal incidence. The opposite side of the mirrors are wideband anti-reflection coated (R < 0.01 for λ = 450 – 650 nm). In the present geometry, standing wave anti-nodes exist midway through the dye solution layer when the thickness of the top SiO_2 film (in contact with the dye solution) is chosen as a multiple of λ/2. The space made by the Ti film was filled by capillary action with a Rhodamine 6G (Rh6G) ethanol solution of 5 x 10^{-3} mol/l. For comparison with emission properties without the cavity effect, a similar structure without highly reflective dielectric layers has also been formed (hereafter we will refer to this as the no–cavity structure).

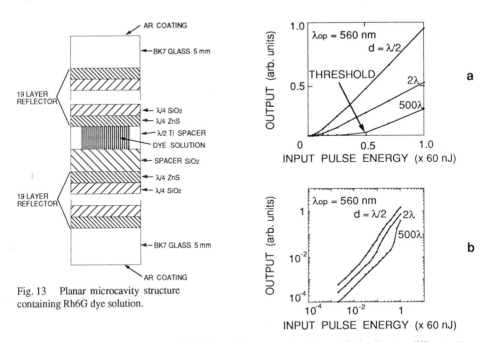

Fig. 13   Planar microcavity structure containing Rh6G dye solution.

Fig. 14   Light output versus excitation laser pulse energy for microcavities having two different mirror distances: linear plots, (b) logarithmic plots. The cavity resonance wavelength is 560 nm for all the microc... or comparison, the curves for a rather long cavity are also shown. Excitation conditions: λ = 480... width ~ 100 ps, pulse repetition = 30 Hz, beam diameter ~ 3 mm, focused by a 50 mm focal length... e detection solid angle is ~10^{-2} π.

As an excitation source for laser oscillation of microcavities, a low pulse repetition rate nitrogen laser pumped Coumarine 102 dye laser (λ_0 : ~480 nm, pulse width: ~100 ps, pulse energy: 60 nJ Max., pulse repetition rate: ~30 Hz) has been mainly used in order to avoid heating and degradation of the dye solution. An excitation laser beam 3 mm in diameter is tightly focused into the dye solution portion of the microcavity by a f = 50 mm single lens. The theoretical limit of the excitation diameter at the focal point is approximately 3 μm. Thus, the maximum irradiation light energy density was ~1 J/cm^2 (however, only 1 % of the irradiation energy was absorbed). Note that this small diameter excitation geometry can avoid unintentional amplified spontaneous emission (ASE) along the dye solution layer. The output from the microcavity is collimated and focused into a 200 mm monochromator, and detected by a photo-multiplier and single photon-counting electronics. The combination of a short pulse excitation scheme and a single photon counting electronics enabled emission lifetime measurements. A low power Ar ion laser (λ: 488 nm, cw power: 1 mW, beam diameter: 1 mm), and a diode laser pumped Q-switched YLF laser 's second harmonic (λ: 527 nm , pulse repetition: 20 kHz Max., pulse energy: 20 μJ Max.) have also been used as excitation sources to measure the spontaneous and laser emission spectra of the microcavities. In optical measurements, the detection solid angle was fixed to ~10^{-2} π around the cavity axis (10 mm diameter aperture in front of 50 mm focal length collimation lens).

Figure 14 shows the linear and logarithm plots of output versus input curves for microcavities having two different F-P cavity mirror distances (L=λ/2 means no spacer SiO$_2$ layer). In these microcavities, the cavity axis resonance wavelength was chosen to be 560 nm. For comparison, the curve for a rather long mirror distance cavity (200 μm thick SiO$_2$, e.g. ~ 300 μm net cavity length) was also shown. As the figure reveals, this "long" cavity laser has a clear threshold feature. Utilizing the rate equations (9) and (10), the coupling ratios of spontaneous emission within the measurement spectral window (~ 10 nm for this laser) was evaluated to be ~10^{-3}. On the other hand, lasing thresholds of microcavities are quite fuzzy, which indicates nearly threshold-less laser action. This shows that the ratios of spontaneous emission coupled into the laser oscillation modes are very large in the microcavities. The coupling fraction of spontaneous emission into the lasing mode were evaluated to be ~0.2 for the L = λ/2 cavity, and ~0.1 for both the 2λ and 4λ cavities in the present experiment. It should be noted that the threshold in input, which is now defined as a transition region from the spontaneous emission dominant regime to the stimulated emission dominant regime, increased with increases in cavity length. This is, as discussed in **4**, qualitatively consistent with the feature guided by the Ujihara's theory [1991].

Although it is not easy to clearly determine laser oscillation (stimulated emission dominant) regimes for the microcavities from the input–output curves, light emission response time measurements are useful to distinguish laser oscillation and spontaneous emission. Figure 15 shows the emission decay data for the λ/2 microcavity under three different excitation energies of 100 ps optical pulses. The decay with the lowest excitation energy corresponds to the spontaneous emission lifetime (2.2 ns). An increase in the excitation energy resulted in faster decay due to stimulated emission. The decay with the 60 nJ excitation is limited by the measurement system response (~300 ps).

The emission spectra of the λ/2 microcavity and the rather long cavity are shown in Fig. 16. The spontaneous emission spectra and the laser emission spectra were respectively measured under excitation with weak Ar ion laser light and the second harmonic of a Q-switched YLF laser. In contrast with the long cavity, there are no drastic changes between spontaneous and laser emission spectra for the $\lambda/2$ cavity, although narrowing is recognized at the foot of the peaks when stimulated emission is dominant. This is due to the fact that even spontaneous emission is restricted by the single F-P cavity resonance curve. (The intensities of laser spectra were not compared to each other because of the optical pass alignment problem in the present experiment.)

When the resonance wavelength in the cavity axis direction was chosen to be 600 nm, thresholds in the input-output curves became clearer than the those at the 560 nm oscillation wavelength. The evaluated coupling ratios of spontaneous emission into the lasing mode decreased to 0.06 for the L = $\lambda/2$ cavity. The higher spontaneous emission coupling efficiency at 560 nm oscillation may be attributed to the larger extraction of spontaneous emission by the cavity resonance curve as understood from the emission spectrum of Rh6G shown in Fig.16.

We also performed a series of similar experiments for the microcavities containing Sulf-Rhodamine 640 solution. The results were quite similar to those of the Rh6G microcavities.

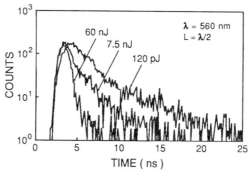

Fig. 15   Emission decay curves for the $\lambda/2$ microcavity with three different excitation pulse energies. Excitation conditions are the same as those in Fig. 14.

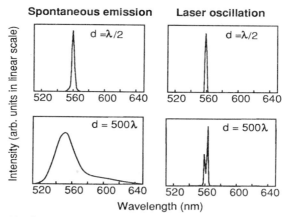

Fig. 16   Spontaneous and stimulated emission spectra of the $\lambda/2$ microcavity and the rather long cavity structure. The detection solid angle is $\sim 10^{-2}\ \pi$.

6. SUMMARIES

In summary, we have theoretically and experimentally studied spontaneous emission and laser oscillation properties of optical microcavity structures having dimensions of the emitted light wavelength. Particular attention has been paid to planar microcavity structures. We have shown that the presence of a cavity causes great modifications in the light emission spectrum and spatial emission intensity distribution, accompanied by changes in the spontaneous emission lifetime. Spontaneous emission experiments have been carried out with planar microcavities containing GaAs MQWs. Regarding the laser oscillation properties, an analysis have shown that a very large coupling of spontaneous emission into a laser mode makes the lasing threshold in the input-output curve unclear, and a threshold-less laser can be achieved if all the spontaneous emission is confined in a single cavity mode. Nearly threshold-less laser operation was demonstrated using planar optical microcavities confining an organic dye solution. Differences between the spontaneous emission dominant regime and the stimulated emission dominant regime have also been observed.

Although changes in the static laser oscillation property have been discussed in the present paper, it should be noted that the microcavity laser will offer the ultrafast response (over 100 Gbps) which cannot be achieved in a conventional diode laser [Y. Nambu and H. Yokoyama 1991]. One reason is due to the extremely short photon lifetime in an appropriately designed microcavity accompanied with the extremely short cavity length. The other reason for the response speed increase is the cavity enhanced spontaneous emission rate; two- or three-dimensionally confined microcavity structures should be fabricated for this purpose.

Another interesting subject of the microcavity optical device is the photon statistics of light output. For example, other authors have pointed out the possibility of generating number state light from a microcavity semiconductor diode laser with a very low excitation power [Yamamoto et al 1989]. This may be possible in the near future with further progress in the VCSEL devices. The application of cavity QED for optical devices will hopefully produce novel kinds of light sources beyond conventional lasers.

References

Baba T, Hamano T, Koyama F and Iga K 1991 IEEE J. Quantum Electron. **27** 1347

Björk G, Machida S, Yamamoto Y and Igeta K 1991 Phys. Rev. A **44** 669

Brorson S D, Yokoyama H and Ippen E P 1990 IEEE J. Quantum Electron. **26** 1492

DeMartini F, Innocenti G, Jacobovitz G R and Mataloni P 1987 Phys. Rev. Lett. **59** 2995

DeMartini F and Jacobovitz J R 1988 Phys. Rev. Lett. **60** 1711

Drexhage K H 1974, in *Progress in Optics*, Wolf E, ed. North Holland, Amsterdam Vol.XII, p.165.

Fouquet J E and Burnham R D 1986 IEEE J. Quantum Electron. **22** 1799

Geels R L and Coldren L A 1990 Appl. Phys. Lett. **57** 1605

Goy P, Raimond J M,Gross M and Haroche S 1983 Phys. Rev. Lett. **50** 1903

Haroche S and Kleppner D 1988 Phys. Today **42** 24

Heinzen D J, Childs J J, Thomas J E and Feld M S 1987 Phys. Rev. Lett. **58** 1320

Heinzen D J and Feld M S 1987 Phys. Rev. Lett. **59** 2623

Jewell J L, McCall S L, Lee Y H, Schere A, Gossard A C and English J H 1989 Appl. Phys. Lett. **54** 1400

Jhe W, Anderson A, Hinds E A, Meschede D, Moi L and Haroche S 1987 Phys. Rev. Lett. **58** 666

John S 1991 Phys. Today **45** 32

Kleppner D 1981 Phys. Rev. Lett. **47** 233

Kobayashi T, Segawa T, Morimoto A and Sueta T 1982 Tech. Dig. of *43th Fall Meeting of Japanese Applied Physics Society*, Sep., paper 29a-B-6 (in Japanese)

Kobayashi T, Morimoto A and Sueta T 1985 Tech. Dig. of *46th Fall Meeting of Japanese Applied Physics Society*, Oct., paper 4a-N-1 (in Japanese).

Koyama F, Kinoshita S and Iga K 1988 Trans. IECE Jpn., **E71** 1089

Meschede D, Walther H and Müller G 1985 Phys. Rev. Lett. **54** 551

Milonni P W and Knight P L 1973 Opt. Commun. **9** 119

Nambu Y and Yokoyama H 1991 Tech. Dig. of 1991*Quantum Electronics and Laser Science Conference*, Baltimore, May, paper JThB1.

Philopott M R 1973 Chemi. Phys. Lett. **19**, 435

Purcell E M 1946 Phys. Rev. **69** 681

Rempe G and Walther H 1987 Phys. Rev. Lett. **58** 353

Schere A, Jewell J L, Lee Y H, Habrison J P and Florez L T 1989 Appl. Phys. Lett. **55** 2724

Soda H, Iga K, Kitahara C and Suematsu Y 1979 Jpn. J. Appl. Phys.**18** 2329

Stehle P 1970 Phys. Rev. A, **2** 102

Suzuki M,Yokoyama H, Brorson S D and Ippen E P 1991 Appl. Phys. Lett. **58** 998

Ujihara K 1991 Jpn. J. Appl. Phys.

Vaidyanathan A G, Spencer W P and Kleppner D 1981 Phys. Rev. Lett. **47** 1592

Yablonovitch E, Phys. Rev. Lett. **58** 2059

Yablonovitch E, Gmitter T J and Bhat R 1988 Phys. Rev. Lett. **61** 2546

Yamamoto Y, Machida S, Igeta K and Horikoshi Y, 1989 Tech. Digest of *6th Rochester Conference on Coherence and Quantum Optics*, Rochester, June

Yamamoto Y, Machida S, Igeta K, Horikoshi Y and Börk G 1991Opt. Commun. **80** 337

Yokoyama H, Nishi K, Anan T and Yamada H 1989 Tech. Digest of *Topical Meeting on Quantum Wells for Optics and Optoelectronics*, March, Salt Lake City, paper MD4

Yokoyama H and Brorson S D 1989 J. Appl. Phys. **66** 4801

Yokoyama H, Nishi K, Anan T, Yamada H, Brorson S D and Ippen E P 1990 Appl. Phys. Lett. **57** 2814

Yokoyama H, Suzuki M and Nambu Y 1991 Appl. Phys. Lett. **58** 2598

Part II

III–V Compound Semiconductors

Chapter 6

Radiative and nonradiative recombination in AlGaAs and InGaAsP heterostructures and some features of the corresponding quantum well laser diodes

A F Ioffe Physico-Technical Institute, Russian Academy of Sciences,
26 Polytechnicheskaya st. 194021 St Petersburg, Russia

Abstract. Studies of MOCVD- and MBE-grown AlGaAs/GaAs SCH SQW heterostructures at low pumping levels have revealed nonradiative channels which tend to saturate with increasing pumping but nevertheless can limit the efficiency of emission from a quantum well near the threshold pumping levels. Such channels are not observed to exist in the InGaAsP/GaAs or InGaAsP/InP SCH SQW structures investigated.

In all these three types of structures the efficiency of emission from the quantum well undergoes at high pump densities ($J > 10^2$ A/cm^2) a falloff caused by a redistribution of nonequilibrium carriers between the QW and the barrier layers. In the case of current excitation and low barrier height at the cladding–waveguide interface the main mechanism responsible for the redistribution is the leakage of nonequilibrium carriers to the p-cladding. This effect results in an anomalous decrease of the differential efficiency in short cavity high output loss SCH SQW lasers.

1. Introduction

The present review is an attempt at summing up the results of photoluminescence (PL) and electroluminescence (EL) studies of InGaAsP- and AlGaAs-based laser structures carried out in the recent five years at the Ioffe Physico-Technical Institute Academy of Sciences of the USSR. It also compares the parameters of the laser diodes based on these structures.

In mid 1970s, soon after the pioneering works dealing with the development and investigation of AlGaAs/GaAs heterostructures and the corresponding laser diodes (Alferov 1989a), studies were started of two other types of heterostructures based on quaternary solid solutions in the InGaAsP/InP and InGaAsP/GaAs systems.

The InGaAsP solid solutions lattice-matched with InP and GaAs substrates represent a universal material for the development of optoelectronic devices operating from the infrared ($\lambda = 1.8$ μm) to the visible ($\lambda = 0.7$ μm). The ideas bearing on the possibility of using these materials to prepare heterostructures were first formulated by Alferov *et al* (1971). The first InGaAsP/InP double heterostructures and laser diodes based on this system were produced by Bogatov *et al* (1974). Our experimental studies in this area started soon after that were summarized in a review article by Alferov *et al* (1984). As for the preparation and investigation of the wide-gap InGaAsP solid solutions, our work fell initially somewhat behind the developments made in this area in other countries (Burnham *et al* 1970, Stringfellow 1972). Later, however, we were apparently the first to grow by liquid phase epitaxy (LPE) the GaAs–InGaP–InGaAsP–InGaP double heterostructures and to use them to construct first pulsed, and, subsequently, CW laser diodes (Alferov *et al* 1975a, b).

Later studies revealed a number of attractive properties of quaternary InGaAsP compounds which distinguish them from the traditional AlGaAs/GaAs structures. Among these properties is the lower rate of surface oxidation which permits preparation of high quality layers of quaternary compounds in the simplest slide version of LPE.

The possibility of using the simplest slide version accounts apparently for the fact that it is these materials that were chosen by Prof. N Holonyak and his colleagues to work on the preparation of ultra thin epitaxial layers by short-time LPE in a specially designed cassette with a rotating substrate (Rezek *et al* 1977).

In mid '80s, we continued and developed further this line of research by applying the short-time LPE technique to the production of multilayered InGaAsP/GaAs and InGaAsP/InP laser quantum well structures (Alferov and Garbuzov 1986). The luminescence characteristics of these structures and the parameters of the laser diodes made of them will be one of the main topics of the present review. In the late '80s, after the

MOCVD and MBE techniques reached a high enough level at Ioffe Institute, similar studies were performed on traditional AlGaAs/GaAs quantum well heterostructures and laser diodes fabricated with their use (Alferov *et al* 1988a, b). A comparison of the characteristics of the as-grown structures and of the laser diodes prepared using the AlGaAs and InGaAsP systems will be the second major topic of this review.

2. The Structures and Their Preparation

The structures based on quaternary compounds were grown by the slide version of LPE in which layers more than 0.1 μm thick were prepared by the traditional technique while thinner layers were deposited during the motion of the substrate under the corresponding melts. In the growth cell intended for the preparation of thin layers the melt contacts the substrate within a narrow slit, its width in the direction of substrate motion (1–2 mm) being an order of magnitude less than the substrate length. Using the growth cell of this configuration permitted obtaining sufficiently thin layers ($\simeq 10^2$ Å) at not excessively high substrate motion velocities. Our studies (Alferov and Garbuzov 1986), as well as the results reported by Brunemeier *et al* (1985), show that by this method one can fabricate quantum well structures with interface abruptness not over a few lattice constants.

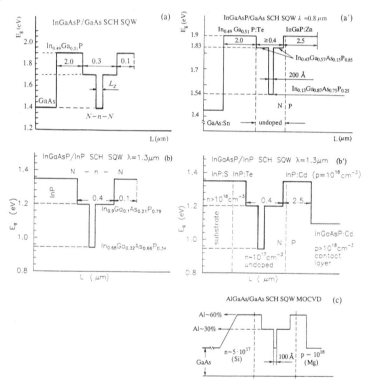

Figure 1. Band diagrams for heterostructures with (a',b',c) and without (a, b) p-n junctions used, respectively, in electro- and photoluminescence studies.

Figure 1 presents schematically the band diagrams of the InGaAsP/GaAs and InGaAsP/InP structures designed for photoluminescence studies (a, b) and the fabrication of laser diodes (a',b'). In the case of the InGaAsP/GaAs structures with all layers lattice-matched with GaAs the wide-band gap cladding layers represented the ternary compound $In_{0.49}Ga_{0.51}P$ ($E_g = 1.9$ eV), the active region being made up of quaternary solid solutions of a composition close to GaAs with emission wavelength of 0.86 or 0.78–0.81 μm. In the narrower band gap structures lattice matched with the InP substrate the composition of the quaternary compound acting as material for the active region ($In_{0.68}Ga_{0.32}As_{0.66}P_{0.34}$) corresponded to the position of the edge emission maximum at $\lambda = 1.3$ μm. In addition to the active region and cladding layers, the SCH SQW structures studied included waveguide layers made of quaternary solid solutions with compositions intermediate between those of the active region and the cladding layers. Besides the SCH SQW devices, we used in the photoluminescence studies also structures without waveguide layers (SQW structures).

In the InGaAsP/GaAs laser structures the p-InGaP and n-InGaP cladding layers were doped by Zn and Te, accordingly. The contact layer and the p-InP cladding in the InGaAsP/InP structures were doped by Cd. When doping both types of the laser structures with the acceptor impurity, particular attention was focused on the localization of the p-n junction near the p-cladding-waveguide interface, with the background impurity concentration in the active region and the waveguide layers maintained as low as possible (figure 1(a', b')).

Just as in the case of the InGaAsP heterostructures, the MBE- and MOCVD-grown AlGaAs/GaAs devices under investigation (figure 1(c)) represented SCH SQW structures with a stepped distribution of composition. (Alferov *et al* 1988a, b). The thickness of their quantum well varied from 50 to 300 Å, the parameters x and y which characterize the waveguide and cladding composition lying, accordingly, in the ranges $0.2 \leq x \leq 0.35$ and $0.35 \leq y \leq 0.6$. The n-type claddings in MBE- and MOCVD-grown structures were doped by Si, and the p-claddings and GaAs contact layers, by Be and Mg, respectively. Just as in the fabrication of the InGaAsP heterostructures, the doping conditions were chosen such as to leave the waveguides and the active region undoped. Similar to the case of the InGaAsP devices, the AlGaAs/GaAs structures grown for photoluminescence studies were isotype, undoped and without contact layers, the upper cladding being about 500 Å thick.

3. Quantum Well Photoluminescence Efficiency

The question of the internal quantum efficiency (η_i) of spontaneous radiative recombination in quantum wells of laser heterostructures cannot be considered totally clear. For the LPE grown direct-gap compounds in the AlGaAs and InGaAsP systems this question reduces primarily to the problem of the effect of nonradiative recombination at interfaces. Indeed, studies of the corresponding double heterostructures with thick ($L_z \geq 0.1$ μm) active region carried out in the '70s revealed that the room temperature rate of radiative recombination in the epitaxial layers of these compounds can exceed by at least an order of magnitude that of bulk nonradiative processes involving deep centers (Garbuzov 1982). At the same time the available estimates for the rate of interface recombination in AlGaAs/GaAs heterostructures (Nelson and Sobers 1978) suggested that interface recombination can compete with radiative processes in the quantum wells where the rate of radiative transitions remains practically the same as in the bulk material (Khalfin *et al* 1986).

In connection with the significance of this problem for the device performance, we started our investigation of InGaAsP QW structures with the determination of the value of η_i for quantum wells of various thicknesses. We developed for this purpose photoluminescence techniques with the luminescence of the QWs excited by different lines of He–Ne, Kr^+ and Ar^+-lasers (Alferov and Garbuzov 1986). These techniques included preparation of special samples which would permit measurements in the transmission geometry, determine the fraction of the absorbed pump radiation, and, in this way, obtain absolute values of the external quantum efficiency (η_e) for emission from the quantum well.

The most suitable for such measurements are structures of the type shown in figure 1(a, b) with the pump radiation absorbed practically completely in the waveguide layers. A comparison with the luminescence efficiency obtained under pumping with shorter wavelength radiation which becomes absorbed in the top cladding showed that due to the low rate of interface recombination in InP and InGaP such method of pumping can also be used to determine the values of η_i provided the thickness of the top cladding does not exceed 0.1 μm. The corresponding experiments were performed under conditions excluding photon recycling, so that the only factor determining the quantity η_e for a given value of η_i would be the reflection processes accompanying the emergence of spontaneous radiation created in the quantum well. It is well known that the values of $1.5 \div 2\%$ are under these conditions the upper limit for η_e corresponding to the values of η_i in the QW close to 100%.

Figure 2 presents the values of η_e for InGaAsP/InP structures differing in QW thickness (Alferov and Garbuzov 1986). These results suggest that in structures with L_z down to 100 Å, more than 70% of photoexcited carriers are captured by the well and undergo in it radiative recombination. Recently we have carried

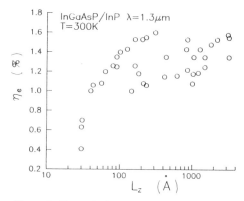

Figure 2. External photoluminescence efficiency of the active region as a function of its thickness for InGaAsP/InP structures with $\lambda = 1.3$ μm (for figure 1(b)).

out more comprehensive studies of the η_e vs L_z dependence for InGaAsP/GaAs SCW SQW structures with band gaps in the active region and the waveguide layers 1.42 and 1.8 eV wide, respectively (Bejanishvilli *et al* 1990). In contrast to the previous studies of Alferov and Garbuzov (1986), we measured, besides the efficiency of the QW photoluminescence, also that of the red PL band due to recombination of a fraction of nonequilibrium carriers in the waveguide layers. As seen from figure 3(a), for $L_z < 80$ Å the decrease of η_e for the QW emission band is accompanied by an increase of the waveguide peak intensity, so that the total luminescence intensity falls off by not more than 20% even for structures with the thinnest wells ($L_z \approx 40$ Å). In structures without waveguide layers (figure 3(b)) the active region PL efficiency falloff and the rise of the InGaP cladding emission intensity occur only for $L_z \approx 40$ Å, i.e. at lower values than in the case of structures with waveguides.

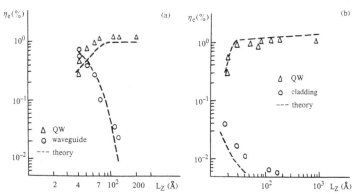

Figure 3. External photoluminescence efficiency vs quantum well thickness for InGaAsP/GaAs structures. (a) luminescence efficiency of the quantum well and waveguide for the structures with the band diagram of figure 1(a); (b) luminescence efficiency of quantum well and cladding in InGaAsP/GaAs structures without waveguide layers.

These results suggest that even in structures with ultrathin quantum wells the decrease of η_i for the QW emission is connected not with nonradiative recombination at interfaces but rather with a redistribution of nonequilibrium carriers between the QW and the adjoining barrier layers (Vavilova *et al* 1982).

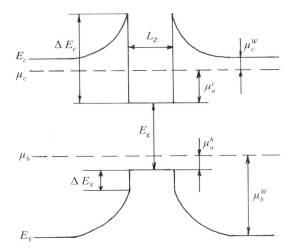

Figure 4. Heterostructure band diagram for the case of high excitation level. One can see the electron and hole quasi Fermi level positions and band discontinuities at the active region-waveguide interfaces.

Calculations show that this process of nonequilibrium carrier redistribution can be described quantitatively in terms of a simple model based on the following assumptions:

(i) In accordance with the absence of interface recombination, the quasi Fermi levels do not have discontinuities at the interfaces and can be approximated with two horizontal straight lines, their spacing being determined by background doping and the pump level (figure 4).

(ii) As a result of modulated doping, the concentration of the majority carriers (electrons) in QW exceeds the background donor concentration ($N_d \simeq 10^{17}$ cm^{-1}).

(iii) The regions where the nonequilibrium carriers recombine are considered quasineutral, recombination in the space charge regions being neglected.

(iv) Radiative transitions in the well occur between the states in the subbands satisfying the momentum selection rules.

The calculations based on these assumptions were carried out by the following scheme:

(i) For a set of quantum wells with L_z from 20 Å to 0.1 μm the energy positions of the subbands were found.

(ii) For each QW, calculations were made for a set of nonequilibrium carrier concentrations (Δn) of the quasi Fermi level position and the radiative transition rate in the QW and the waveguide, after which the value of Δn where the total recombination rate becomes equal to the given pump rate was determined.

(iii) The recombination rate in the waveguide was calculated under the assumption that the nonequilibrium carrier concentration in the quasineutral part of the waveguide layers corresponds to the thermal equilibrium condition. Note that, as seen from figure 4, the height of the barrier which determines the concentration of nonequilibrium carriers (holes) in the waveguide layers, is comparable with the total difference of the band gaps of the waveguide and the quantum well ($\Delta E_g = \Delta E_c + \Delta E_v$) and, thus, it is the value of ΔE_g that affects primarily the relative magnitudes of the recombination rates in the QW and the waveguide layers. As for the values of ΔE_c and ΔE_v, they can influence only the position of the subbands in the wells and of the quasi Fermi levels, which affects the carrier redistribution to a much lesser extent than the total height of the barrier for minority carriers.

The calculated dependences presented in figure 3(a, b) agree satisfactorily with experimental data, which argues for the validity of the model. A similar effect accounts apparently for the reduced efficiency of luminescence also in the thinnest InGaAsP/InP QW studied earlier (Alferov and Garbuzov 1986).

As for structures with thicker quantum wells ($L_z > 100$ Å), the data presented in figures 2 and 3 permit a conclusion that both in InGaAsP/GaAs and in InGaAsP/InP QW structures with QW thickness down to 100 Å there are no processes that could reduce the radiative recombination efficiency in the region of low pump densities (1–10 A /cm^2) considered here.

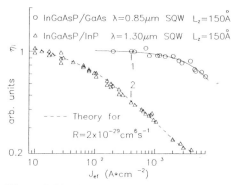

Figure 5. Photoluminescence efficiency vs optical pumping density (in A × cm^{-2} units) for In-GaAsP/GaAs and InGaAsP/InP QWs with $L_z =$ 150 Å. The dashed curve is the result of theoretical calculations with $R = 2 \times 10^{-29}$ cm^6/s

The difference in the properties of InGaAsP/GaAs and InGaAsP/InP heterostructures becomes noticeable as one goes over to higher photoexcitation densities (Alferov and Garbuzov 1986, Garbuzov *et al* 1988a). As seen from figure 5, in a 150 Å-thick InGaAsP/InP quantum well an increase of the pump level up to 200 A/cm^2 reduces the value of η_e by one half, while a InGaAsP/GaAs quantum well of the same thickness reveals only a weak falloff of η_e at 20–30 times higher levels of excitation. Experiments show that the pump level corresponding to a fixed decrease of η_e in InGaAsP/InP quantum wells is proportional to their thickness, the decrease of the luminescence efficiency with increasing excitation level following the relation

$$\eta_e = \eta_e^0 \cdot \frac{\tau_{nr}}{\tau_r + \tau_{nr}} \tag{1}$$

where η_e^0 is the external quantum efficiency at low excitation levels; τ_r is the radiative lifetime in the quantum well which depends on the pump level, $\tau_r \approx (B\Delta p)^{-1}$; τ_{nr} is the nonradiative time of the Auger process which falls off still faster with the excitation level, $\tau_{nr} = R^{-1}\Delta p^{-2}$.

In the calculation of the η_e vs J relation the absolute values of the coefficient B and its dependence on the concentration of nonequilibrium carriers were determined in the way described by Garbuzov *et al* (1983). The Auger recombination coefficient R should depend on pump level considerably weaker than the coefficient B does. The experimental $\eta_e = f(J)$ plot presented in figure 5, just as similar data for InGaAsP/InP structures with a thicker QW (Alferov and Garbuzov 1986), agrees well with the calculation assuming R to be constant and equal to 2×10^{-29} cm^6/s. One can thus suggest that in InGaAsP/InP SQW lasers Auger recombination should result in an increase of the lasing thresholds at 300 K, just as this was found to occur earlier in similar lasers with 3D-active regions.

The reason for the decrease of η_e in InGaAsP/GaAs structures at high pump levels will be considered later when discussing the results of the electroluminescence efficiency studies.

Let us turn now to the results of PL studies into the emission efficiency of AlGaAs/GaAs heterostructures with the GaAs QW active regions. The studies carried out in the '70s of the radiative recombination efficiency in AlGaAs/GaAs DH devices with 3D-GaAs active regions showed unambiguously that LPE-grown GaAs has a close to 100% internal efficiency of radiative recombination ($\eta_i > 80\%$ at $T \geq 300$ K) over a wide range of the majority carrier concentrations, $10^{16} - 10^{18}$ cm^{-3} (Garbuzov 1982). The upper limit for the recombination rate at the interface in such structures does not exceed 2×10^2 cm/s, and interface recombination does not affect the lifetime of nonequilibrium carriers at least for the active region thicknesses of 10^{-4}–10^{-5} cm (Abdulaev *et al* 1979).

AlGaAs/GaAs quantum wells represent a much more complex subject for studies of the radiative recombination efficiency than the AlGaAs/GaAs DH structures. The value of η_i in AlGaAs/GaAs QWs may depend both on such physical parameters as well thickness and AlAs content in the barrier layers, and on the growth conditions whose effect in the MOCVD and MBE methods of fabrication manifests itself much more strongly than it does in liquid epitaxy. This apparently accounts for the strong scatter in the literature data for the value of η_i in AlGaAs/GaAs quantum wells. In some studies quoting very low threshold currents for AlGaAs/GaAs SCH SQW lasers the discussion of the experimental data and calculations are based on the assumption of 100% internal quantum efficiency of radiative recombination (Chen *et al* 1987, Alferov *et al* 1990). Conversely, other publications report on a high rate of interface recombination in QWs grown by MBE (Dawson and Woodbridge 1984) and MOCVD (Ahrenkiel *et al* 1991), as well as on the existence in such structures of nonradiative channels which do not saturate up to the threshold current densities (Blood *et al* 1991).

Our studies of the radiative recombination efficiency in MOCVD- and MBE-grown AlGaAs/GaAs SCW SQW structures started in 1987 (Alferov *et al* 1988a,b) also revealed that growth conditions affect strongly the values of η_i of the structures under investigation. The work on improving the growth technology is still being continued, and therefore we will present here the data from the abovementioned publications relating to the best samples available at the time.

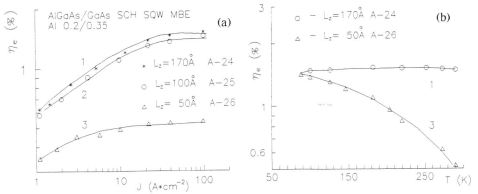

Figure 6. (*a*) Photoluminescence efficiency as a function of pumping level for AlGaAs/GaAs MBE-grown SCH SQW with different active region thickness. Optical pumping level are expressed in units of equivalent current density. (*b*) Temperature dependencies of photoluminescence efficiency for two structures from figure 6(*a*).

Figures 6(*a*) and 8 show the dependences of η_e on optical pump density (expressed, just as in figure 5, in units of equivalent current density) for several MBE- and MOCVD-grown AlGaAs/GaAs SCH SQW structures differing in well thickness and AlAs content in the barrier layers. The main parameters of the layers making up these structures are listed in table 1. As seen from figure 6(*a*), for MBE-grown structures with $L_z \geq 100$ Å the external quantum efficiency increases with pump intensity and reaches for $J > 20$ A /cm^2 a steady-state level ($\simeq 1.5\%$) corresponding to close to 100% values of η_i. In structures with $L_z = 50$ Å the maximum value of η_e at 300 K is nearly three times smaller. In accordance with the 100%-value of η_i, a decrease of temperature down to 80° K does not practically increase the luminescence efficiency of the structure with $L_z = 100$ Å, whereas for the structure with $L_z = 50$ Å we see a threefold increase of the internal efficiency of radiative recombination at such a decrease of temperature (figure 6(*b*)). The structure A–26 with $L_z = 50$ Å had a comparatively small AlAs content in the waveguide layer and the low values of η_e and η_i are partially due to the fact that at 300 K part of the nonequilibrium carriers recombine in the waveguide. Accordingly, the emission spectra of this structure displayed in figure 7 exhibit a relative enhancement with increasing temperature of the intensity of the luminescence band originating from recombination in the waveguide.

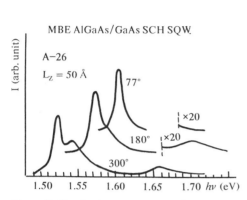

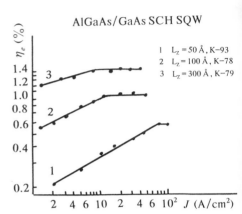

Figure 7. Photoluminescence spectra of MBE-grown AlGaAs/GaAs structure (A-26) with $L_z = 50$ Å for different temperatures.

Figure 8. Same as in figure 6(a), but for MOCVD-grown AlGaAs/GaAs SCH SQW structures with different L_z.

A trend towards a decrease in luminescence efficiency with decreasing L_z is clearly seen also in the case of AlGaAs/GaAs SCW SQW structures grown by low-pressure MOCVD (figure 8). A low value of η_e in the thinnest QW structure ($L_z = 50$ Å, curve 3) was observed here as well, despite the specially increased AlAs content in the barrier layers (table 1). Note that η_e continues to increase with pumping up to pump densities close to the level corresponding to the lasing threshold. Just as in the case of MBE-grown structures, a decrease of temperature affects only weakly the luminescence efficiency of thick QW structures while producing a strong increase of η_e for structures with small L_z.

Table 1.

Wafer No.	L_z (Å)	Doping	AlAs (%)	
			Waveguide	Cladding
A–24 MBE	170	N–n–N	20	35
A–25 MBE	100	N–n–N	20	35
A–26 MBE	50	N–n–N	20	35
K–79 MOCVD	300	N–n–N	17	40
K–78 MOCVD	100	N–n–N	17	40
K–93 MOCVD	50	N–n–N	35	60
K–275 MOCVD	100	N–p–P	30	60
K–267 MOCVD	60	N–p–P	30	60
A–182 MBE	130	N–p–P	20	38

Summing up the results presented in this section one may conclude that our photoluminescence studies of InGaAsP/GaAs and InGaAsP/InP isotype structures did not reveal nonradiative recombination channels associated with deep centers or interface recombination which could affect the threshold current densities in the region $J \geq 10^2$ A cm^{-2}. The probability of an influence of such channels on the lasing thresholds is higher in the case of AlGaAs/GaAs SCH SQW structures, particularly for quantum well thicknesses of less than 100 Å.

For diodes using InGaAsP/InP SCH SQW structures, just as for similar diodes with 3D-active regions, our studies of the photoluminescence efficiency suggest that nonradiative Auger recombination should affect noticeably the lasing threshold.

4. Spontaneous Emission of Laser Diodes

The measurements described in this and subsequent sections were performed on the three types of diodes shown in figure 9. The width of the stripe in broad-contact laser diodes was, as a rule, 100 μm. The In-GaAsP/InP and InGaAsP/GaAs-based buried diodes were fabricated by chemical etching of mesas and subsequent regrowth by two or four blocking layers with alternating conduction types. The active region width in the buried laser diodes varied from 3 to 12 μm. The diodes of the third type represented a modification of the broad stripe contact devices while differing from the latter in the presence of a narrow slit in the stripe contact through which spontaneous radiation from the well and the waveguide layers could be coupled out perpendicular to the plane of the structure. In AlGaAs/GaAs SCH SQW diodes of the third type the p-GaAs contact layer was removed from under the slit by selective etching.

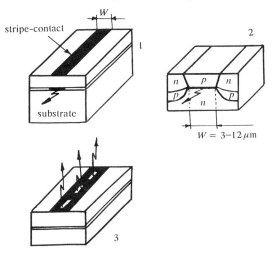

Figure 9. Schematic representation of the three diode types used in the experiments under consideration.

To avoid the effect of stimulated radiation in studies of the spectra and efficiency of spontaneous emission over a broad current density range, we used very short-cavity diodes with roughened mirror facets. The suppression of the effects associated with the decrease of self-absorption and amplification of radiation with increasing current density in these diodes is favored by the fact that the optical confinement factor (Γ) in them does not exceed 0.05 and the main fraction of spontaneous radiation propagates in the plane of the structure along the waveguide layers. Our estimation yields for the fraction of spontaneous radiation self-absorbed in the quantum well, when it is coupled out from a 100 μm-long diode through a mirror facet, a value not in excess of 10% for low excitation levels, which implies that the effects associated with the onset of transparency and gain at high current densities in diodes with roughened mirror facets can be only of the same order of magnitude. This conclusion is supported by the data presented in figure 10 which compares the current density dependences of the efficiency of spontaneous emission from the well when coupling it out through the mirror facet of a short cavity diode and through the slit in the stripe contact.

It should be pointed out that due to the absence of self-absorption, measurements of the emission efficiency with the radiation coupled out through the mirror facet of a short-cavity diode makes it possible not only to find the current dependence of the radiation efficiency but also to determine the absolute values of η_e and to derive the internal efficiency (η_i) of radiative recombination under current pumping. The corresponding evaluation shows that if the step of the refraction index at the waveguide-cladding interface is small ($n_w^2 - n_c^2 \ll 1$), the coefficient k_e determining the ratio of the quantities η_e and η_i ($\eta_e = k_e \eta_i$) may be considered in a first approximation as product of two factors, k_1 and k_2, the first of them characterizing the fraction of radiation trapped in the plane waveguide, and the second, the probability for the radiation trapped in the waveguide to escape through the mirror facet. Under the above assumption, the coefficient k_2 is determined by the absolute value of the waveguide refractive index which depends relatively weakly on chemical composition of the materials in question, and the coefficient k_1, by the difference between the refractive indices of the cladding and waveguide (Δn_{wc}), i.e. by a quantity which is strongly dependent on the solid solution composition and the bandgap difference between the corresponding materials. Since the value of Δn_{wc} for the AlGaAs/GaAs diodes is about three times that for the InGaAsP/GaAs devices, our estimates of the coefficient k_e yield, accordingly, 1.5×10^{-2} and 0.7×10^{-2}.

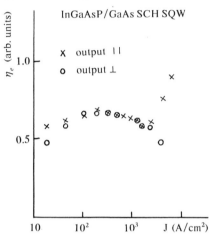

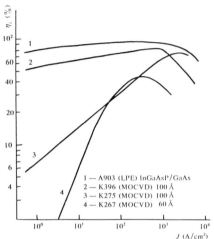

Figure 10. A comparison of current density dependences of the quantum well emission efficiency for a short-cavity InGaAsP/GaAs diode with the radiation coupled out through the mirror facet and through a slit in the stripe contact.

Figure 11. External quantum efficiency of radiative recombination vs current density for short-cavity laser diodes ($L \leq 100\ \mu m$) with roughened facets. Curve 1 relates to typical InGaAsP/GaAs diodes, curve 2—to a diode, prepared from the best MOCVD-grown AlGaAs/GaAs structure ($L_z = 100$ Å), curve 3—to a diode fabricated from a Al-GaAs/GaAs structure (K–275, $L_z = 100$ Å) referred also in figures 15 and 19; and curve 4—to a diode fabricated from a MOCVD-grown Al-GaAs/GaAs structure (K–267) with $L_z = 60$ Å.

Our measurements of the external quantum efficiency of short-cavity diodes, together with these estimates, were used to calculate the internal efficiency of radiative recombination and to draw the $\eta_i = f(J)$ plots displayed in figure 11. Curve 1 in this figure represents the dependence typical of most InGaAsP/GaAs diodes fabricated from different wafers. The other curves relate to MOCVD-grown AlGaAs/GaAs SCW SQW diodes whose characteristics in question could vary considerably from one structure to another. The possible scatter of these characteristics is demonstrated by curves 2 and 3 corresponding to diodes with $L_z = 100$ Å fabricated from two different wafers. Curve 2 refers to the structure on which one observed the highest efficiency of the quantum well radiation, and curve 3, to the structure to which the data of figures 15 and 19 relate. Despite the large difference in the η_e vs J characteristics for AlGaAs/GaAs SCW SQW diodes associated with the various uncontrollable technological factors, an analysis of all the results obtained permits a conclusion that here, just as in the photoluminescence experiments, a decrease of well thickness leads to a decrease of the maximum values of η_e and a stronger dependence of η_e on current. This trend is illustrated by curve 4 relating to AlGaAs/GaAs SCW SQW structures with $L_z = 60$ Å.

Thus the measurements of the η_e vs J dependences for InGaAsP/GaAs SCW SQW diodes are in agreement with the photoluminescence studies of isotype structures, and support the suggestion that nonradiative recombination channels should not affect the threshold current densities in the corresponding laser diodes. As for the AlGaAs/GaAs SCH SQW structures, we believe that fabrication of diodes with $\eta_i \simeq 100\%$ under current pumping would be a still more complex problem than the preparation of high efficiency isotype structures and, therefore, the effect of nonradiative channels on the thresholds in AlGaAs/GaAs SCH SQW diodes appears quite probable.

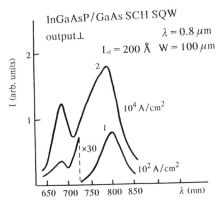

Figure 12. Emission spectra for a short-cavity InGaAsP/GaAs diode with the radiation coupled out through a slit in the stripe contact. Curves 1 and 2 relate to the current densities 10^2 and 10^4 A $\times$ cm^{-2}.

Consider now the behavior of the η_e vs J dependences in the high current density range. In these measurements on InGaAsP/GaAs and AlGaAs/GaAs diodes (figures 12, 13), the radiation was coupled out through a slit in the stripe contact. Integral measurements of the radiation intensity were complemented by spectral studies permitting one to obtain the pump current dependences of radiation efficiency separately in the quantum well band (η_a) and in the short wavelength band due to the nonequilibrium carrier recombination in the waveguide (η_w). As seen from the spectra given in figure 12, the relative intensity of the waveguide emission band grows dramatically as one approaches high pump levels. The general pattern of the η_a vs J and η_w vs J dependences for InGaAsP/GaAs and AlGaAs/GaAs SCH SQW diodes is shown in figure 13. These graphs argue convincingly for a falloff of the QW radiation efficiency at high current densities in both types of the diodes, which is accompanied by an increase of radiation efficiency from the waveguide.

Obviously enough, the reason for this phenomenon, just as in the case of decreasing QW thickness (figures 2 and 3), lies in a redistribution of injected carriers between the quantum well and the barrier layers. Besides the experimental data, figure 13(a, b) presents the results of the calculations of these effects (Garbuzov *et al* 1991c, Khalfin *et al* 1991). The scheme of the corresponding calculations and the major assumptions used coincide with the model employed in the analysis of the PL experiments. When considering the diode measurement data, however, the following essential alterations and additions were introduced:

(i) It was assumed that an increase of pump density leads to an enhancement of recombination not only in the waveguide layers but in the p-cladding as well, which appears due to the oppositely directed injection of electrons from the waveguide to the p-cladding. Thus the total pump current was considered to be a sum of the recombination currents in the quantum well, J_{QW}, the waveguide, J_w, and the p-cladding J_c. Estimates show that the leakage of holes to the n-cladding may be neglected because of their larger effective mass.

(ii) It was assumed that the concentration of excess holes in the waveguide at the QW interface is determined by the equilibrium termal distribution, their concentration increasing as one moves away from this interface because of the gradient associated with diffusive hole transport from the cladding to the well. The magnitude of this gradient can be calculated from the ambipolar diffusion equations with the inclusion of

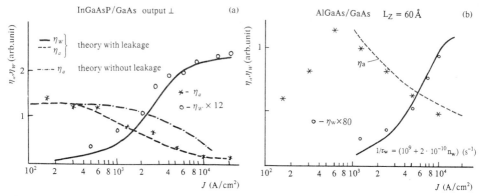

Figure 13. (*a*) Efficiency of spontaneous emission from active region (η_a) and waveguide (η_w) vs current density for very short-cavity ($L \simeq 50$ μm) non-lasing InGaAsP/GaAs diode. Stars and circles are experimental values of η_a and η_w. Solid and dashed curve represent the result of calculation taking into a account electron leakage. Broken curve corresponds to the result of calculation in which the leakage was neglected. The scaling factors for experimental and theoretical values of η_w are 12 and 6 correspondingly. (*b*) Same as in figure 13(*a*) but for AlGaAs/GaAs SCH SQW diodes fabricated from a structure with $L_z = 60$ Å (K–267). The scaling factor for the experimental values of η_w is 80, and for the calculated values of η_w it is one half of that. The calculations assume the reciprocal lifetime of nonequilibrium carriers in the waveguide to be described by the relation $1/\tau_w = (10^9 + 2 \times 10^{-10} n_w)$ s^{-1} where $n_w = n_{ow} + \Delta n_w$ is the full electron concentration in the waveguide.

recombination in the waveguide. By virtue of quasineutrality the electron concentration in the waveguide should also increase from the quantum well to the p-cladding.

(iii) Waveguide recombination in the InGaAsP/GaAs and AlGaAs/GaAs SCH SQW structures was calculated in the framework of different models. Just as this was done in the photoluminescence studies, it was assumed, in accordance with experimental data, that only radiative recombination occurs in the InGaAsP waveguide layers (Garbuzov 1982). On the contrary, measurements carried out on MOCVD-grown AlGaAs/GaAs structures reveal a high rate of nonradiative recombination in their waveguide layers (Ahrenkiel 1991, Blood *et al* 1991). Therefore for the waveguide of the InGaAsP/GaAs diode we took into consideration only the radiative recombination, and for that in the AlGaAs/GaAs diodes, linear-in-concentration nonradiative and quadratic radiative recombination (Khalfin *et al* 1991).

(iv) Electron leakage to the p-cladding was calculated by the conventional expression taking into account the diffusion of nonequilibrium carriers and their drift in the Ohmic field across the p-cladding of a finite thickness.

Turning back to a consideration of figure 13(*a*,*b*)) it should be pointed out first of all that in the construction of the experimental and theoretical curves the maximum experimental and calculated values for the quantum well emission efficiency were matched, however, after this normalization the scaling factors that had to be used to match the experimental and theoretical values of waveguide efficiency occur to be slightly different. Nevertheless, there is no doubt that the calculations yield a correct description of the shape of the experimental curves and predict correct current densities at which the intensity redistribution of the emission bands in question should occur.

In the case of the InGaAsP/GaAs diodes (figure 13(*a*)) the QW emission efficiency calculations were made in two versions, namely, with and without taking into account the leakage to the p-cladding. A comparison of the corresponding curves with experiment shows persuasively that the increase of the leakage current to the p-cladding affects the decrease of the QW emission efficiency with increasing current to a greater extent than the increase of recombination in the waveguide does. At a current density $\approx 5 \times 10^3$ A/cm^2 the effect of the leakage becomes so pronounced that its growth results in a slowing down of the increase of efficiency of the waveguide emission band as well.

Because of the higher barriers in the AlGaAs/GaAs structures (the contents of AlAs in the waveguide and cladding layers are typically 30 and 60%), the redistribution of radiation intensity in the corresponding diodes takes place at higher current densities than is the case with the InGaAsP/GaAs devices. Note that, in contrast to the InGaAsP/GaAs diodes, the falloff of the QW emission efficiency occurs here primarily as a result of the waveguide recombination enhancement. The effect of leakage to the p-cladding manifests itself only at the highest obtainable current densities.

As will be shown in the next section, the above phenomena of redistribution of the current components affects the characteristics of short-cavity laser diodes with large output losses (Garbuzov *et al* 1991 c, Khalfin *et al* 1991).

5. Threshold Current Densities

Figures 14 and 15 present the dependences of the threshold current densities on output losses for In-GaAsP/GaAs and AlGaAs/GaAs SCH SQW diodes. The measurements of J_{th} were carried out in the pulsed mode on broad stripe diodes ($W = 100~\mu$m) with the cavity length varied over a wide range. In the case of InGaAsP/GaAs diodes we used both samples with uncoated mirror facets and devices with a high reflective coating on one of the facets.

The experimental $J_{th} = f(\alpha)$ dependences presented in figure 14 are typical of the InGaAsP/GaAs diodes whose active region composition corresponds to the lasing wavelength $\lambda = 0.81~\mu$m (figure 1(a')). A large number (a few hundreds) of such structures were fabricated to be used in the pumping systems of YAG:Nd^{+3} lasers, the threshold current densities of 200–300 A/cm^2 being typical of the present level of growth technology (Garbuzov *et al* 1990b).

The experimental values of the threshold current densities for InGaAsP/GaAs SCH SQW diodes with low output losses exceed by not more than two times the lowest limit calculated in the approximations discussed by Garbuzov *et al* (1987). This difference appears only natural since the corresponding calculations illustrated by solid curve in figure 14 did not take into account such factors which reduce the gain at a given pump current density as intraband relaxation and band tailing caused by microfluctuations in the solid solution composition.

The calculations mentioned above (Garbuzov *et al* 1987) assumed a 100% current pump efficiency while not taking into account nonradiative recombination in the active region and current leakage. Nevertheless, the calculated dependence $J_{th} = f(\alpha)$ is superlinear which is associated with the effect called frequently 'gain saturation' typical of QW lasers (Zory *et al* (1986), Alferov *et al* 1989b). The experimental values of J_{th} reveal a still faster growth of J_{th} with increasing output losses than this is predicted by the theoretical model in question. The data presented in figure 13(a) provide a convincing argument for this fast growth of J_{th} in low-barrier InGaAsP/GaAs SCH SQW diodes under investigation to be due primarily to electron leakage to the p-cladding.

In the case of AlGaAs/GaAs SCH SQW structures, the best values of J_{th} for samples with small α were also found to be about 200 A/cm^2 (figure 15), despite the thinner well ($L_z = 100$ Å) and substantially better conditions of optical confinement than those typical of the InGaAsP/GaAs lasers. As mentioned in the previous section, in the region of the threshold current densities the QW emission efficiency in most of the AlGaAs/GaAs diodes tested still continues to grow with current, and the values of η_e at $J = J_{th}$ do not yet reach the level corresponding to the 100% internal quantum efficiency of radiative recombination in the active regions. The presence of the nonradiative current component is apparently one of the reasons for the increase of the threshold current densities in the AlGaAs/GaAs diodes considered here, just as in similar devices studied by other authors (Blood *et al* 1991, Abrenkiel *et al* 1991).

As seen from figure 15, a reduction of the cavity length in AlGaAs/GaAs SCH SQW diodes also results in a superlinear growth of the threshold current densities. Due to the higher barriers and a weaker influence of the leakage to the p-cladding, however, this growth is not so strong as is the case with the InGaAsP/GaAs diodes, and for AlGaAs/GaAs diodes with the shortest cavities the value of J_{th} turns out to be four to five times smaller than that for the InGaAsP/GaAs diodes with the same output losses (figure 14 and 15). The results presented in the preceding section suggest that the superlinear growth of J_{th} with increasing output losses originates predominantly from redistribution of carriers with their transfer to the AlGaAs waveguide layers where nonradiative recombination primarily takes place.

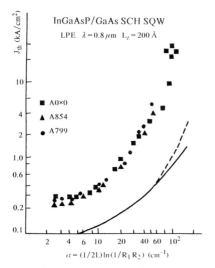

Figure 14. The experimental data illustrate typical dependences of the threshold current density on output losses for InGaAsP/GaAs SCH SQW diodes fabricated from three different wafers. The solid and dashed curves are calculated J_{th} vs α plots obtained without and with the inclusion of leakage to the p-cladding, respectively.

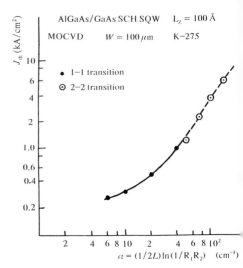

Figure 15. J_{th} vs α plot for diodes fabricated from MOCVD-grown AlGaAs/GaAs SCH SQW structures with $L_z = 100$ Å (K–275). The points and circles refer, respectively, to the 1e–1h and 2e–2h transitions.

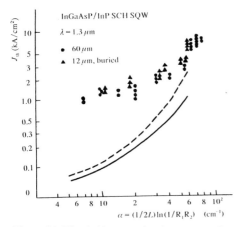

Figure 16. Threshold current density vs output loss for InGaAsP/InP SCH SQW diodes with a broad contact and for buried diodes. The solid and dashed curves are calculated J_{th} vs α plots obtained without and with the inclusion of Auger recombination, respectively.

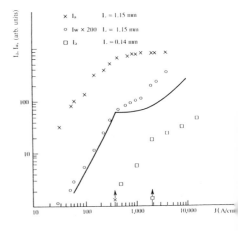

Figure 17. Spontaneous emission intensity from active region (J_a) and waveguide (J_w) vs current density for AlGaAs/GaAs diodes with cavity length 1.15 mm (○ and ×) and 140 μm (□). Diodes were prepared from AlGaAs/GaAs SCH SQW structures A–182 (see table 1).

It should be pointed out that the dashed part of the J_{th} vs α dependence, as shown by an analysis of the lasing spectra, relates to transitions not between the lowest quantum well subbands but rather to those involving the second electron and hole subbands (Zory *et al* 1986, Garbusov *et al* 1987). The lasing wavelength shifts shortward in short-cavity InGaAsP/GaAs diodes too, however here the shift occurs in a smoother way, and one not always succeeds in determining accurately the value of α which corresponds to the change of the transition mechanisms.

Figure 16 presents experimental data and calculated $J_{th} = f(\alpha)$ plots for InGaAsP/InP laser diodes (Garbuzov *et al* 1988b, Garbusov *et al* 1991b). These measurements were performed on broad-stripe lasers ($w = 60$ μm), as well as on buried mesa diodes with stripes 12 μm wide. These diodes were fabricated from structures with active regions 200–300 Å thick. It is remarkable that the difference between the experimental and calculated threshold current densities for the InGaAsP/InP lasers is substantially greater than that for the InGaAsP/GaAs and AlGaAs/GaAs diodes discussed earlier.

A comparison of the calculations performed for laser diodes with $In_{0.68}Ga_{0.32}As_{0.66}P_{0.34}$ and GaAs active regions shows that a decrease of the band gap of the active region material should result in a decrease of the threshold current densities in both the 3D- and 2D-cases (Alferov *et al* 1989b). This decrease is associated both with the increase of the spontaneous radiative lifetime and the decrease of the m_e/m_{hh} ratio with decreasing E_g. This relates to the ideal radiative model of calculations which does not include nonradiative recombination channels. According to these calculations (solid curve in figure 16), the threshold currents J_{th} for InGaAsP/InP laser diodes ($\lambda = 1.3$ μm) with low output losses should be less than 50 A/cm^2 even for $L_z = 200$ Å. Neither the addition of the current component associated with Auger recombination, nor the inclusion of the leakage currents are capable of accounting for the observed difference between the calculated and experimental values of the threshold current density for the InGaAsP/InP diodes. There exists apparently another factor (e.g., a density-of-states tailing caused by fluctuations in the composition of the $In_{0.68}Ga_{0.32}As_{0.66}P_{0.34}$ solid solutions) which results in a three-to fourfold decrease of the gain at a given current density compared with the calculations. The associated increase of the threshold current density leads to an additional enhancement of the current components due to the Auger recombination and leakage to the p-cladding, so that the experimental values of J_{th} reflect apparently the resultant action of all three aforementioned factors.

6. Differential Efficiency

Prior to discussing the experimental data on the differential efficiency of the laser diodes under consideration, we shall present the results of an experimental investigation (Khalfin *et al* 1991) into the efficiency of spontaneous emission of laser diodes representing a continuation of the studies described in section 4. Just as before, the spontaneous radiation was coupled out through a slit in the p-stripe contact, however here the mirror facets were not roughened, and the purpose of the experiments was an investigation of the dependence of the emission efficiency from the active region and waveguide on current below and above the lasing threshold.

Figure 17 displays the results of the corresponding measurements for two AlGaAs/GaAs SCH SQW diodes with cavities differing strongly in length. These diodes were fabricated from AlGaAs/GaAs SCH SQW structures with $L_z = 130$ Å and AlAs content in the waveguide and p-cladding layers 20 and 38% respectively, that is lower than in the other structures. The relatively low content of AlAs in the waveguide layers permitted a reliable detection of the radiation emitted from them. As seen from figure 17, in the case of long cavity diodes the growth of the intensity of the quantum well spontaneous radiation with increasing current observed to occur practically stops as one approaches the lasing threshold, in agreement with the assumption of the quasi Fermi levels becoming pinned above the lasing threshold. At the same time the waveguide radiation intensity continues to grow with current even above the threshold.

Similar features were observed in other experiments performed both on AlGaAs/GaAs and on InGaAsP/GaAs laser diodes with long cavities.

As for diodes with cavities 100–200 μm long and threshold current densities ≥ 1 kA/cm^2, we have not succeeded in finding among them samples in which the intensity of spontaneous radiation from the active region would stop completely above the lasing threshold. A typical dependence of the intensity of the QW spontaneous emission on current for a short cavity diode is illustrated in figure 17.

Putting aside for the time being the aforementioned features of the short-cavity SCH SQW diodes, we will try to explain the reasons for the growth of the waveguide radiation intensity J_w above the lasing threshold which is observed to occur on all samples including the long cavity diodes. Obviously enough, this pattern of the $J_w = f(J)$ dependence is a direct consequence of the model developed in section 4. Indeed, the pinning of the quasi Fermi levels in the active region above the lasing threshold implies only a fixed concentration of nonequilibrium carriers in the part of the waveguide layers adjoining the active region. As for the increase of the current density, it should be accompanied by an increase of the carrier concentration gradient in the waveguide, an increase of the average carrier concentration and, hence, a growth of the waveguide radiation intensity.

The results of a calculation of the J_w vs J dependence in the framework of this model shown by the solid line in figure 17 is in a qualitative agreement with experiment.

Thus the increment of the total current through a diode above the lasing threshold (ΔJ) should be equal to the sum of the increments of the recombination currents in the active region and waveguide (ΔJ_{QW} and ΔJ_W) and of the additional electron leakage current to the p-cladding (ΔJ_e).

Obviously, the role of the current components ΔJ_W and ΔJ_e which reduce the deferential efficiency, will be the grater, the larger are the absolute values of the threshold current density. The quantity $\eta_u = \Delta J_{QW}/\Delta J$ representing a function of the total current density through the diode ($\eta_u = f(J)$) is an upper limit of the differential efficiency for a laser diode with a threshold current density $J_{th} = J$. To determine η_d, the values of η_u should be multiplied by the quantity η_{ex} characterizing the differential efficiency of long cavity diodes in which carrier leakage is negligible. The value of η_{ex} was calculated by means of the expression

$$\eta_{ex} = \frac{\alpha}{\alpha + \alpha_f + \alpha_{QW} + \alpha_W} \tag{2}$$

where α is the output losses; α_f is the current-independent losses due to the lasing mode scattering and absorption by the equilibrium free carriers; α_{QW} and α_W are the current-dependent losses related with the absorption by excess carriers injected into the quantum well and waveguide layers.

Figures 18–20 present experimental values of η_d and the calculated dependences of the differential efficiency on output losses (or reciprocal cavity length) for the three types of the SCH SQW laser diodes under consideration. When constructing the theoretical curves for η_d, the values of η_u were derived from the experimental threshold current densities for the corresponding laser diodes (figures 14–16), the parameter α_f being chosen such that the calculated values of η_{ex} be equal to the maximum experimental values of the differential efficiency for diodes with the lowest output losses.

A remarkable feature of the data presented in figures 18–20 is the anomalous decrease of the differential efficiency with decreasing cavity length which is indeed observed in the case of the short-cavity laser diodes of all the three types studied (Wilcox *et al* 1989, Garbuzov *et al* 1990). For the InGaAsP/GaAs and InGaAsP/InP diodes (figures 18–19) whose barriers are lower than those of the AlGaAs/GaAs devices the efficiency η_d drops substantially with cavity length decreasing down to 100 μm despite the comparatively large thickness of their active region ($L_z \geq 200$ Å). A substantial decrease of the differential efficiency of AlGaAs/GaAs SCH SQW diodes with the cavity length decreasing down to 100 μm was observed to occur only for devices with the thinnest quantum wells ($L_z = 60$ Å).

As seen from the experimental data shown in figures 18–20, in all the three diode types studied the region of the anomalous decrease of η_d is adjoined by a domain of moderate output losses where the differential efficiency varies only weakly with α. Only in the region of the lowest output losses the expected normal falloff of η_d with decreasing α becomes noticeable.

The calculated dependences shown by the solid lines in figures 18–20 fit satisfactorily to the experimental data for the InGaAsP/GaAs and InGaAsP/InP diodes. The comparatively weak growth of η_d with α observed for these diodes at medium values of α can be accounted for within the theoretical model used by the super-linear increase of the threshold current densities that corresponds to the increase of the threshold concentration of nonequilibrium carriers in the quantum well and waveguide and the associated enhancement of absorption by free carriers. Further increase of α leads also to a growth of the leakage current which results, as already pointed out, in the appearance of region of anomalous decrease of η_d with increasing α. It should be noted, however, that a comparison of the experimental data with calculations in figures 18–19 indicates that the experimental values of η_d decrease with increasing α faster than this is predicted by theory.

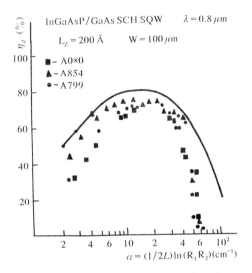

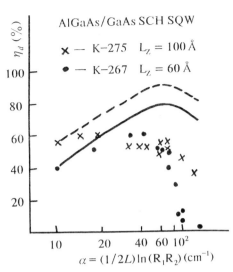

Figure 18. The experimental η_d vs α dependence obtained on the InGaAsP/GaAs diodes to which figure 14 relates. The calculated η_d vs α plot assuming the current-independent internal losses to be $\simeq$ 1 cm^{-1}.

Figure 19. Experimental and calculated dependences of differential efficiency on output losses for AlGaAs/GaAs diodes fabricated from two structures with L_z = 100 Å (K–275) and L_z = 60 Å (K–267). The current-independent internal losses were assumed to be 6.5 and 13.5 cm^{-1}, which provides a good agreement between the calculated and experimental values of η_d for diodes with the lowest output losses.

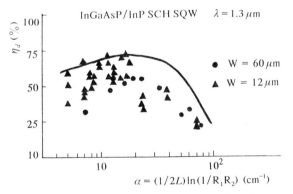

Figure 20. Experimental values of η_d as a function of output losses for the InGaAsP/InP diodes to which the data of figure 16 relate. The calculated η_d vs α plot was constructed assuming the background internal losses to be $\simeq$ 1 cm^{-1}.

Much more serious disagreement is revealed when examining the data relating to the AlGaAs/GaAs SCH SQW laser diodes. Because of the high barriers the losses associated with the nonequilibrium carrier absorption should increase with J_{th} for these devices slower and, in contrast to the aforementioned InGaAsP diodes, theory predicts here a normal growth of η_d in the range of α where the experimental values of η_d practically do not increase with increasing α. As for the region of the largest α, here the experimental values of η_d drop noticeably, whereas theory predicts only an insignificant falloff of the differential efficiency. It thus becomes obvious that in these diodes there are factors, besides those included in the theoretical model, which reduce η_d as the output losses grow. One of such factors is the dependence of the near-field distribution pattern on cavity length revealed by us earlier (Garbuzov *et al* 1990a) and confirmed, in particular, in an analysis of the near-field pattern for the samples relating to figures 18–19. For the broad contact diodes studied, the spatial uniformity of the near-field distribution deteriorates dramatically as one crosses over to diodes with cavities less than 0.5 mm long. An enhanced effect of current leakage through the nonlasing parts of the stripe area is another obvious reason which increases J_{th} and reduces η_d in broad-contact diodes as their cavity length decreases. Our preliminary experiments performed on single-mode laser diodes with stripes a few microns wide suggest that one can obtain here $\eta_d = f(\alpha)$ experimental dependences fitting better to the theoretical curves shown in figures 18–19.

7. Light-Current Characteristics

In this, last section we are going to discuss some results bearing on the light-current characteristics of laser diodes fabricated from InGaAsP/GaAs ($\lambda = 0.8$ μm) and InGaAsP/InP ($\lambda = 1.3$ μm) SCH SQW structures. These results are of interest since the corresponding diodes seem to offer a possibility of obtaining CW power outputs at 0.8 μm and 1.3 μm in excess of the figures quoted in earlier publications for diodes with a similar aperture. As for the AlGaAs/GaAs SCH SQW diodes, the highest CW output powers obtained on them in these studies (2–3 W for a 100 μm-wide stripe) were typical for such devices which would make the discussion of their light-current characteristics here unnecessary.

Turning back to the preceding section, we should draw attention to the high values of η_d obtained for the InGaAsP/GaAs and InGaAsP/InP diodes with very long cavities. The possibility of using long cavity diodes is a condition essential for the production of high CW power outputs, since the electric and thermal resistance of diodes decreases proportionately with their length, while the local temperature rise of a diode active region for a given emission power decreases, respectively, proportional to the square of cavity length. As shown by our experiments, in order to produce diodes with low internal losses and a high differential efficiency for long cavities, the corresponding SCH SQW structures must have an undoped active region and waveguide (figure 1). A comparison of the experimental and calculated dependences (figures 18 and 20) supports the conclusion that in long-cavity, low-threshold diodes fabricated from such structures the losses due to the absorption by nonequilibrium carriers in the waveguide layers and the QW may be reduced below 1 cm^{-1} and are comparable with those due to scattering and absorption by the majority carriers in the doped cladding layers.

Figure 21 demonstrates the best result obtained thus far in experiments aimed at reaching the highest possible radiation power outputs on InGaAsP/GaAs SCH SQW laser diodes with $\lambda = 0.8$ μm (Garbuzov *et al* 1991a). The light-current characteristic displayed in this figure relates to a laser diode with a 100 μm-wide stripe and a cavity length in excess of 1 mm which had a high reflective and a low reflective coatings on the front and rear mirror facets. The highest radiation power obtained on this diode in CW operation is 5.3 W which is among the best quoted figures for 100 μm-aperture AlGaAs/GaAs diodes (Welsch *et al* 1988). Due to the high values of η_d and a low series resistance the conversion efficiency of electric power to light for such diodes was in excess of 50% for a CW power in the range of 1–4 W.

The temperature rise in the active region in high power laser diodes is an essential factor determining both their maximum power and device lifetime. The temperature of the active region in InGaAsP/GaAs diodes was determined by making use of the short wavelength emission band associated with recombination in the waveguide and discussed already in the preceding sections (figure 12). Since the diodes in question were bonded to the heatsink with the stripe down, this band could be studied only with the radiation coupled out through the diode mirror facet. Nevertheless, due to the fact that this band intensity reaches saturation with increasing current neither below nor above the lasing threshold (figures 13 and 17), it can be spectrally selected even with such a method of detection without any difficulty. Since the concentration of nonequilibrium carriers in the waveguide layers remains comparatively low throughout the current density range studied, the position

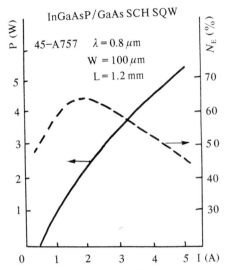

Figure 21. Output power (solid curve) and power conversion efficiency (dashed curve) plotted against driving current for InGaAsP/GaAs laser diode. Average heatsink temperature was near 5 °C.

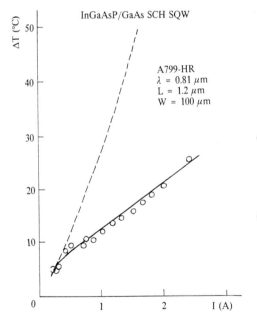

Figure 22. Local temperature rise near the active region of an InGaAs/GaAs diode as a function of driving current. The solid and dashed curves are the calculations with and without taking into account of the diode optical output.

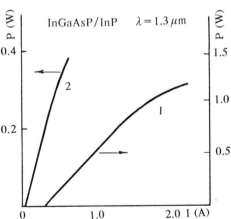

Figure 23. Light vs current characteristics for two InGaAsP/InP diodes ($\lambda = 1.3\ \mu$m). Curve 1 relates to a 12 μm wide-stripe buried diode, and curve 2 to a broad-contact ($w = 100\ \mu$m) diode. Both diodes have a high reflectivity coating on rear mirror facet.

of the waveguide band maximum depends only on the temperature in the domain of the diode where the waveguide emission is generated. Basing oneself on the known values of the self-absorption coefficient at the edge emission maximum (Dyment *et al* 1977), one may conclude that the region of the waveguide layer adjoining the mirror facet which contributes to the detected radiation is not less than 10 μm long. Estimates show that for the region of such a size the additional contribution to the local temperature rise associated with the mirror facet heating should not exceed 10%, and that the temperature variations derived from the shift of the waveguide peak maximum should reflect the local temperature rise in the bulk of the active region (Garbuzov *et al* 1991a).

Figure 22 presents the current dependence of the local temperature rise determined from the long wavelength shift of the waveguide peak for one of the InGaAsP/GaAs diodes studied. The part of this dependence below the threshold was used to find the thermal resistance of the diode

$$R = \frac{\Delta T}{\Delta (VI)} \tag{3}$$

where I and V are the current and the voltage applied to the diode. The dashed curve in figure 22 shows how the local temperature rise in the active region should have grown if equation (3) was met also above the lasing threshold (for instance, if the front mirror facet had also a high reflective coating). The solid curve presents the dependence

$$\Delta T = R(VI - P_{opt}) \tag{4}$$

where P_{opt} is optical power coupled out of the diode.

The agreement between equation (4) and the experimental ΔT values evidences that the local temperature rise in the active region of high power diodes depends not only on the thermal characteristics but on the diode efficiency as well. In the case in question the local temperature rise of the active region ($\simeq 20°$ for a CW optical power of 2 W) is almost one half that expected for a non-emitting diode (dashed curve in figure 22).

Estimates show that the thermal and series resistances of these diodes exceed nevertheless by a few times the calculated values, so that improvements in fabrication technology (e.g. reduction of contact resistance) can broaden the power range within which the temperature rise will not influence the diode operation.

High CW power outputs have been obtained also for the InGaAsP/InP ($\lambda = 1.3$ μm) SCH SQW laser diodes. Figure 23 displays light-current characteristics for a broad-contact diode and a buried diode with a stripe 12 μm wide. Curve 1 relating to a broad-contact diode shows that CW power outputs of 1 W can be reached with 100 μm-aperture diodes not only in the 0.86–0.8 μm range but at $\lambda = 1.3$ μm as well, thus making possible their matching to optical fibers (Garguzov *et al* 1991b). Curve 2 reveals that one can achieve CW power outputs in excess of 380 mW from a diode whose stripe width provides efficient (80%) output coupling to a 50 μm dia. fiber (Garbuzov *et al* 1988).

8. Conclusion

The results of our studies of the three types of heterostructures and of the corresponding laser diodes can be summed up in the following way.

8.1. Heterostructures and lasers in the InGaAsP/GaAs system

8.1.1. Nonradiative recombination involving deep levels in the active region or interface processes do not affect the total recombination rate in InGaAsP/GaAs SCH SQW structures even for $L_z \simeq 25$ Å and pump levels ≥ 10 A/cm^2.

8.1.2. The falloff of the radiative recombination efficiency for quantum wells with $L_z \leq 100$ Å, as well as for high pumping levels ($\geq 5 \times 10^2$ A/cm^2), is due to a redistribution of nonequilibrium carriers between the active region and the barrier layers. In the case of high current densities the spatial redistribution of the recombining carriers is connected primarily with an enhancement of the electron leakage current to the p-cladding.

8.1.3. Recombination in the waveguide and leakage to the p-cladding continue to grow with increasing pump level after the onset of lasing in the quantum well.

8.1.4. The threshold current densities in low output loss SCW SQW laser diodes are close to calculated values.
8.1.5. The superlinear increase of the thresholds and the decrease of differential intensity with increasing output losses are caused by the factors specified in items 8.1.2 and 8.1.3

8.2. Heterostructures and lasers in the InGaAsP/InP system

8.2.1. All the statements made in items 8.1.1, 8.1.2, 8.1.3 and 8.1.5 with respect to the InGaAsP/GaAs heterostructures are applicable here too.
8.2.2. An additional factor increasing the threshold current densities in both long-and short-cavity diodes is the Auger recombination.
8.2.3. The factors considered in items 8.2.1 and 8.2.2 cannot account fully for the experimentally observed excess of the thresholds above the calculated values in the case of low output loss, InGaAsP/InP SCH SQW laser diodes. Band tailing in the $In_{0.68}Ga_{0.32}As_{0.66}P_{0.34}$ solid solutions is apparently the dominant reason for the comparatively high threshold current densities observed.

8.3. AlGaAs/GaAs heterostructures

8.3.1. In contrast to the quaternary laser structures in the AlGaAs/GaAs SCH SQW heterostructures the nonradiative recombination channels in the active region, some of which at least are connected with interface recombination, affect strongly the efficiency of radiative recombination at low pump densities ($J < 100$ A/cm^2).
8.3.2. The rate of nonradiative recombination in the active region and the values of pump densities required to reach 100% efficiency of radiative transitions depend on growth conditions hard to control. In the case of thin quantum well structures the effect of the nonradiative recombination can be strong enough to increase by a few times the threshold current densities in lasers with low output losses.
8.3.3. Due to the higher barriers, the effects associated with the redistribution of nonequilibrium carriers start to influence the efficiency of radiative recombination in the active region at higher pump levels than is the case with the InGaAsP/GaAs structures. Note that both under photo- and current-pumping the decrease of the radiation efficiency in the active region is primarily due to the increasing fraction of the nonequilibrium carriers which recombine nonradiatively in the waveguide layers.
8.3.4. Redistribution of the nonequilibrium carriers produces the same features in the characteristics of the AlGaAs/GaAs lasers with high output losses as the ones described in item 8.1.5 for InGaAsP/GaAs diodes. In the case of the AlGaAs/GaAs diodes, however, these features are observed to occur at shorter cavity lengths and only on diodes fabricated from structures with the thinnest quantum wells.

Acknowledgments

In conclusion the authors express their gratitude to V.B.Khalfin and Yu.Shernyakov for the calculations and experiments included in the present review, as well as to the coauthors of the cited references making up the bulk of the review.

References

Abdulaev A, Garbuzov D Z, Ermakova A M and Trukan M K 1979 *Fiz. Tekh. Poluprovodn.* **14** 1744
Ahrenkiel R K, Deyes B M and Dunlavy D V 1991 *J. Appl. Phys.* **70(1)** 225
Alferov Zh I, Andreev V M, Konnikov S G, Nikitin V G and Tretyakov D N 1971 *Proc. Intern. Conf. Phys. Chem. Semicond heterostructures* Budapest, Hungary vol. 1 p 93
Alferov Zh I, Arsent'ev I N, Garbuzov D Z and Rumyantsev V D 1975a *Sov. Tech. Phys. Lett.* **1** 191
Alferov Zh I, Arsent'ev I N, Garbuzov D Z, Konnikov S G and Rumyantsev V D 1975b *Sov. Tech. Phys. Lett.* **1** 147
Alferov Zh I, Garbuzov D Z and Gorelenok A T 1984 *InGaAsP heterostructures: Growth, Recombination Processes and Application for Optoelectronics Devices*, in: Advances in Science and Technology in the USSR. Physics Series, ed A M Prokhorov (Moscow: Mir Publishers.)
Alfereov Zh I, Garbuzov D Z, Arsent'ev I N, Ber B I, Vavilova L S, Krasovskii V V and Chudinov A V 1985 *Sov. Phys.-Semicond.* **19** 679
Alferov Zh I and Garbuzov D Z 1986 *18th Conf. on the Physics of Semicond.* Stockholm, Sweeden p 136

Alferov Zh I, Garbuzov D Z, Denisov A G, Evtikhiev V P, Komissarov A B, Senichkin A P, Skorokhodov V N and Tokranov V E 1988a *Sov. Phys.-Semicond.* **22** 1331

Alferov Zh I, Garbuzov D Z, Zhigulin S N, Kyz'min I A, Orlov B B, Sinitchin M A, Strugov N A, Tokranov V E and Yavich B S 1988b *Sov. Phys.-Semicond.* **22** 1334

Alferov Zh I, Garbuzov D Z, Evtikhiev V P, Komissarov A B, Sokolova Z N, Ter-Martirosjan A L, Tokranov V E, Chalyi V P and Khalfin V B 1988c *19th Int. Conf. on the Physics of Semiconductor* Warsaw, Poland vol. 1, p 271

Alferov Zh I 1989a In: *Semiconductor Heterostructures: Physical Processes and Applications* (Moscow: Mir Publishers) pp 9–11

Alferov Zh I, Garbuzov D Z and Khalfin V B 1989b *in Semiconductor Heterostructures: Physical Processes and Applications* (Moscow: Mir Publishers) pp 126–158

Alferov Zh I, Ivanov S V, Kopjev P S, Ledentzov N N, Lutzenko M E, Meltzer B Yu, Nemenov M I, Uxtjugov K M and Schaposchnikov S C 1990 *Sov. Phys.-Semicond.* **24** 92

Bejanishvilli G R, Buinov P P, Garbuzov D Z, Juravkevitch E V and Khalfin V B 1990 *Proc. of 5th All-Union Conf. on Physics of Semicond. Heterostructures.* Kaluga, USSR p 5

Blood P, Kucharska A I, Jacobs J P and Criffith K 1991 *J. Appl. Phys.* **70(3)** 1144

Bogatov A P, Dolginov L M, Druzhinina L V, Eliseev P G, Sverdlov B N and Shevchenko E G 1974 *Kvant. Electron. (Moscow)* **1** 2294

Brunemeir P E, Hsieh K C, Deppe D G, Brown J M and Holonyak N J 1985 *J. Cryst. Growth* **71** 705

Burnham R D, Holonyak N, Keune D L, Scifres D R and Dopkus P D 1970 *Appl. Phys. Lett.* **17** 430

Chen H Z, Ghaffari A, Morkoc H and Yariv A 1987 *Electron. Lett.* **23** 1334

Dawson P and Woodbridge K 1984 *Appl. Phys. Lett.* **45** 1227

Dyment J C , Kapron F P and Springthrope J 1977 *IEEE Trans. Electron. Devices* **ED–24** 995

Garbuzov D Z 1982 *J. Lumin.* **27** 109

Garbuzov D Z, Agaev V V, Khalfin V B and Chalyi V P 1983 *Sov. Phys.-Semicond.* **17** 1052

Garbuzov D Z, Arsent'ev I N, Chalyi V P, Chudinov A V, Evtikhiev V P and Khalfin V B 1984 *Sov. Phys.-Semicond.* **18** 1272

Garbuzov D Z and Khalfin V B 1987 *Proc. Third Binational USA–USSR Symp. on Laser Optics of Condensed Matter.* Leningrad, USSR ed Birman *et al* (New York and London: Plenum Press) p 103

Garbuzov D Z, Chalyi V P, Svelokuzov A E, Khalfin V B and Ter-Martizosyan A J 1988a *Sov. Phys.-Semicond.* **22** 410

Garbuzov D Z, Arsent'ev I N, Ovchinnilov A V and Tarasov I V 1988b In: *Technical Digest, Conf. on Lasers and Electro-Optics Washington* (Opt. Soc.of America, Washington DC, USA) pp 396–398

Garbuzov D Z, Kochergin A V and Rafailov E U 1990a *Proc. of 5th All-Union Conf. on Physics of Semicond. Heterostructures* Kaluga, USSR p 60

Garbuzov D Z, Gulakov A V, Kochergin A V, Scurko A P, Strugov N A, Ter-Martirosyan A L and Chalyi V P 1990b In: *Technical Digest, Conf. on Lasers and Electro-Optics.* Washington DC, USA vol. 7 p 468–469

Garbuzov D Z, Antonishkis N Yu, Bondarev A B, Zhigulin S N, Katsavets N I, Kochergin A V and Rafailov E V 1991a *IEEE J. Quantum Electron* **QE–27** 1531

Garbuzov D Z, Goncharov S E, Il'in Yu V, Ovchinnikov A V, Pikhtin N A and Tarasov I S 1991b *First Intern. Soviet Fiber-Optics. Conf. Leningrad USSR.* vol. 1 p 144

Garbuzov D Z, Ovchinnikov A V, Pikhtin N A, Sokolova Z N, Tarasov I S and Khalfin V B 1991c *Fizika and Teknika Poluprovodn.* **25** 928

Khalfin V B, Garbuzov D Z and Krasovskii V V 1986 *Sov. Phys.-Semicond.* **20** 1140

Khalfin V B, Gulakov A B, Kochnev I V Rafailov E U, Shernyakov Yu M, Yavich B S and Garbuzov D Z 1991 *Joint Soviet-American Workshop on Physics of Sem. Lasers* Leningrad, USSR pp 49–57

Nelson R J and Sobers K G 1978 *J. Appl. Phys.* **49** 6103

Rezek E A, Holonyak N and Vojak B A 1977 *Appl. Phys. Lett.* **31** 288

Strigfellow G B 1972 *J. Appl. Phys.* **43** 3455

Vavilova L S, Garbuzov D Z, Tulashvili A V, Trukan M K, Arsent'ev I N 1982 *Sov. Phys.-Semicond.* **16** 989

Welsch D F, Chen B, Streifer W and Scifres D R 1988 *Electron. Lett.* **24** 113–115

Wilcox J Z, Ou S, Yang J J, Jansen M and Peterson G L 1989 *Appl. Phys. Lett.* **55** 825

Zory P S, Reisinger A R, Waters R G, Mawst L J, Zmudzinski O A, Emanuel M A, Givens M E and Coleman J J 1986 *Appl. Phys. Lett.* **49** 16

Chapter 7

Far-infrared cyclotron resonance of a two-dimensional electron gas in III–V semiconductor heterostructures

M. O. Manasreh

Electronic Technology Directorate (WL/ELRA), Wright Laboratory,
Wright-Patterson Air Force Base, OH 45433-6543

ABSTRACT: Far-infrared cyclotron resonance (CR) measurements of a 2-dimensional electron gas (2DEG) in InAs/AlSb and InAs/AlGaSb single quantum wells as well as GaAs/AlGaAs heterostructures are obtained at liquid helium temperature. The electron effective mass (m*) calculated from the CR transmission peak position energy, effective g_o^* factor of the 2DEG, relaxation time (scattering time), and full width at half maximum (FWHM) of the CR spectra as a function of the magnetic field strength (B) will be presented. An oscillatory behavior of these parameters is observed as a function of B which can be explained by Landua levels filling factors.

1. INTRODUCTION

The InAs/AlSb and related heterostructure combinations are a potentially very interesting system due to their large conduction band offset of about 1.30 - 1.35 eV (Nakagawa 1989) providing a good confinement of carriers in the quantum well. In addition, this system offers a high room temperature mobility 2-dimensional electron gas (2DEG) which makes it an ideal material for heterostructure field effect transistors (Tuttle 1989). The high mobility 2DEG is an excellent system in which to study electron-electron interactions (Goldberg et al 1889; Nicholas et al 1988; Ando 1975). These interactions cause an enhancement of the electronic Landé g factor. Ando and Uemura (1974) were the first to point out that the Landé g factor should be an oscillatory function of the filling of Landau levels. The Landé g factor for electron spin splitting in bulk InAs is large as compared to that of GaAs and negative due to spin-orbit interactions (Palik and Stevenson 1963). In this chapter we report on the far-infrared cyclotron resonance (CR) of the 2DEG in InAs/AlSb and InAs/AlGaSb single quantum wells (SQWs) as well as GaAS/AlGaAs heterostructes. The results show oscillatory behavior as a function of magnetic field strength (B) in the electron effective mass (m*) calculated directly from the peak position energy of the CR spectra, relaxation time (τ) extracted from fitting the CR spectra with a transmissivity equation based on Drude-type conductivity, and full width at half maximum (FWHM) measured directly from the CR spectra. The effective g factor (g_o^*) of the 2DEG calculated from m* also shows oscillation as a function of B.

The molecular beam epitaxial sample investigated here consisted of a 500 Å GaAs buffer layer grown on a semi-insulating GaAs substrate, a 3 μm undoped AlSb layer, a 150 Å undoped InAs layer, a 150 Å undoped AlSb layer, and a 100 Å undoped GaSb cap layer.

The quantum well was modulation-doped by deep donors formed in AlSb. The sample had InSb-like interfaces which were grown according to Tuttle *et al* (1990) with RHEED patterns being 4x2 and 1x3 for InAs and AlSb, respectively. The GaAs/AlGaAs heterostructures investigated consisted of a GaAs buffer layer grown on a semi-insulating GaAs substrate, an undoped GaAs channel layer , an $Al_{0.28}Ga_{0.72}As$ undoped spacer layer, an $Al_{0.28}Ga_{0.72}As$ donor layer doped with Si with a typical concentration of $\sim 1 \times 10^{18} cm^{-3}$, and an undoped GaAs cap layer. These samples were grown by molecular beam-epitaxial (MBE) and metal-organic chemical vapor deposition (MOCVD) techniques. The far-infrared measurements were made at 4.2 K with a Bomem DA3 Fourier-transform spectrometer in conjunction with a Si-bolometer and a 7 T superconducting magnet cryostat. The substrate was wedged and the reference file was taken with sample at 4.2 K and zero magnetic field in order to eliminate any interference effects. Both the magnetic field and the incident radiation were parallel to the growth axis.

2. InAs/AlSb SINGLE QUANTUM WELL

Figure 1 shows selected CR spectra taken at different magnetic field values for the InAs/AlSb single quantum well. According to this figure the CR spectra labeled (a)-(c) are composed of two peaks while spectra (d)-(g) exhibit only one peak. The vertical arrows in the top panel of Fig. 1 indicate the two peak positions as obtained from fitting the spectra with two Gaussians. The separation between the two peak positions is $\sim 16 \, cm^{-1}$ at B=2.6 T, which is decreased to $\sim 8 \, cm^{-1}$ at B=5.2 T. Results similar to spectra (a)-(c) were reported for bulk InAs at higher magnetic fields (Palik and Stevenson 1963) (B>10 T). Palik and Stevenson (1963) interpreted the structure observed in their CR spectra as being due to spin splitting and variation of the effective g_o^* factor of the conduction band with Landau levels. Similar structure and interpretations were also reported for bulk InSb (Palik et al 1961). However, the current results are different from Palik and Stevenson (1963) results in two ways. First, the present CR spectra were obtained at low magnetic fields ($1.6 \leq B \leq 7T$). Second, the present CR spectra provide a strong evidence that g_o^* exhibits an oscillatory behavior in InAs/AlSb single quantum well as a function of B; only one peak is observed in the CR spectra when g_o^* is approximately the same for different Landau levels [see Fig. 1 (d)-(g)] while two peaks can be clearly resolved [see Fig.1 (a)-(c)] when g_o^* is different for different Landau levels.

In Fig. 2 we plot the relative electron effective mass (full squares) obtained from $\omega_c = eB/m^*c$, where ω_c is the cyclotron resonance frequency obtained directly from the CR spectra. In case of CR spectra with two peaks, m* was calculated from the average cyclotron frequencies of the two peak position energies. The data in this figure show oscillatory behavior as a function of B. The calculated effective mass of the InAs quantum well is $\sim 0.043 \, m_0$ at B=7 T, which is larger than that obtained for bulk InAs (for example, Palik and Stevenson (1963) reported m*= 0.024 m_0 and Litton *et al* (1969) reported m*=0.023 m_0) by a factor of 1.8. However, the present value of m* at B=7 T is in good agreement with the value reported for an InAs/GaSb superlattice (Bluyssen et al 1979) at B=6.39 T. The g_o^* is also calculated from the following expression

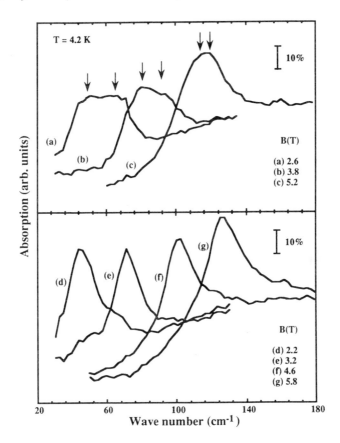

Fig. 1. Selected cyclotron resonance spectra taken at certain values of B. The top panel show spectra with two peaks and the bottom panel show spectra with only one peak. The vertical arrows in the top panel indicate the peak positions obtained from fitting these spectra with two Gaussians.

$$g_0^* = 2[1 - \{(1-x)/(2+x)\}\{(1-y)/y\}], \tag{1}$$

which is derived from the fourth order effective mass theory (Palik et al 1961) at the center of the Brillouin zone. In this expression $x = 1/(1+\Delta/E_g)$, $y = m^*/m_0$, $\Delta = 0.43$ eV, and $E_g = 0.41$ eV. The results are shown in Fig. 2 (open squares). Both g_0^* and m^*/m_0 exhibit similar oscillations which is not surprising since g_0^* is directly proportional to m^*. However, Fig. 2

indicates that m* is larger when the absolute value of g_o^* is smaller. In addition, the spin-splitting energy decreases when g_o^* is smaller. This can be clearly seen in the Fig.1(a)-(c) where the separation between the two peaks decreases as B increases. This behavior is also observed in bulk InAs (Palik and Stevenson 1963) and bulk InSb (Palik et al 1961). It should pointed out that Eq. (1) is derived from single particle calculations and g_o^* may be enhanced due to many-body effects.

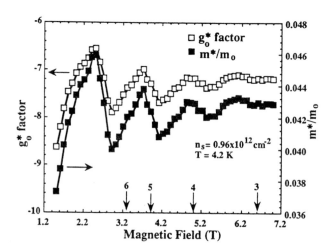

Fig. 2. The relative effective mass, m*/m_0, of the InAs 2DEG obtained directly from the peak position energies of the CR spectra (closed squares) and the g_o^* factor (open squares) calculated from Eq. (1). The vertical arrows indicate the positions of a few even integral values of ν.

The vertical arrows in Fig. 2 represent even integer filling factor (ν) calculated from the expression $\nu = n_s h/2eB$ where n_s is the density of the 2DEG. The n_s value was retrieved from fitting the CR transmission spectra with the following transmissivity equation based on a Drude-type conductivity (Chiu et al 1976):

$$T(\omega) = 2\beta^2 \left[\frac{1 + \Omega^-}{(1 + \delta n_s \tau/m^*)^2 + \Omega^-} + \frac{1 + \Omega^+}{(1 + \delta n_s \tau/m^*)^2 + \Omega^+} \right], \quad (2)$$

where $\Omega^\pm = (\omega \pm \omega_c)^2 \tau^2$, ω is the infrared radiation frequency, τ is the relaxation time, $\delta = 4\pi\beta e^2/c$, c is the speed of light, and $\beta = 1/(1+n)$ where n is the refractive index of InAs and is taken to be 3.52. In order to minimize the number of fitting parameters, m* values were taken from Fig. 2. Thus, τ and n_s are the only fitting parameters. Small amplitude oscillations were observed in n_s as a function of B with an average value of (0.96 ±0.01)x10^{12} cm^{-2}. If the value of n_s =0.96x10^{12} cm^{-2} is to be considered accurate enough and if the peak positions of m* vs. B are corresponding to half-integer values of ν, then one should expect to observe more oscillations in m* for B < 3.0 T (see Fig. 2). One possible explanation for the absence of many oscillations for B< 3.0 T is that the resolution and stability of the superconducting magnet are not high enough to resolve further oscillations especially when the energy separation between Landau levels is small.

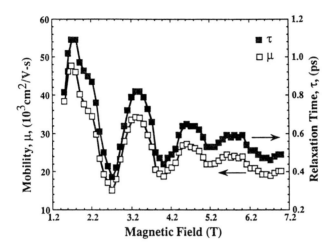

Fig. 3. The relaxation time, τ, retrieved from fitting the CR spectra with Eq. (2) [closed squares]. The mobility, μ, obtained from τ is also shown (open squares).

The relaxation time, τ, was also extracted from fitting the CR transmission spectra with Eq. (2). The results exhibit oscillatory behavior similar to that of m* as shown in Fig. 3. This time can be converted into the 2DEG mobility (μ) as shown in Fig. 3 using the relationship $\mu=\frac{e}{m^*}<\tau>$ where $<\tau>$ is the average relaxation time which is equivalent to τ obtained from the CR measurements. The Hall effect mobility on the other hand is measured and found to be 33.6x10^3 cm^2/V-s at 78 K. Note, however, that the Hall mobility involves a different relaxation time and different energy averaging. The relaxation time in Fig. 3 can also be converted into the FWHM (Γ) using the relationship (Ando and Uemura 1974)

$$\Gamma = \sqrt{\frac{2}{\pi}\hbar\omega_c\frac{\hbar}{\tau}} \ , \tag{3}$$

which was derived from the self-consistent Born approximation where short-ranged potentials were assumed for the main scatterers. The results are shown in Fig. 4 (open squares) along with the FWHM measured directly from the CR spectra (closed squares). However, Eq. (3) was derived in terms of τ under zero magnetic field. Since τ is not known for the present quantum well under zero magnetic field, we calculated τ from the Hall mobility which was measured at B < 0.5 T. The result of Eq. (3) with Hall τ is shown in Fig. 4 as the solid line. Both measured and calculated FWHM show similar oscillatory behavior on top of a slow rise as B increases. Despite these good agreements, there are two problems. First, the oscillation amplitudes of the measured FWHM are much larger than those of the calculated FWHM. Second, the calculated FWHM, in particular the solid line, is smaller than the measured FWHM. These disagreements suggest that other effects such as screening (Hopkins et al 1989), impurities and carrier-carrier scattering (Kobori et al 1990b), phonon scatterings (Kobori et al 1990a), and scattering from long range potentials (Tsui 1982) are not negligible. The agreement between theory and experiment is improved by calculating FWHM from Eq. (3) using the CR relaxation time extracted from Eq. (2). This improvement indicates that the CR τ reflects some of the above effects [assuming that Eq. (3) is still valid for τ obtained at B $\neq$ 0 T]. In addition, the g_o^* factor enhancement at certain filling factors (even integers) seems to play an important role (see Fig. 1) in determining the FWHM in the present single quantum well. Further analysis will be reported elsewhere.

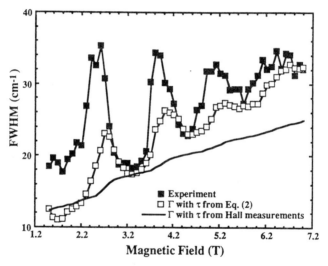

Fig. 4. The FWHM measured directly from the CR spectra (solid squares), and FWHM calculated from Eq. (3) using Hall relaxation time (solid line) and using the CR τ of Fig.3.

Linewidth oscillations of CR spectra in GaAs 2DEGs have been observed previously with maxima at even integral values of ν (Ando et al 1981; Englert et al 1983; Heitmann et al 1986; Sarma 1980; Lassnig and Gornik 1983; Ando and Murayama 1985; Manasreh 1991a) which seem to reflect the absence of a contribution to screening from intra Landau level excitations whenever a Landau level is full. In addition, it was speculated that the same effect should produce linewidth maxima at odd integral values of ν when the exchange enhancement of the g_o^* factor is large enough (Shlesinger et al 1987). The g_o^* factor in InAs (see Fig. 2) is larger than that observed in GaAs, and therefore one would expect to observe oscillations at even and odd integral values of v. The oscillations in the present FWHM and other parameters, namely, m^*, g_o^*, and τ, seem to correlate with even integral values of v for $B > 4.2$ T (as shown in Fig. 2). However, for $B < 4.2$ T one should observe many oscillations, but as pointed out earlier the absence of these oscillation is perhaps due the low resolution and stability of the superconducting magnet.

It should be pointed out the the CR technique cannot measure g_o^* directly, but definitely can probe the changes in this factor. Figure 1 shows strong evidence that g_o^* oscillates as a function of B and Fig. 2 provides additional evidence to support this oscillatory behavior. Equation (1) however is oversimplified because it does not take into account the electron-electron exchange interaction effect which produces enhancement in the g_o^* factor (Goldberg et al 1989; Nicholas et al 1988; Ando and Uemura 1974; Englert et al 1982). Further analysis regarding this issue will be presented elsewhere. In addition, the spin splitting in an InAs/AlSb single quantum well similar to ours has been observed recently at gentetic field as low as 2 T by using Shubnokov-de Haas technique (Hopkins et al 1991).

3. InAs/AlGaSb SINGLE QUANTUM WELL

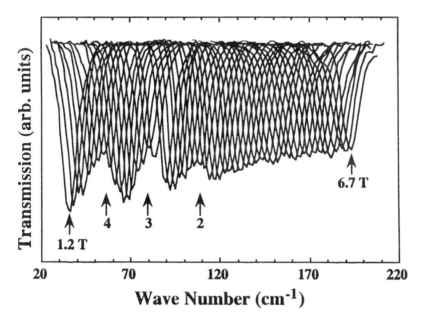

Fig. 5. Cyclotron resonance transmission spectra InAs/AlGaSb single quantum well taken at 4.2 K in the magnetic field range of 1.2 - 6.7 T at an increment of 0.1 T. The vertical arrows with the integers represent the position of integral values of the filling factor. The spectra were taken after a red LED illumination.

Figure 5 shows several CR transmission spectra taken at 4.2 K in the B range of 1.2 - 6.7 T at a constant increment of 0.1 T after illuminating the sample with a red LED light.   It is clear from this figure that the amplitude as well as the linewidth vary (oscillate) as a function of B. The vertical arrows with the integer numbers indicate the positions of the integral values of the filling factors. The filling factors v = 2, 3, and 4 are calculated using n_s = 3.6x10^{11} cm^{-2} obtained directly from Shobnikov - de Haas measurements. The amplitude minima of the CR spectra correlate very well with v. However, it should be noted that the amplitude minima at v = 5 and 6 are not clear in Fig. 5 because LL's separations are too small for CR measurements to resolve.

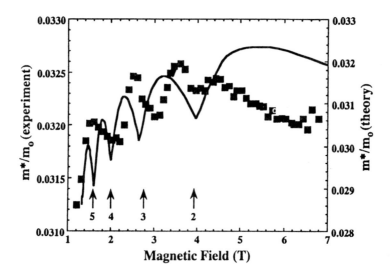

Fig. 6.   Comparison between the measured (solid squares) and calculated (solid line) electron effective mass. The measured data were obtained directly from the cyclotron resonance spectra of Fig. 6 as a function of the magnetic field strength. The calculated electron effective mass was obtained from Manasreh et al. (Ref. 28).

The electron effective mass, m^* = $eB/\omega_c c$, can be obtained directly from the peak position frequencies (ω_c) of the CR transmission spectra of Fig. 5 or by fitting CR spectra with Eq. (2). The results of m^* as a function of B are shown in Fig. 6.   The fitting is excellent for all CR spectra.   An example is shown in Fig. 7.   The solid curve in this figure is the experimental CR spectrum taken at B = 6.0 T and the dashed curve is the theoretical fit. Small amplitude oscillations were observed in n_s as a function of B with an average value of (3.70 ± 0.07)x10^{11} cm^{-2}.   This value is in excellent agreement with SdH measurements (obtained after LED illumination) of n_s = 3.60x10^{11} cm^{-2}. On the other hand, the scattering time (τ) obtained from fitting the CR spectra of Fig. 5 using Eq. (2) exhibits oscillatory behavior with minima at integral values of v. A detailed discussion regarding τ and the linewidth will be presented elsewhere (Manasreh 1992). It should also be noted that τ is about an order of magnitude larger than the scattering time (τ_s) obtained from Shubnikov - de Haas measurements. This discrepancy arises because τ is obtained at a fixed B while τ_s is extracted at different values of B (Manasreh et al 1991b). Therefore, these parameters represent two different scattering mechanisms. The results of m* obtained directly from the CR spectra or from Eq. (2) are identical and are presented by the solid squares in Fig. 6. The data exhibit oscillatory behavior with three clear oscillations at $v = 2, 3$, and 4, but the maxima do not coincide with integral filling factors, in disagreement with the results obtained for AlGaAs/GaAs heterostructures (Richter et al 1989, 1990). The maxima appear to be shifted to half-integer filling factors.

In order to explain the oscillatory behavior of m* as a function of B, Manasreh *et al* (1991b) and Gumbs *et al* (1991) developed an electron self-energy model based on the Hartree-Fock approximation in which the LL broadening (due to disorder) is simulated by a Lorentzian lineshape. The result is shown as a solid line in Fig. 7. Both experimental and theoretical data in Fig. 7 exhibit similar oscillatory behavior. The theoretical m* minima occur at integral values of v while the maxima occur at half-integral values of v. The experimental m* minima at $v = 2$ and 4 are in excellent agreement with the theoretical results. On the other hand, the experimental m* minimum at $v = 3$ is within 0.4 T from the theoretical minimum. The m* oscillation around $v = 5$ is not obvious as shown in Fig. 7 because the separation between LLs is very small and therefore it is difficult for the CR technique to probe this oscillation.

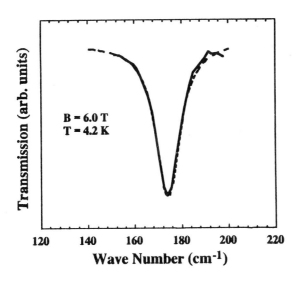

Fig. 7. Comarison between the measured (solid line) and fitted (dashed line) cyclotron resonace transmission spectra. The measured spectrum was taken at 4.2 K and B = 6.0 T and the fitted spectrum is Eq. (2) with n_s, m*, and τ being the fitting parameters.

4. GaAs/AlGaAs HETEROSTRUCTURES

Two samples of GaAs/AlGaAs heterostructures are taken here as an example and their characteristics are summarized in table I. In Fig. 8 we show selected CR spectra obtained at different magnetic field values for sample No.1. The transmission scale is the same for all spectra, but the spectra were shifted vertically with respect to each other for the purpose of comparison. The amplitudes, FWHM, and total integrated area of these spectra show a similar oscillatory behavior as a function of the magnetic field for all samples investigated.

The effective mass, m^*, is calculated from $\omega_c = eB/m^*c$ where ω_c is the cyclotron resonance frequency obtained directly from the spectra in Fig. 8 and the results are shown in Fig. 9. The data in this figure show anomalous behavior around B ~ 6, 4, and 3 T. The FWHM and amplitude of the CR spectra also oscillate as the magnetic field is increased.

Table I.   Characteristics of GaAs/AlGaAs heterostructure samples used in the current study. Hall measurements were made at 77 K with white light on. Cyclotron resonance measurements were performed at 5 K and B=7.0 T. n_s was obtained from fitting Eq. (1) to the CR spectra.

	Sample number	
	1	2
Channel layer (Å)	5000	10 000
Spacer layer (Å)	20	200
Donor Layer (Å)	500	400
Cap layer (Å)	500	200
Hall Mobility (10^3 cm^2/V-s)	40	146
CR mobility (10^3 cm^2/V-s)	73	204
Hall concentration (10^{11} cm^{-2})	5.64	4.92
n_s (10^{11} cm^{-2})	4.20	2.97
m^*/m_0	0.0727	0.0699

The theoretical fit of Eq. (2) to the CR spectra is excellent and the extracted n_s values are shown in Figs. 10. The n_s values obtained from fitting Eq. (1) to the CR spectra are in good agreement with those obtained by Hall effect measurements as shown in table I. For example, the value of n_s obtained from fitting Eq. (2) to the CR spectrum of sample No. 1 at B=7 T is 4.20×10^{11}cm^{-2}, while n_s is 5.64×10^{11}cm^{-2} from Hall effect measurements at 77K. The relaxation time obtained from the CR can be converted into the 2DEG mobility using the relationship (Look 1989) $\mu = \dfrac{e}{m^*} <\tau>$ where $<\tau>$ is the average relaxation time which is equivalent to τ obtained from the CR measurements. We obtain μ ~ 73 000 and 204 000 cm^2/V-s at B=7T for samples No.1 and 2, respectively. These values are comparable with the mobility of 40 000 and 146 000 cm^2/V-s obtained from Hall effect measurements at 77K for samples No.1 and 2, respectively. The main reason for the large difference between the Hall and CR mobilities is that the two experiments were performed at different temperatures. However, preliminary results (see table I) indicate that the thicker the spacer layer the higher the mobility in agreement with Ando and Murayama (1985). This finding may have an important implication on electron-ion interaction in HEMT structures.

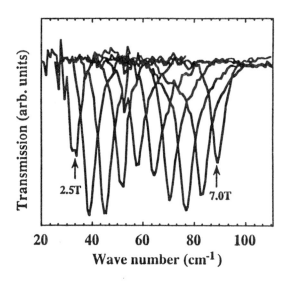

Fig. 8. Selected CR transmission spectra of GaAs/AlGaAs heterostructure taken at 5.0 K and at a constant increment (0.5T) of the magnetic field for sample No. 1 of table I. The spectra are shifted vertically with respect to each other for comparison purposes, but all spectra have the same vertical scaling.

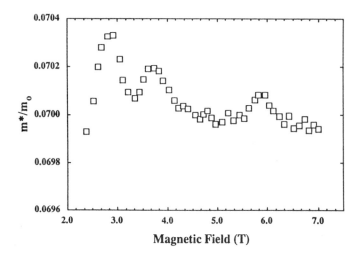

Fig.9. Relative CR effective mass m^*/m_0 of GaAs/AlGaAs heterostrucure taken for sample No. 2 of table I, where m_0 is the free electron mass, as a function of the magnetic field.

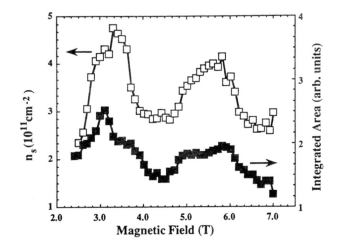

Fig. 10. The total integrated absorption recorded directly from the CR spectra of GaAs/AlGaAs heterostructure and n_s obtained from fitting Eq. (2) to the CR spectra of sample No. 2 of table I.

One striking feature of the present results is that the concentration of the 2DEG, n_s, varies as a function of the magnetic field. Additional supporting evidence for this finding is that the total integrated area, which is directly proportional to the concentration of electrons which undergo Landau transitions, of the CR spectra varies with magnetic field strength in a fashion similar to that of the n_s (the results are polotted along n_s in Fig. 10). The oscillatory behavior of n_s is in excellent agreement with that of the total integrated area measured directly from the CR spectra. In addition, the magnitude of n_s variation ($\sim\frac{4.6}{2.8} \sim 1.6$) is remarkably in good agreement with that of the integrated area ($\sim\frac{2.5}{1.4} \sim 1.8$). It should be emphasized that the variation of n_s with the magnetic field does not necessarily imply that the carriers are moving into or out of the triangular quantum well formed at the channel-spacer interface; it simply indicates that the number of electrons involved in transitions between Landau levels as measured by the CR depends on the magnetic field strength. The oscillatory behavior of n_s as well as the integrated area is in apparent agreement with the self-consistent calculations reported by Sanchez-Dehesa *et al* (1987) for a GaAs/AlGaAs selectively doped single quantum well. Sanchez-Dehasa *et al* based their interpretation of the oscillatory behavior of n_s on the existence of charge oscillation between the quantum well and the doped barrier. This interpretation is in agreement with the explanation of the quantum Hall effect in heterojunctions (Baraff and Tsui 1981; Toyoda et al 1984) which was based on the existence of an electron reservoir outside the 2DEG system. According to Baraff and Tsui (1981), the donor impurities in the barrier may act as a reservoir which in turn offers a possibility of changing n_s over wide limits in order to observe the quantum Hall effect in the SdH experiment. This idea is supported by the experimental measurements (Nizhankovskii et al 1986) of the chemical potential in GaAs/AlGaAs heterostructure which also shows oscillatory behavior as a function of the magnetic field. Nizhankovskii *et al*[35] found that the assumption of a constant n_s does not agree with the magnetic field dependence of the chemical potential of the 2DEG and therefore n_S should have an oscillatory character. Additional support is found in the reported inhomogeneity in n_S near the filling factor $v = 2$ in gated AlAs/GaAs heterostructures (Thiele et al 1989) and in the anomalous behavior of m* as a function of n_S near integer values of v in a tilted magnetic field experiment (Demin and Malyavkin 1990).

The above argument is in conflict with the localized state interpretation for the quantum hall effect (Laughlin 1981; Halperin 1982; von Klitzing 1986; Prange 1981; Aoki and Ando 1981; Chalker 1983). The conduction according to this interpretation is produced by electrons with energy lying near the center of the Landau levels (extended states), while the rest of the electrons are lying on the wings of the Landau levels (localized states). One important aspect of the localized state model is that n_S remains constant (Anenko and Lozovik 1985). This is, however, in disagreement with the present results of n_S as shown in Fig. 10. Our intention here is not to provide a microscopic interpretation for the quantum Hall effect, but to demonstrate that the pronounced variation of n_S as a function of the magnetic field may reflect a strong electron-ion interaction. This premise is supported by the dramatic influence of impurities, both donor and acceptor, on the CR measurements (Richter et al 1989, 1990) in particular m* and FWHM.

The anomalous behavior of m* near even numbers for the filling factor is in good agreement with the previously published results (Richter et al 1989, 1990) for δ-doped and n-type (Thiele et al 1987) GaAs/AlGaAs heterostructures. In addition, a sudden jump in m* was observed (Ensslin et al 1987) to occur near $v=1$. The FWHM also exhibits anomalous behavior near integral values of the filling factors (Englert et al 1983; Richter et al 1989, 1990). The correlation between the FWHM and the filling factor of the Landau levels is not well understood. However, Schlesinger *et al* (1987) observed a dramatic resonance narrowing and shift as the 2DEG density was reduced below the point at which the lowest spin-split Landau level was filled. They provided a qualitative explanation for such behavior. The 2DEG loses its ability to screen at very low temperature when $v=1$, because a large gap opens up at the Fermi surface as a result of the exchange-enhanced spin splitting. Hence, the interaction between electrons and the finite-wavelength magnetoplasmons becomes stronger causing a broadening in the FWHM at $v=1$ and narrowing when $v<1$. In other words, the screening effect is minimum when Landau levels are empty or full (v=integer) and maximum when Landau levels are half-full. Additional effects that may strongly influence the anomalous behavior observed in m*, FWHM, n_S, and τ as a function of magnetic field strength near integer values of the filling factor include dynamic screening (Englert et al 1983; Heitmann et al 1986; Sarma 1980, 1981), finite temperature (von Klitzing 1986), inhomogeneity (Woltjier 1989), electron-electron interaction (Shlesinger et al 1987), energy renormalization (Ensslin et al 1987), band structure nonparabolicity (Thiele et al 1987; Hopkins et al 1987), polaron (Larsen 1984; Horst 1985; Wei et al 1988, 1989; Wu et al 1989), electron-phonon interaction (Larsen 1984; Horst 1985; Wei et al 1988, 1989; Wu et al 1989; Chang et al 1988), and electron-impurity interaction (see for example Refs. 29, 33-34, 54-56).

5. CONCLUDING REMARKS

Cyclotron resonance measurements were performed for the first time in a InAs/AlSb single quantum well structure grown by the molecular beam epitaxial technique. The electron effective mass, m*, in the InAs quantum well was calculated from the peak position energy of the CR spectra and found to be $0.043m_0$ at B=7 T. This value is about a factor of 2 larger than that observed in bulk InAs. We also calculated the effective g_0^* factor using an expression derived for single particle calculations from the fourth order effective mass theory (Palik et al 1961). The CR spectra provided strong evidence that g_0^* exhibits an oscillatory behavior as a function of B and that this factor is changing with Landau levels. In addition, similar oscillations were observed in m*, the relaxation time obtained by fitting the CR spectra with a Drude-type conductivity model, and the FWHM obtained either directly from the CR spectra or calculated from the self-consistent Born approximation linewidth. The CR spectra show clear splitting at certain values of the magnetic field (see Fig. 1). This splitting is explained in terms of Landau levels spin splitting. Further analysis regarding electron-electron exchange enhancement of g_0^* and the effects of various contributions to the FWHM are needed. The effective mass of the 2DEG in an $Al_{0.6}Ga_{0.4}Sb/InAs$ SQW is also determined to be $\sim 0.032m_0$. The cyclotron resonance results exhibit an oscillatory behavior as a function of the magnetic field with minima occuring near half-integral values (both even and odd) of the filling factors. An electron self-energy model based on the Hartree-Fock approximation developed by Gumbs *et al* (1991) seems to explain the oscillatory behavior of the cyclotron resonance effective mass as a function of the magnetic field. The theoretical calculations of the effective mass show a trend similar to that of the measured effective mass. Spin-splitting of Landau levels in the above structure is inferred indirectly from the cyclotron resonance measurements where the oscillations of the spectral amplitudes (as well as the linewidth) were found to occur at both even and odd integral values of the filling factors. A persistent photoconductivity effect opposite to that of the DX center in AlGaAs is observed in this single quantum well after red LED illuminations. Further analysis is in progress to understand the mechanisms of Landau levels broadening.

Far-infrared cyclotron resonance has also been used to study the 2DEG system in GaAs/AlGaAs heterostructures grown by either MBE or MOCVD techniques. The relaxation time and concentration of the 2DEG were obtained from fitting the CR transmission spectra with a transmissivity equation based on the dynamical conductivity of a Drude type model. The estimated values of the 2DEG mobility and concentration obtained by the CR technique have the same trend as those obtained from the conventional Hall effect measurements. A pronounced anomalous behavior was observed in the total integrated absorption of the CR transmission spectra, and the concentration of the 2DEG system near integer values of the filling factor obtained from SdH results. The most important result of the latter study is that n_s varies as a function of the magnetic field in support of recent theoretical and experimental reports. The implication of this finding lies in the microscopic interpretation of the quantum Hall effect. This result does not, however, rule out the localized and extended states model for the quantum Hall effect, but it shows a strong indication that electron-impurity interaction is not negligible.

ACKNOWLEDGMENTS

This work was partially supported by the Air Force Office of Scientific Research. I would like to thank Mr. G. R. Landis of University of Dayton Research Institute and T. Cooper of Wright State University for performing the Hall effect measurements, I. Lo of the National Research Council and W. C. Mitchel of Wright Laboratory for providing Shubnokov - de Haas results, Mr. R. E. Perrin for his help in the computer analyses, and C. E. Stutz, K. R. Evans, E. Taylor, and J. Ehret of Wright Laboratory for the MBE growth.

References

Ando T and Uemura Y 1974 J. Phys. Soc. Jpn. **37** 1044
Ando T 1975 J. Phys. Soc. Jpn. **38** 989
Ando T, Fowler A B and Stern F 1981 Rev. Mod. Phys. **54** 437
Ando T and Murayama Y 1985 J. Phys. Soc. Jpn. **54** 1519
Anenko S M and Lozovik Yu E 1985 Zh. Eksp. Teor. Fiz. **89** 573 [1985 Sov. Phys. JETP **62** 328].
Aoki H and Ando T 1981 Solid State Commun. **38** 1079
Baraff G A and Tsui D C 1981 Phys. Rev. B **24** 2274
Bluyssen H, Maan J C, Wyder P, Chang L L and Esaki L 1979 Solid State Commun. **31** 35
Chalker J T 1983 J. Phys. C **16** 4297
Chang Y -H, McCombe B C, Mercy J -M, Reeder A A, Ralston J and Wicks G A 1988 Phys. Rev. Lett. **61** 1408
Chiu K W, Lee T K and Quinn J J 1976 Surf. Sci. **58** 182
Demin A A and Malyavkin A V 1990 Fiz. Tverd Tela **32** 309 [1990 Sov. Phys. Solid State **32** 178].
Englert Th, Tsui D C, Gossard A C and Uihlein Sh 1982 Surf. Sci. **113** 295
Englert T, Maan J. C, Uilein C, Tsui D C and Gossard A C 1983 Solid State Commun. **46** 545
Ensslin K, Heitmann D, Sigg H and Ploog K 1987 Phys. Rev. B **36** 8177
Glaser E, Shanabrook B V, HJawkins R L, Beard W, Mercy J -M, McCombe B D and Musser D 1987 Phys. Rev. B **36** 8185
Huant S, Stepniewski R, Martinez G, Thierry-Mieg V, and Etienne B 1989 Superlattices and Microstructures **5** 331
Goldberg B B, Heiman D and Pinczuk A 1989 Phys. Rev. Lett. **63** 1102
Gumbs G, Zhang C and Manasreh M O 1991 Phys. Rev. B (submitted).
Halperin B I 1982 Phys. Rev. B **25** 2185
Heitmann D, Ziesmann M and Chang L L 1986 Phys. Rev. B **34** 7463
Hopkins M A, Nicholas R J, Brummell M A, Harris J J and Fox C T 1987 Phys. Rev. B **36** 4789
See for example Hopkins M A, Nicholas R J, Barnes D J, Brummell M A, Harris J J and Foxon C T 1989 Phys. Rev. B **39** 13302
Hopkins P F, Rimberg A J, Westervelt R M, Tuttle G and Koemere H 1991 Appl. Phys. Lett. **58** 1428
Horst M, Merkt U, Zawadzki W, Mann J C and Ploog K 1985 Solid State Commun. **53** 403
Huant S, Stepniewski R, Martinez G, Thierry-Mieg V, and Etienne B 1989 Superlattices and Microstructures **5** 331
von Klitzing K. 1986 Rev. Mod. Phys. **58** 519
See for example Kobori H H, Ohyama T and Otsuka E 1990a J. Phys. Soc. Jpn. **59** 2141
See for example Kobori H, Ohyama T and Otsuka E J 1990b Phys. Soc. Jpn. **59** 2164
Larsen D M 1984 Phys. Rev. B **30** 4595
Lassnig R and Gornik E 1983 Solid State Commun. **47** 959
Laughlin R B 1981 Phys. Rev. B **23** 5632
Litton C W, Dennis R B and Smith S D 1969 J. Phys. C **2** 2146
Look D C 1989 *Electrical Characterization of GaAs Materials and Devices*, (Wiley, New York) p. 65.
Manasreh M O, Fischer D W, Stutz C E and Evans K R 1991a Phys. Rev. B **43** 9772
Manasreh M O, Gumbs G, Zhang C, Stutz C E, Evans K R, Bozada C A, Lo I and Mitchel W C 1991b Phys. Rev. B (submitted).
Manasreh M O *et al* . 1992 (unpublished).

Mercy J -M, Chang Y -H, Reeder A A, Brozak G and McCombe B D 1988 Superlattices and Microstructures **4** 213

Nakagawa A, Kroemer H and English 1989 J Appl. Phys. Lett. **54** 1893

Nicholas R J, Haug R J, Klintzing K V and Weimann G 1988 Phys. Rev. B **37** 1294

Nizhankovskii V I, Mokerov V G, Medvedev B K and Shaldin Yu V 1986 Zh. Eksp. Teor. Fiz. **90** 1326 [1986 Sov. Phys. JETP **63** 776].

Palik E D, Picus G S, Teitlee S and Wallis R F 1961 Phys. Rev. **122** 475

Palik E D and Stevenson J R 1963 Phys. Rev. **130** 1344

Prange R E 1981 Phys. Rev. B **23** 4802

Richter J, Sigg H, Klitzing K V and Ploog K 1989 Phys. Rev. B **39** 6268; 1990 Surf. Sci. **228** 159

Sanchez-Dehesa J, Meseguer F, Borondo F and Maan J C 1987 Phys. Rev. **B** 5070

Sarma S Das 1980 Solid State Commun. **36** 357 ; 1981 Phys. Rev. B **23** 4592

Shlesinger Z, Wang W I and MacDonald A H 1987 Phys. Rev. Lett. **58** 73

Thiele F, Merkt U, Kotthaus J P, Lommer G, Malcher F, Rössler U and Weimann G 1987 Solid State Commun. **62** 841

Thiele F, Batke E, Dolgopolov V, Kotthaus J P, Weimann G and Schlapp W 1989 Phys. Rev. B **40** 1414

Toyoda T, Gudmundsson V and Takahashi Y 1984 Phys. Lett. **A 102** 130

See for example Tsui D C, Störmer H L and Gossard A C 1982 Phys. Rev. Lett. **48** 1559; Ref. 21 and references therein.

Tuttle G, Kroemer H and English J H 1989 J. Appl. Phys. **65** 5239

Tuttle G, Kroemer H and English J H 1990 J. Appl. Phys. **67** 3032

Wei C -W, Kong X. -J, Gu S -W 1988 Phys. Rev. B **38** 8390; 1989 ibid **39** 3230

Woltjer R 1989 Semicond. Sci. Technol. **4** 155 and references therein

Wu X, Peeters F M and Devreese J T 1989 Phys. Rev. B **40** 4090

Part III

II–VI   Compound Semiconductors

Chapter 8

Optics in lower-dimensional quantum confined II–VI heterostructures

A.V. Nurmikko[1] and R.L. Gunshor[2]
[1]Division of Engineering and Department of Physics
Brown University, Providence, Rhode Island 02912
[2]School of Electrical Engineering,
Purdue University, West Lafayette IN 47907

ABSTRACT: Recent progress in wide-gap II-VI heterostructures is reviewed with emphasis on optical properties ensuing from quantum confinement in multilayer structures. The results include blue and green diode lasers and light emitting diodes in the ZnSe-based quantum wells configurations. An important element in these structures are quasi-2 dimensional exciton effects which are likely to be of relevance as practical optoelectronic devices emerge at short visible wavelengths.

1. INTRODUCTION

Considerable enthusiasm envelopes today much of the research in the field of wide-gap II-VI compound semiconductors. While the field has a venerable past, with much of the work on ZnSe, ZnTe, ZnS, CdTe, CdSe and their compounds traced to the 1960's, a big "second wind" blew in about eight years ago. At that time a few adventurous groups began to apply modern epitaxial instrumentation, especially molecular beam epitaxy (MBE), to the fabrication of II-VI heterostructures, both in the wide-gap [Gunshor et al 1990] and the narrow-gap [Farrow et al 1987] materials. A rather rapid rate of progress has in fact ensued, leading to profound improvement in material quality and purity, as well as a broad physical understanding of a number of II-VI quantum well (QW) and superlattice systems. Consequently, as illustrated in this article, guidelines for designing quantum confining structures for optoelectronic device applications can now be discussed. In another seminal specific development, useful p-type doping of ZnSe has recently been realized [Park et al 1990], and, while still not fully optimized, has nonetheless led to p-n heterojunction proof-of-concept devices for light emitters in the blue and the green. The point of culmination of this work was the achievement in 1991 of diode laser operation in the (Zn,Cd)Se/ZnSe [Haase et al 1991], [Jeon et al 1991(a)] and (Zn,Cd)Se/Zn(S,Se) QWs [Jeon et al 1992(a)].

The purpose of this paper is to focus on selected aspects of wide-gap II-VI heterostructures, primarily ZnSe-based QWs. In addition to their intrinsic electronic properties, these structures are here also considered from the perspective of light emitter applications at short visible wavelengths, both for diode lasers and LEDs. Given the breakthroughs in the past year, there is now a real possibility that the II-VI semiconductors will find a special niche for such applications in the blue and the green portions of the spectrum. This follows from the fact that no practical semiconductor optoelectronics is yet present in this wavelength range. By way of comparison, so far SiC LEDs of modest efficiency (10^{-4}) and broad spectrum are commercially available as blue emitters; however, the indirect bandgap and polytype nature of SiC makes it less attractive as a laser material. Elsewhere, recent work on bulk GaN and some efforts at heterostructure design with GaN/(Al,Ga)N have yielded promising initial results for LED devices [Akasaki et al 1992]. The optical transition responsible for blue emission in this wide gap material is associated with an impurity/defect level. Finally, GaAsP QW diode lasers have been operated in the red at room temperature and in the green at T=77 K [Valster and Acket 1991], however it is rather unlikely that the green wavelength regime can be reached at room temperature with this heterostructure.

When contrasted with "mature" III-V semiconductors such as GaAs, it is important to recognize the large ionic component in the chemical bonding which enters into defining the optical and other physical properties (e.g. mechanical) of the wide-gap II-VIs. One consequence is the strong excitonic character which literally "colors" the photoresponse of QW structures so that both absorptive and emissive processes are likely to contain a substantial excitonic component up to device temperatures in these quasi-2-dimensional (2D) systems. The excitonic component is clearly present in absorption and spontaneous emission to room temperature in (Zn,Cd)Se QWs and appears to be important also in stimulated emission processes in laser devices.

Among the wide-gap II-VI heterostructures which have been explored through optical investigations are superlattice and QW configurations composed of ZnSe, ZnTe, ZnS, CdTe, MnSe, and MnTe, including their ternary compounds. This is illustrated in **Figure 1**, which shows the bandgap vs. lattice diagram for the wide gap II-VI semiconductors in the cubic phase (room temperature). The solid lines indicate some of the heterostructures studied so far. Below, a selected sampling of this work will be presented, with emphasis on results obtained in the authors' laboratories at Brown University and Purdue University in collaboration with other groups, especially the important contributions at the University of Notre Dame. Consequently, many other important results will not be covered here and we apologize for these omission; we especially wish to note work on the

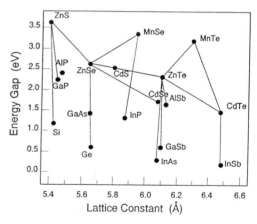

Fig.1. Bandgap vs lattice constant for cubic II-VI wide-gap semiconductors. The solid lines indicate some of the strained heterostructures studied. Some choices for III-V buffer layers and substrates are indicated.

ZnSe/Zn(S,Se) system. A good cross section of contemporary work can be found in the proceedings of the last conference on II-VI semiconductors held in 1991 in Tamano, Japan.

This chapter is organized as follows. In section 2 we consider choices for II-VI heterostructures from the standpoint of electronic confinement for possible optoelectronic applications at short visible wavelengths. We show earlier examples of QW physics in the CdTe/MnTe and ZnSe/(Zn,Mn)Se systems. In contemporary context this leads to the introduction of a particular QW system of specific focus here, based on the (Zn,Cd)Se/ZnSe [and (Zn,Cd)Se/Zn(S,Se)] heterostructure. We underline the chief physical reasons why excitons can remain stable beyond cryogenic temperatures in such a type I system (i.e. both electron and hole confinement in the same well), in spite of the large Fröhlich interaction in ZnSe. In section 3 recent work with pn-junctions and light emitting diodes work is described. In section 4 we first discuss optical pumping of the QW lasers, especially from the standpoint of physical mechanism for gain. This is followed by a review of recent results of blue-green laser diode action in these structures which exploits both the pn-junction development and the QW aspect of the problem. It is important to note that the diode laser development is proceeding at a considerable pace that state-of-the-art devices are not a well defined concept. In anticipation of novel laser geometries, we describe briefly in section 5 work which has produced patterned nanostructures. Finally, we offer some concluding remarks in section 6.

2. Electronic Confinement and Two Dimensional Electron-Hole Systems in Wide-Gap II-VI QWs

An important goal of wide gap II-VI heterojunction research from its beginning has been the realization of a useful type I QW system. A number of early choices (e.g. ZnSe/(Zn,Mn)Se [Kolodziejski et al 1985], [Hefetz et al 1985] and CdTe/ZnTe [Miles et al 1986], [Hefetz et al 1986(a)] were found to possess rather small valence band offsets, while the opposite case of a small conduction band offset was found in the ZnSe/Zn(S,Se) system [Olego et al 1988], [van de Walle 1989]. In case of the Mn-containing diluted magnetic semiconductor (DMS) heterostructures, magneto-optical spectroscopy could be used to directly obtain information about the degree of penetration of the QW electron-hole states into the adjacent DMS layers [Chang et al 1988]. There is certainly no valid a priori justification for referring to the often disparaged "common anion rule" to estimate bandoffsets in the wide-gap II-VI heterostructures, yet from a purely pragmatic viewpoint it nonetheless appears that this rule of thumb operates fairly effectively in most of the systems studied so far, at least for low concentrations of the substitutional alloy component in a ternary case. However, since all the ternary II-VI heterostructures are also strained by lattice mismatch, it follows that allowances must be made for the possibility of using "strain-engineering" to enhance the electronic confinement. In this case, there are severe constrains present in practical structures by the critical thickness considerations, in order to avoid the nucleation of misfit dislocations. Revisiting Figure 1, it is evident from the figure that lattice mismatch strains on the order of few percent will readily occur with random choices of heterostructure constituent materials. While not addressing the problem of critical thickness and dislocation formation in this paper, as a very rough rule of thumb we have found that a mismatch strain of about 1% corresponds to a critical thickness of some 300-400 Å in the ZnSe-based QWs and superlattices.

In this section we first give an example of a highly strained, strongly confining type I QW, namely the CdTe/MnTe binary heterostructure, to demonstrate very large confinement induced electronic energy changes which can occur in the wide gap II-VI materials. We then illustrate a very different aspect of the DMS based materials, namely the conversion of electron-hole pair excitations into the d-electron states of the Mn-ion. The d-electron states are in use today in electroluminescence phosphors in (especially yellow) flat panel display devices.

Following these examples of QW physics we then introduce and concentrate in section 2 and in the balance of this article on the (Zn,Cd)Se/ZnSe (and (Zn,Cd)Se/Zn(S,Se)) QWs which have been found to exhibit particularly useful quasi-2 dimensional electron-hole behavior. This characteristic brings along the very strong excitonic effects in these quasi-2 dimensional systems.

2.1 Strong Electronic Confinement in a Metastable System: The CdTe/MnTe QW

Possibly the strongest type I confinement so far in a semiconductor heterostructure has been demonstrated in the highly strained CdTe/MnTe QWs where the effective optical bandgap was shown to be "tunable" across much of the visible spectrum by varying the CdTe QW thickness in the ultrathin well limit [Ding et al 1991]. This system offers the further example of the interesting "metastable" configuration in that MnTe does not crystallize in the bulk in zincblende cubic form. Once the synthesis of cubic zincblende MnTe was accomplished by MBE [Durbin et al 1989], optical measurements indicated that the s-p bandgap occurs at 3.2 eV at low temperature, thus roughly following extrapolation from previously available data for (Cd,Mn)Te or (Zn,Mn)Te at low Mn composition. Such wide bandgaps were also reported for MnS in evaporated polycrystalline zincblende films [Goede et al 1988].

The anticipated very large lattice mismatch in a CdTe/MnTe heterostructure (a = 6.48 Å and 6.28 Å, respectively) implies immediately that very thin layered structures must be employed in order to obtain pseudomorphic (coherently strained) heterostructures. The strong confinement effects are illustrated in **Figure 2** which shows photoluminescence spectra (PL) taken at T= 10K for three SQW samples with approximate well thickness 22, 15, and 10 Å. Note the remarkable degree of confinement which shifts the QW bandgap from the infrared, characteristic of bulk CdTe, across the visible spectrum into the blue. The quantum efficiency in samples emitting in the red, yellow and green is high up to 77K - clearly visible to the eye under modest laser excitation even below the bandgap of MnTe! On the

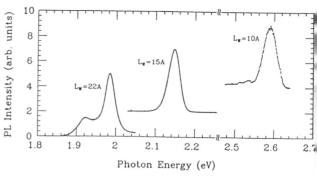

Fig. 2. PL spectra at T= 10 K of three CdTe/MnTe SQW's of different well widths (given in the figure). The bulk exciton bandgap is 1.595 eV [from Ding et al 1991].

other hand, the efficiencies for well thicknesses on the order of 10 Å or less have so far been rather low. Results of the optical studies give values for the conduction and valence band offsets as $\Delta E_c \approx 1.28$ eV, $\Delta E_{HH} \approx 0.34$ eV, and $\Delta E_{LH} \approx 0.16$ eV [Pelekanos et al 1990]. Despite of the demonstration of such large confinement effects, it may be difficult to use the CdTe/MnTe QWs in light emitting applications, given the severe limitations to the layer thicknesses imposed by the lattice mismatch strain (>3%).

While the CdTe/MnTe QW presents a case of a particularly dramatic electronic confinement, work in the past two years has also identified the (Zn,Cd)Se/ZnSe and (Zn,Cd)Te/ZnTe heterojunctions as a good choice for a type I QW, without the requirement of a huge built-in strain. The growth of the (Zn,Cd)Se/ZnSe layered structure was begun at the University of Notre Dame by MBE, with rapid initial progress in terms of the crystalline quality [Samarth et al 1990]. That reasonable confinement of electrons, and especially, of holes was indeed occurring came from direct evidence in optical studies at Brown University on these QW structures, initially through the use of DMS (Zn,Mn)Se barrier layers. From magneto-optical studies, the spatial penetration of the electron-hole wavefunction into the barrier layers could be directly determined [Walecki et al 1990]. The (Zn,Cd)Te/ZnTe QWs have been pioneered at AT&T Bell laboratories and shown also to exhibit optical properties commensurate with type I confinement [Lee et al 1990]. One principal manifestation of such confinement on linear optical properties are enhanced excitonic effects as discussed next.

2.2 Electronic Energy Relaxation in DMS Quantum Wells: The Role of Mn-ion d-Electron States

Apart from increasing the direct bandgap at $k=0$, the incorporation of Mn into ZnSe leads to additional features in the optical spectrum due to the d-electron transitions internal to the Mn-ion. The lowest (crystal field split) transition 6A1→4T1 corresponds to absorption at about 2.1 eV and the zero-phonon line in corresponding luminescence is occurs at about 2.0 eV. This emission is used in thin film insulating (ac) electroluminescence display devices (e.g. with ZnS:Mn) where the d-electron states are excited by very hot electrons injected into the phosphor. In heterostructures such as QWs which contain (Zn,Mn)Se, these 'yellow' resonances can compete for electronic excitation with the 'blue' resonances which are associated with band edge transitions. In particular, there are efficient energy transfer paths from the bandedge exciton states in (Zn,Mn)Se directly into the Mn-ion internal excitation, something which can be graphically demonstrated by comparing thin alloy films of (Zn,Mn)Se with ZnSe/(Zn,Mn)Se quantum wells through time-resolved luminescence spectroscopy [Hefetz et al 1986(b), 1986(c)].

As an example, **Figure 3** shows luminescence spectra from a ZnSe/(Zn,Mn)Se MQW sample at T=2K under cw excitation above the barrier bandgap. The ZnSe well thickness was 67 Å and the Mn-ion concentration in the $Zn_{1-x}Mn_xSe$ barriers $x=0.23$. The spectrally sharp (blue) exciton recombination dominates the broad yellow Mn-ion recombination from the barriers; the ratio of spectrally integrated intensities being about 17 to 1. In strong contrast, the blue recombination in a thin film of(Zn,Mn)Se is dwarfed by the now dominant yellow contribution so that the blue/yellow intensity ratio is about 0.004. Thus, the quantum well structure is efficient in collecting electron-hole pairs prior to any significant energy transfer of such bandedge excitation to the d-electron states of the Mn-ion.

Once captured in a ZnSe quantum well, the photoenergetic electrons and holes relax by optical phonon emission to the n = 1 confined particle states. This relaxation step is fast (probably well below one psec in such a polar material). The subsequent exciton formation and further energy relaxation (localization by quantum well width fluctuations) is a slower process which can be time-resolved through the use of picosecond pulsed laser excitation and a streak camera [Hefetz 1986(b),(c)]. **Figure 4** (left panel) shows the transient luminescence (dotted line)

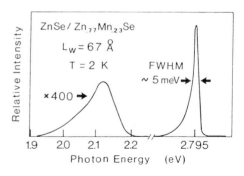

Fig.3. Comparison of blue (≈ 2.8 eV) and yellow (≈ 2.1 eV) luminescence emitted from a ZnSe/(Zn,Mn)Se MQW sample. Note the differences in amplitude and energy scale for each (after Hefetz et al 1985).

from the ZnSe/(Zn,Mn)Se MQW sample when excited above the n = 1 exciton resonance. The risetime is approximately 90 psec; this time constant shortens to approximately 20 psec under resonant excitation, thus yielding a direct measure of the exciton formation step. (Recombination lifetime is, of course, also obtained from the data here $\tau \approx 200$ psec). In contrast, the 'blue' exciton decay in a (Zn,Mn)Se thin film is very fast (≈ 15 psec) as shown in **Figure 4** (right panel); this gives a direct measure of the rate of energy conversion from the exciton state to the d-electron excitation of the Mn-ion.

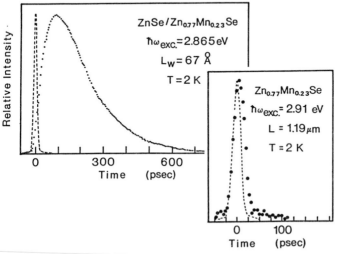

2.3 Quasi-Two Dimensional Exciton Effects in (Zn,Cd)Se QWs

The wide-gap II-VI semiconductors possess naturally strong excitonic effects, derived from the substantial ionic component in their chemistry of bonding. However, until recently, the (very large body of) observations of excitons in absorption,

Fig.4. Time-resolved exciton luminescence for (left) a ZnSe/ZnMnSe MQW and (right) a ZnMnSe film. The dashed lines show time response of monochromator/streak camera system.

reflection, or luminescence has been almost entirely limited to low temperatures (near liquid helium). This is chiefly due to the fact that the exciton-LO phonon (Fröhlich) interaction is very strong and leads to rapid dissociation of the exciton into the free electron-hole pair states at elevated temperatures. For example, in an absorption experiment, this leads to lifetime broadening of the exciton resonance. Two potentially important advantages in a good type I QW are (a) an increase in the exciton oscillator strength (a factor of eight is expected for the case where the QW thickness equals the exciton Bohr radius, $L_w = a_B$), and (b) the possibility that the binding energy E_x can be made to exceed the LO-phonon energy. One or both of these advantages have been identified in ZnTe and ZnSe based heterostructures, respectively, with the addition of Cd to form a ternary QW.

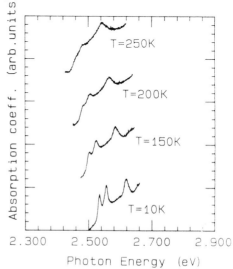

Lee et al [1990] were the first to demonstrate clear exciton absorption features at room temperature in (Zn,Cd)Te/ZnTe MQWs in MBE grown structures. In this case one remains with the usual condition that $E_x < \hbar\omega_{LO}$. Apart from well defined n = 1 heavy-hole exciton resonance in linear absorption, these authors were also able to demonstrate absorption saturation at intensities roughly commensurate with the expected quasi-2D exciton densities for phase-space filling.

Fig.5. Absorption coefficient of a (Zn,Cd)Se/ZnSe sample containing three single quantum wells of thicknesses $L_w = 200$ Å $L_w = 90$ Å, and $L_w = 35$ Å.

The case where, in addition to the advantage in enhanced oscillator strength, a reduction in the Fröhlich interaction is also realized occurs in the (Zn,Cd)Se/ZnSe QWs. Clear room temperature exciton absorption was observed in MQW samples by Ding et al [1990(a)]. Furthermore, exciton absorption even from single QW samples can be resolved. **Figure 5** shows results from a structure into which three (Zn,Cd)Se single QWs of thicknesses 30 Å, 90 Å, and 200 Å were imbedded. Given the approximately 90 Å bulk exciton Bohr diameter in ZnSe, the widest well sample approximates the bulk limit, while the narrowest well sample should correspond to a quasi-2D exciton, provided that the electron-hole confinement is substantially larger than the bulk exciton Rydberg (~17 meV in ZnSe). That this is indeed the case is shown by the dramatically different temperature dependences of the two structures. While all show a strong n = 1 heavy-hole (HH) exciton absorption peak at low temperature ($\alpha \sim 10^5$ cm^{-1}), the linewidth broadens very rapidly with temperature for the widest well sample (similar to bulk ZnSe). In strong contrast, the sample with $L_w = 30$ Å displays the dominant HH exciton absorption peak to well above room temperature.

By using spectroscopic data as input in model calculations in which the excitons are treated variationally, we have shown that the bandoffset values e.g. for the strained $Zn_{0.76}Cd_{0.24}Se/ZnSe$ QW are approximately as follows: $\Delta E_c \approx 180$ meV and $\Delta E_v \approx 80$ meV. Furthermore, the exciton binding energy has been determined to be approximately $E_x \approx 40$ meV, including evidence derived from high magnetic field studies [Pelekanos et al 1992]. Such a large binding energy is a key aspect in reducing the rate of thermally driven dissociation of the exciton into free electron-hole pair states by LO-phonon scattering: in the quasi-2D case we now have that $E_x > \hbar\omega_{LO}$, where the LO-phonon energy is $\hbar\omega_{LO} = 31$ meV. It is the very strong Fröhlich interaction which prevents the observation of excitons at room temperature in bulk ZnSe.

The quasi-2D excitons have been further characterized by magnetoabsorption experiments in fields up to 24 Tesla [Pelekanos 1992]. In addition to the main 1S $n = 1$ HH exciton resonance, the absorption peak due to the 2S HH excited exciton state was also employed. The magnetic field induced shifts of the 1S and 2S peaks are graphed in **Figure 6** for a quasi-2D QW (inset shows the absorption spectrum at B=23.5 Tesla). Although at low magnetic fields the shift of the 2S state is small and comparable to the 1S exciton diamagnetic shift, for B > 10 Tesla the shift becomes almost linear with the magnetic field. For B = 23.5 T this blueshift is six times larger than the one for the 1S exciton. Such a contrasting 1S-2S behavior is consistent with the results of Akimoto and Hasegawa [1966] for 2-dimensional excitons. A portion of such a calculation is included as a solid line in Figure 6 which matches very well with the data points. The identification of the 2S-exciton state enables a determination of the quasi-2D exciton binding energy. Specifically, the difference between the zero-field 1S and 2S exciton energies is measured to be 31 meV. For 3D excitons $|E_{1S}-E_{2S}| = 0.75 E_x^{3D}$, whereas for 2D excitons $|E_{1S}-E_{2S}| = 0.88 E_x^{3D}$. Hence it can be concluded that the binding energy of the quasi-2D exciton in the typical narrow well ZnCdSe/ZnSe QWs is approximately 40 meV, i.e. a value which is considerably larger than the optical phonon energy of 31.7 meV in ZnSe.

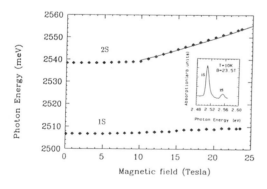

Fig.6. Energy of the 1S and 2S n=1 exciton states as a function of magnetic field. The solid line is a theoretical fit. Inset shows relevant portion of absorption spectrum at B= 23.5 Tesla [from Pelekanos 1992].

Current work into futher details of the excitons in the quasi-2D wide gap QWs includes ultrafast time-resolved spectroscopy. For example, Lee et al [1992] have shown in femtosecond pulsed experiments in (Zn,Cd)Te/ZnTe MQWs how in the presence of a higher density of electron-hole pairs, the exciton absorption saturation is dominated by a gas of cold excitons as opposed to those of thr free particle pairs. This reintroduces problems regarding many-body interactions in circumstances distinct from GaAs QWs where such effects have been actively studied in the past few years. In another different context of time resolved work, Pelekanos et al have demonstrated pronounced hot exciton luminescence effects in ZnTe/MnTe SQWs with thin tunneling barrier layers [Pelekanos et al 1991].

The strong, spectrally well isolated exciton absorption features can in principle be used for devices such as Stark effect modulators. Lee et al have constructed a QW waveguide intensity modulator using (Zn,Cd)Te/ZnTe MQWs [Lee et al 1991]. Work at Brown with the (Zn,Cd)Se/ZnSe QWs has also shown that the quantum confined Stark effect is observable up to room temperature, although the relatively large linewidths make the effect less dramatic than in III-V materials.

3. (Zn,Cd)Se QWs as the Active Region in p-n Heterojunctions and LEDs

Undoubtedly the most frustrating problem in II-VI semiconductor research over the past decades has been the difficulty in their electrical control. The failure to dope e.g. ZnSe both n- and p-type at practical levels was in part responsible for relegating these semiconductors to basic research within a small community. In hindsight, it is clear that much of the difficulties in sorting out the "defect chemistry" which invariably led to charge compensation originated from extrinsic problems with materials. The rapid evolution of the material quality with MBE growth made it finally possible to identify intrinsic issues in the doping process. While successful demonstration of n-type doping of ZnSe, especially by chlorine [Ohkawa et al 1987], [Cheng et al 1988] within MBE-grown epilayers was accomplished some time ago, efforts with p-doping have made the most important leaps only within the past year. In particular, with recent developments in the p-type doping of ZnSe and Zn(Z,Se) by atomic nitrogen in an RF plasma during MBE growth [Park et al 1990], prospects for device quality p-n homo- and heterojunctions suddenly improved dramatically. Contemporary discussion about p-type doping can be found elsewhere, e.g in the II-VI Tamano Conference proceedings [1992] and will not be detailed it here. Suffice it to say that the intrinsic microscopic ideas pointing to the choice of nitrogen as the "best" choice for an acceptor-like dopant in ZnSe have crystallized (in one interpretation) with the recognition that the other heavier and more electropositive column V candidate dopants would likely undergo strong lattice relaxation, henceforth generating deep, compensating (doubly occupied) centers. The insight of Chadi who has modelled this in terms of local bondbreaking and lattice rearrangement around a heavier element such as As or P has been especially important in this viewpoint [Chadi and Chang 1989]. Such lattice relaxation phenomena are not unexpected in the relatively ionic wide-gap II-VIs. We mention in passing studies related to isoelectronic centers of Te in ZnSe QWs which have shown that trapping of (photo)holes is associated with large renormalization energies (up to several hundred meV) at the capture site [Kolodziejski et al 1988], [Fu et al 1989, 1991].

In the work summarized here, (Zn,Cd)Se/ZnSe, (Zn,Cd)Se/ZnSe/Zn(S,Se), and (Zn,Cd)Se/Zn(S,Se) QW p-i-n heterojunctions were grown at Purdue University and fabricated into devices at Brown. Schematic illustration of the principal layered geometries is given in **Figure 7 (a)-(c)**, including doping levels and layer thicknesses. Typically, the intrinsic (Zn,Cd)Se QW layers are thin ($L_w < 100$ Å) and separated by doped ZnSe or Zn(S,Se) barrier layers ($L_b < 500$ Å), while the doped outer layers defining

Principal diode laser and LED layered configurations:

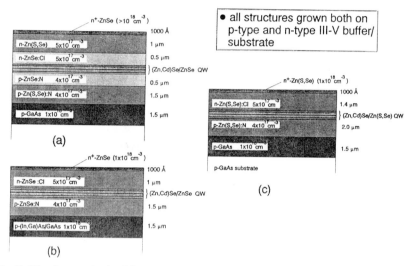

Fig. 7. *The three principal layering configurations used by the authors in LED and diode laser p-n heterojunction devices.*

the pn-junction are relatively thick (~1 μm). Figures 7(a) and 7(b) both house a (Zn,Cd)Se/ZnSe QW active region but differ in that the former structure contains an extra Zn(S,Se) cladding layer. Figure 7(c), in turn, is based on the (Zn,Cd)Se/Zn(S,Se) QW/heterojunction. An important additional point with our work is that the multilayer structures have been grown on both p-type and n-type III-V epilayer/substrates (referred to as "opposite polarities of growth"). These layered geometries have also been used to realize blue and green injection diode lasers as reviewed in section 4.3. A key advantage in the growth of all such structures has been the two chamber MBE machine at Purdue where the III-V epilayer can be grown separately prior to the deposition of the II-VI heterostructure. It has been shown previously that the MBE-grown ZnSe/GaAs epilayer/epilayer interface has excellent electronic properties [Gunshor et al 1987], [Qiu et al 1990]. Generally our measurements of I-V characteristics of such structures indicate that the effective turn-on voltage $V_T \approx 10$ V (for significant forward conduction) is low when compared with those reported previously in ZnSe-based p-n junctions [Xie et al 1992(a)]. Nonetheless, such voltage levels are at this writing still unacceptably high for practical devices, and reflect the problems with the metal/II-VI and other heterojunction impedances. In terms of systematic study, the overall understanding of contact problems in these multilayer ZnSe-based p-n heterojunctions is still at a very rudimentary stage and much active work is required to facilitate e.g. the reproducible fabrication of the low barrier height contacts.

As an example of the heterojunction 'engineering' in such structures, the combination of heavy n-doping within the top ZnSe layer and the finite potential energy barrier for holes at the p-GaAs/p-Zn(S,Se) heterointerface suggests that considerable lateral transport can be expected in broad area structures which are electrically only point contacted. We have made use of this effect by fabricating a "display structure" shown in the upper panel of **Fig. 8** [Jeon et al 1992(b)]. The letters were first defined photolithographically and then wet etched through the II-VI portion of the structure. The overall sample size is about 4x7 mm^2 in the figure, but considerably larger devices could have been made. The combination of lateral current spreading within the top ZnSe layer and the transparent top layers gave raise to a high external quantum efficiency ($\eta_{ext} \approx 10^{-2}$ at T= 77K); for example in photographing the LED emission a 100 W lightbulb was placed next to the sample in order to provide a reasonably compensated background illumination while about 15 mW electrical power was fed through the device. Bandstructure analysis shows the possibility of a quasi-two dimensional hole gas to exist at the ZnSe/GaAs heterointerface and to be an important factor in such parallel conduction.

Fig.8. (Top) A LED display, after QW of Fig. 7(a) [Jeon et al 1992(b)]. (Bottom) 7-segment blue LED after QW of Fig. 7(c) [Hagerott et al 1992].

The shortest wavelengths todate obtained from such LEDs have been achieved with the (Zn,Cd)Se/Zn(S,Se) QW/heterostructure (Fig. 7(c)). This arrangement has also the advantage that lattice matching of the buffer and cap layers to a GaAs substrate/buffer can be achieved, with the (Zn,Cd)Se QW segment pseudomorphically strained, hence drastically reducing the misfit dislocation density to below the 10^5 cm^{-2} range. (Note also, that the structure in Fig. 7(b) has been also grown on (In,Ga)As buffer layers for similar lattice matching and low defect density). **Figure 9** shows a wide area transmission electron cross sectional image to emphasize the scarcity of misfit dislocations in the II-VI portion of such a heterostructure [Xie et al 1992(b)]. For the QW and barrier layer composition of x(Cd)≈0.13 and y(sulfur)≈0.07, room temperature LED emission at 494 nm has been obtained, up to 150 μW output power, in simple planar 2x2 mm^2 devices housing six 75 Å thick QWs [Xie et al 1992(b)]. Room temperature quantum efficiencies approaching 10^{-3} have been reached. **Figure 10** shows the output power from such a LED device as a function of input current; the actual spectrum is presented in the lower half of the figure. Further lithography and processing of this material with a transparent electrode has led to the demonstration of a prototype

Fig.9. Cross-sectional TEM image of structure of Fig. 7(c).

blue numeric display device [Hagerott et al 1992], shown in the bottom panel of **Figure 8**. The EL spectra matched both the photoluminescence and absorption spectra of such structures, indicating that the emission originated from the (Zn,Cd)Se QWs. Spectral redshifts of about 15 meV in the emission with respect to the n = 1 heavy-hole exciton absorption edge are due to the alloy potential fluctuations. These fluctuation are largely responsible for the linewidth at T= 77 K while at higher temperatures the electron- (hole) longitudinal optical phonon interaction broadens the line. Especially at 77 K, the emission is clearly of excitonic in origin and it is likely that the pairwise Coulomb correlations persist up to room temperature in these quasi-2D conditions.

In structures grown on n-GaAs buffer layer/substrates, we have also obtained good junction and LED characteristics. The turn-on voltage is generally higher, typically well in excess of 10 V, most likely because of the higher Schottky barrier at the top p-ZnSe/Au contact, due in part to the lower maximum p-doping level possible at this writing (say $\approx 10^{18}$ cm^{-3}). This presently offsets the advantage in such a structure of the relative small barrier at the n-GaAs/n-ZnSe buried heterointerface for electron injection. Note that in this polarity of growth, the parallel conduction effects are small.

4. Wide Bandgap II-VI Quantum Well Lasers

4.1. Optical Pumping Studies: Overview

Prior to recent success in the p-type doping of ZnSe, laser action in wide bandgap II-VI semiconductors have been demonstrated either by optical or electron beam pumping. By the latter technique, for example, Cammack et al were able to demonstrate pulsed room temperature operation in epitaxial layers of ZnSe and

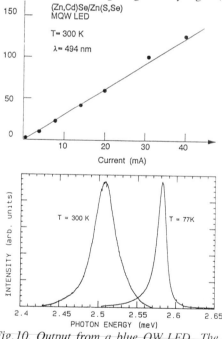

Fig.10. Output from a blue QW LED. The bottom part shows emission spectrum [after Xie et al 1992(b)].

the ZnSe/Zn(S,Se) QW [1987]. In early optically pumped experiments with heterostructures, Bylsma et al employed the ZnSe/(Zn,Mn)Se QW system [1985]. Relatively low threshold operation by optical pumping in the ZnSe/Zn(S,Se) QWs have been obtained recently by optical pumping [Suemune et al 1991], [Sun et al 1991]. Furthermore, Lee et al have demonstrated low threshold room temperature optically pumped lasing in quasi-continuous wave operation from (Zn,Cd)Te/ZnTe lasers [1992(a)] in the red.

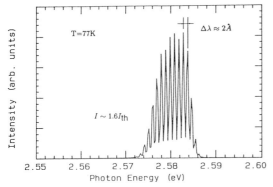

Fig.11. Continuous wave laser emission at T=77K from the (Zn,Cd)Se/ZnSe MQW structure (L_w=35 Å) displaying the longitudinal mode spectrum [Jeon et al 1991(b)].

In this section we focus on the role of (Zn,Cd)Se QWs as active media in both optically and electrically pumped lasers in the blue-green. One point of emphasis is on the issue of lasing mechanism in such strongly quasi-2D excitonic systems. Optical pumping studies in the past two years have given valuable insight about the usefulness of this QW as the key component in a laser structure in the blue green, including requisite carrier injection levels and the physical origin of the lasing mechanism. Following the discussion of the optical pumping studies, we describe recent results by the Brown/Purdue team where (Zn,Cd)Se QW diode laser structures have been fabricated in both polarities of growth (i.e. on n-GaAs and p-GaAs). We also point out that in the recent seminal work by the 3M group on diode lasers [Haase et al 1991], the (Zn,Cd)Se QW system was the active medium of choice.

As a brief summary of the performance of optically pumped (Zn,Cd)Se/ZnSe QWs at Brown University, on structures grown by MBE at the University of Notre Dame (where this heterostructure was first intriduced), we note that apart from pulsed room temperature devices (threshold ≈30 kW/cm²), continuous wave operation has also been demonstrated [Jeon et al 1991(b)]. **Figure 11** shows the emission spectrum from a cw device of 500 μm cavity length, with well defined longitudinal modes present. Six QWs of 30 Å thickness constituted the active medium, but laser action with single QWs has also been demonstrated [Ding et al 1990(b)]. In addition the (Zn,Cd)Se/ZnSe QWs have also been studied as passive waveguides and show good optical confinement characteristics [Walecki et al 1991]. Furthermore, optically pumped laser operation has been realized in structures in which the QW section is fabricated as an ultrashort period CdSe/ZnSe superlattice, including graded index (GRINSCH) configuration in our laboratory. So far, however, problems associated with lattice mismatch strain in these more complex heterostructures have not led to improvement over the simpler ternary MQW arrangement. We also note the work by Kawakami et al who have operated pulsed optically pumped (Zn,Cd)Se/Zn(S,Se) MQW lasers up to 400 C [1991].

As another illustration of the usefulness of the optical pumping approach as a testbed for developing blue-green lasers, we point to recent work in which a distributed feedback (DFB) laser has been realized with the (Zn,Cd)Se/ZnSe QWs, under pulsed conditions at room temperature [Ishihara et al 1992]. The DFB grating structure was fabricated by a combination of two-beam holographic lithography and reactive ion etching methods. **Figure 12** compares the laser emission spectrum from such a DFB laser and a conventional cleaved cavity structure, the latter showing a rich

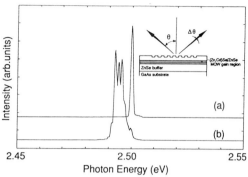

Fig.12. Cleaved cavity laser emission and that from a DFB structure (geometry in inset) in a (Zn,Cd)Se MQW sample at room temperature [after Ishihara et al 1992].

longitudinal mode spectrum. The Bragg condition in this device (5th order diffraction with 4507 Å grating spacing) gives rise to laser emission through the top transparent cap layer of the QW structure with low beam divergence ($\Delta\theta < 1°$), shown in the schematic inset of the figure.

4.2 The Role of Excitons in Stimulated Emission

Apart from being useful in a diagnostic sense of the material quality and helping in the design of actual laser devices, the optically pumped laser studies on the (Zn,Cd)Se/ZnSe QWs have also indicated an interesting likely departure from the conventional laser mechanism associated with other semiconductor lasers. In particular, the extensive spectroscopic studies have accumulated evidence pointing to the important role of excitons in stimulated emission and laser operation in this quasi-2D system. We have already shown above how the 2D excitons in the (Zn,Cd)Se/ZnSe QWs are particularly robust in absorption and spontaneous emission, including room temperature stability. An excitonic element in stimulated emission is, of course, quite contrary to the models of operation of conventional III-V semiconductor lasers where, excepting perhaps very low lattice temperatures, gain in the laser is derived from population inversion supplied by a degenerate electron-hole plasma injected into the active region. The typical

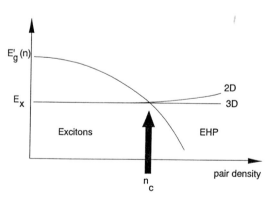

Fig.13. Schematic of an exciton-EHP "phase diagram" showing the bandgap normalization effect on the free e-h pair states and the near constancy of the bound exciton states.

electron-hole pair density at room temperature in a double heterostructure GaAs/(Ga,Al)As laser is approximate 3×10^{18} cm^{-3} and about 1×10^{18} cm^{-3} at T = 77K (in terms of the 3D equivalent density). In terms of many electron (and hole) Coulomb interactions, including exchange and correlation effects, such a density is on the "metallic" side of an insulator-metal transition. This is schematically illustrated in **Figure 13** where the exciton (insulator) energy is seen to be nearly constant with pair density, while the free electron-hole plasma is subject to the bandgap renormalization effect by exchange and correlation effects; the crossing of the two curves defining the phase transition density [Schmitt-Rink et al, 1989], [Zimmerman 1988]. The experimental point in the observation of stimulated emission in the (Zn,Cd)Se QWs is that at the pair densities at which laser action commences, the electron-hole population is still expected to be in the excitonic phase. Gain is possible within the inhomogeneously broadened lowest exciton resonance, starting from the low energy edge, where exciton phase-space filling first bleaches the absorption, followed by gain, while the remainder of the resonance still exhibits (partially saturated) absorption.

An example of the direct observation of the excitonic process is shown in **Figure 14** where optical pumping into the lowest energy resonance of a (Zn,Cd)Se/ZnSe MQW is activating laser emission [Ding 1992(a)]. The work reviewed here was carried out at Brown on structures prepared at the University of Notre Dame. A 1000 Å thick cap layer of ZnSe was grown to complete a waveguide structure, designed in such a way that the lowest order TE$_0$ mode overlapped substantially with the MQW section while not reaching the absorbing GaAs substrate. The optical pumping experiments were carried out in the usual perpendicular (to QW plane) geometry, with stimulated emission measured off the cleaved facets. The effect of varying the pump photon energy is shown by directing some of the pump laser light (from a tunable dye laser) into the spectrometer. Recall that the (electron-hole) bandgap is some 40 meV above this resonance, an energy in excess also of the optical phonon energy. In such an experiment one can precisely measure the pair density at laser threshold, especially if the experiment is performed by picosecond laser pulses delivered in accurate doses. Furthermore, at low lattice temperatures the exciton gas under such resonant conditions is cold such that thermal excitation to the pair states is energetically impossible (especially for the large binding energy). Lasing under resonant excitation conditions has been

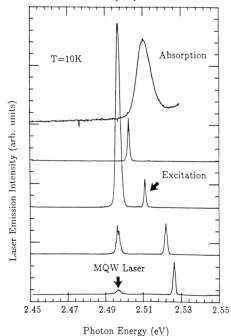

Fig.14. Resonant pumping of the (Zn,Cd)Se/ZnSe MQW laser at T=10 K (top trace shows n=1 HH exciton absorption unexcited sample) [after Ding et al 1992].

observed to near room temperature conditions in this QW system. Other spectroscopic measurements strenghten the argument that the exciton resonance is still present under these conditions, that is, that a transition to the "metallic" regime has not yet occurred.

In another illustration of the origin of the laser action, **Fig. 15** shows how the stimulated emission emerges directly from spontaneous emission spectra in the n = 1 HH exciton region at T = 77K and T = 295 K for 150 μm and 780 μm long QW devices, respectively, pumped under steady-state conditions in the edge emission geometry [Ding 1992(b)]. (In order to

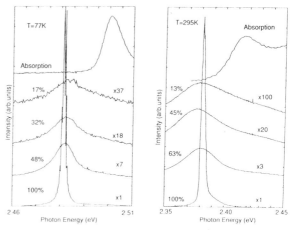

Fig.15. Evolution of laser emission from spontaneous emission at T= 77K and T= 295K (n=1 HH exciton absorption is shown for reference). The maximum excitation level 100% is slightly above the laser threshold [Ding 1992].

maintain a comparable intensity of excitation in both cases, the lower gain at room temperature required the use of a longer resonator). The exciton absorption resonance is included for spectral reference and such experiments have been carried out at photon energies both at and above the exciton resonance. Studies of the emission at excitation levels as low as four orders of magnitude below the laser threshold demonstrate the lack of any substantial spectral shifts which might be indicators of an exciton-EHP transition. Note also that the condition $E_x > \hbar\omega_{LO}$ ensures that quasi-2D excitons are expected to dominate the spontaneous emission. (In the edge geometry, the spontaneous emission is subject to finite self-absorption and hence a spectral redshift due to the long optical pathlength; however, we have verified by simultaneous measurement of the surface emission in the -z direction that the spontaneous emission very nearly coincides with the absorption peak and is clearly excitonic in origin).

Time-resolved pump-probe experiments further show that exciton relaxation into the low energy states of the **inhomogeneously broadened** n = 1 HH resonance can lead to the presence of gain following exciton phase-space filling of these states [Ding 1992(b)]. In simplest terms of such a phase-space filling model, an exciton occupying a particular volume (defined by localization or mean free path) is no longer available for further absorption from the crystal ground state, and is hence automatically inverted with respect to the gorund state. The results of a phenomenological model calculation show an analog to a three level system of the excitonic. Both from the standpoint of the typical electron-hole pair densities at laser threshold ($\sim 9 \times 10^{11}$ cm^{-2} at 77 K) as well as the experimental observation of the remaining presence of (a partially saturated) exciton resonance in pump-probe experiments, we deduce that a free electron-hole plasma is clearly **not** present at cryogenic temperatures (e.g. at T = 77 K), and appears to play a much diminished role even at room temperature conditions. The typical pair densities are in fact well below the exciton screening density in terms of its transition to an electron-hole plasma in the spirit of the Mott transition (screening is, of course, in any case weakened in our quasi-2D conditions); at the same time this density does, however, imply a finite phase-space filling in the inhomogeneous exciton line on its low energy side.

We note that other ideas about excitonic lasing have been previously experimentally explored in bulk II-VI semiconductors under very intense pulsed excitation [Hvam 1973], [Koch et al 1978], [Newbury et al 1991]; however, these have been limited to low temperatures only (T < 10 K) and interpreted in terms of exciton-exciton scattering process to supply gain outside the exciton resonance. In the (Zn,Cd)Se QWs, the quasi-2D enhancement of the binding energy makes excitons relevant at considerably higher temperatures, moreover, the model shows how gain can be available directly within the exciton line itself.

4.3 Blue-green diode lasers.

The MBE-grown p-i-n heterostructures where (Zn,Cd)Se QWs are imbedded within a ZnSe/Zn(S,Se) p-n junctions of varying layering sequences were already introduced in section 3 in connection with LED devices. The three structures of Figure 9 have also been employed in initial demonstration of blue-green diode laser action. The group at 3M first showed laser action at T=77K in structures of Fig. 7(a) [Haase et al 1991], as did the Brown/Purdue team (up to T=250K) [Jeon et al 1991(a)]. Following this, our group has realized diode lasers also in the simpler structures of Figs. 7(b) and 7(c), which offer the important advantage of lattice matching to (In,Ga)As or GaAs buffer layers, respectively [Xie et al 1992(b)], [Jeon et al 1992(a)]. Overall, the rapid recent progress suggests that practical blue-green II-VI laser structures may now be within reach. The Brown/Purdue team has obtained laser diode operation in such multilayer structures grown on both p-type and n-type GaAs, a feature which may add useful flexibility in terms of finding a solution to the difficult (ohmic) contact problem. The contact difficulty presently translates itself into a major electrical dissipation problem which leads to unacceptably high levels of device heating at room temperature. The contact problem notwithstanding, the diode lasers exhibit high peak power and good differential quantum efficiency at T=77K, and pulsed operation has been obtained at this writing up to room temperature in devices with uncoated resonator facets. The group at 3M also reported pulsed room temperature operation with devices where coatings were applied to reduce the requisite threshold current density.

We first touch on results of laser emission from our laboratories with both mesa and broad area devices (the two polarities, respectively), based on the (Zn,Cd)Se/ZnSe/Zn(S,Se) p-n junction geometry [Jeon et al 1991(a)]. In this initial work, the current pulse duration varied from 100 to 500 nsec; at T=77 K a duty cycle up to several percent was reached without special efforts in heatsinking. The onset of lasing was observed as a bright forward emission from the end facets, coupled with expected far-field intensity distribution, and strong spectral narrowing. In mesa structures especially, a rich transverse mode spectrum could be seen. **Figure 16** shows the spectral

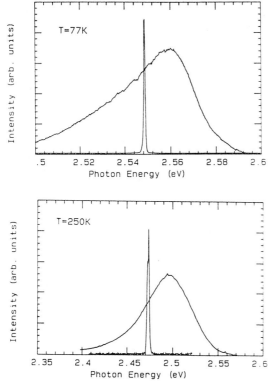

Fig.16. From spontaneous (scale magnified) to stimulated emission in QW laser diode structure of Fig. 7(a) at T=77 K and T=250 K for a mesa device with n^+-ZnSe top contact [after Jeon et al 1991(a)].

narrowing at the onset of stimulated emission at T= 77 K and 250 K for a mesa laser device, compared with the incoherent (LED) emission spectrum below threshold of laser action (the latter multiplied by several orders of magnitude to bring it to the same amplitude scale). Note that the spectral position of the stimulated emission overlaps the spontaneous emission spectrum in the low energy tail of the n = 1 heavy-hole exciton transition of the (Zn,Cd)Se/ZnSe QW, similar to the optically pumped case. Under higher spectral resolution, the laser emission exhibited any number of longitudinal modes depending on temperature and current injection conditions. The peak power from laser devices in both polarities of growth at T=77K has reached in excess of 600 mW, while the differential quantum efficiency for both sets of structures was also rather similar and at least $\eta_{ext} \approx 25\%$ per facet.

One of the major issues which could be readily identified for further improvement of these ZnSe-based QW diode lasers is the control of strain related defects. A key point about the heterostructures in Fig. 7(b)-(c) is that, except for the thin quantum well (QW) region, they allow lattice matched growth onto a GaAs or (In,Ga)As buffer layer so that strain related defects can now in principle be minimized (the QW region itself remaining pseudomorphic). The additional Zn(S,Se) cladding layers (Fig. 9(a)) were first argued to be necessary for optical waveguide purposes [Haase et al 1991] or to contribute favorably to hole injection [Jeon et al 1991(a)]. In fact, both passive waveguide studies [Walecki et al 1991] as well as direct optically pumped laser experiments [Jeon et al 1990] indicate that the (Zn,Cd)Se/ZnSe QW system is quite capable of supporting waveguide action necessary for a diode laser. Furthermore, the level og hole injection at the levels of present p-type doping appear to be adequate in such a simpler structure. Hence lattice matching opportunities to a III-V buffer layer (InGaAs) can be taken advantage of. As before, we have realized these new diode lasers on both n-type and p-type GaAs substrates and buffer layers, an important feature in terms of potential flexibility in addressing the difficult and fundamental electrical contact problem with wide-gap II-VI semiconductors.

For the (Zn,Cd)Se/ZnSe QW structures in Fig 7(b), (In,Ga)As buffer layers with 4.3% indium were used to obtain lattice matching to the ZnSe layers, while GaAs buffer layers were lattice matched to the Zn(S,Se) cladding layers in the second set of devices by incorporating a sulfur fraction of approximately 7% (Fig. 7(c)). The Cd concentration was approximately 20% in the ZnSe-based devices resulting in a lattice mismatch strain of about 1.6%; the Zn(S,Se) devices employed 12% Cd in the wells. Typically six 75 Å thick (Zn,Cd)Se QWs formed the gain medium, separated by 100 Å thick ZnSe or Zn(S,Se) barrier layers. Transmission electron microscope studies show that very low misfit dislocation densities have been achieved in these structures, moreover, for the Zn(S,Se) alloy containing structures X-ray double crystal rocking curves have shown linewidths as narrow as 20 arcsec [Xie et al 1992(b)]. The dopants were also introduced within the 50 Å center region of the barrier layers. In the devices discussed here the doping in the QW barrier region was p-type, although we have also obtained similar results with n-type doping.

In the (Zn,Cd)Se/Zn(S,Se) QW's the increased hole confinement energy (with the addition of sulfur) blueshifts the laser emission spectra. **Figure 17** shows the diode laser output power vs. input current density from T=77 K to T=273 K for a mesa device, again with uncoated facets. The current threshold density at T=77 K was 400 A/cm^2, or 160 mA for a typical device. Without any particular effort at heatsinking, beyond attaching the GaAs substrate to a copper block, these devices have been operated at a duty cycle as high as 20% and at a couple of mW average output power levels. With coated endfacets, the group at 3M has reported cw operation from the longer wavelength structures (Fig. 8(a)) [dePuydt et al 1992]. Typically we find at this stage of our device development that as room temperature conditions are approached (beyond 250K), the heating due to the non-ohmic contacts (dominantly at the top contact) rapidly escalate thermal problems. As an indication of this, note the increasing current threshold density and apparent decrease in the differential quantum efficiency from T=250 K to T= 273 K. At room temperature the threshold current density increased to 1500 A/cm^2 (corresponding to 600 mA actual current) for this device. Furthermore, for the T=273 K data (coldfinger temperature), we find from comparison of the room temperature (low current density) LED spectrum for such devices that the effective junction temperature on the laser diode at threshold is already well above room temperature. Major electrical losses occur at the contacts and other heterojunction barriers so that strong

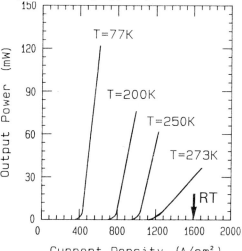

Fig.17. Pulsed laser output power per uncoated facet as a function of current density from a (Zn,Cd)Se/Zn(S,Se) QW laser diode from T=77 K to T=273 K [after Jeon et al 1992].

local heating is unsurprising. Nonetheless useful pulsed diode laser operation under room temperature ambient conditions is expected to readily occur in these devices with facet coating or other forms of increased feedback (the uncoated facet reflection coefficient being approximately $R = 0.21$).

To conclude this section, laser diode emission from (Zn,Cd)Se/ZnSe and (Zn,Cd)Se/Zn(S,Se) QW structures has been obtained up to room temperature conditions. A key feature of our results is the realization of laser devices in which lattice matching to an appropriate III-V buffer layer brings under control a major concern in further development of these devices, namely that of strain induced defects.

5. Fabrication of Small Structures

Because the material and physical properties of the wide-gap II-VI heterostructures differ in many instances substantially from those of the III-V semiconductors, it is likely that a range of device geometries will be explored while advancing the performance of the blue and green diode lasers. Nanometer scale patterned structures are one area of potential interest in this connection. Such small structures provide also possibilities for basic research with photons, excitons, and phonons confined to volumes in which size quantization effects for each are becoming important. In this brief section we preface such work by reviewing a recent example of fabrication techniques in ZnSe-based heterostructures by Walecki et al [1990] who employed short wavelength holographic laser lithography and reactive ion etching to define wire and dot-like patterns in ZnSe thin epitaxial films and heterostructures with spatial feature size to better than 100 nm. In particular, photoluminescence measurements showed that surface damage from etching appears to be much less severe than in III-V semiconductors.

The approach by Walecki et al was by UV laser holographic techniques to expose photoresist on MBE grown ZnSe epilayers and heterostructures either in the form of an interference 'grating' pattern or two such grating patterns crossed at 90° with respect to each other. Electron beam writing techniques provide a complementary and more flexible means to define even smaller structures with arbitrary geometry. **Figure 18** shows scanning electron microscope images of wire and dot structures, produced after etching grating patterns into $Zn_{1-x}Cd_xSe/Zn_{1-y}Mn_ySe$ heterostructures. The individual wires and dots have here a spatial definition of about 100 nm. One advantage of the the holographic approach is to produce the periodic pattern with high uniformity over a macroscopic area of the sample (up to 1 cm^2). The heterostructures were composed of isolated (Zn,Cd)Se quantum wells of different thicknesses so that combined optical (luminescence) and electron microscopic techniques could be used to aid in the depth calibration of the dry etching rate. Clausen et al had earlier reported on the RIE of ZnSe thin epitaxial films, grown by molecular beam epitaxy (MBE) while and studied patterns on the larger spatial scale of about 1 μm [1988].

A key question about the usefulness of such small etched structures for light emitting purposes concerns the electronic quality of the etch-exposed 'vertical walls'. Walecki et al studied the effects of the RIE process in terms of possible surface damage

Fig.18. SEM images of a periodic 'quantum wire' and 'quantum dot' structure etched into the $Zn_{1-x}Cd_xSe/Zn_{1-y}Mn_ySe$ heterostructure [Walecki et al 1990].

by direct comparison of the PL efficiency in both etched and unetched samples, including their temperature dependence. **Figure 19** shows such a comparison for a typical case from T=4 K to T=170 K, where the amplitude from the etched sample corresponds to emission from a particular (Zn,Cd)Se quantum well. Note how for both samples the PL efficiency is nearly the same over a range of temperatures. While at low temperatures (T≤20 K) the recombination radiation is dominated by localized excitons, in the upper edge of the temperature range in Fig. 3, the recombining particles are either free excitons of electron-hole pairs (the overall decrease in the PL efficiency is due to 'bulk' defects in the heterostructures). The quantitatively nearly identical temperature dependence for the etched and unetched samples suggests that possible surface damage induced in the RIE process is not adding significant nonradiative channels under these conditions. This is in striking contrast with PL measurements e.g. with InGaAs/GaAs dry etched quantum well samples where the etched samples of comparable dimensions show a precipitous decrease in their PL efficiency and a very disparate temperature dependence vs. unetched samples. In case of the ZnSe heterostructures for the temperature range above approximately 100 K the photoexcited electron-hole pairs are free from local traps etc., and are subject to normal ambipolar diffusion. An estimate can then be made about the likelihood of the nonequilibrium carriers reaching nonradiative surface states. Although the measurement of hole mobilities in ZnSe has not been historically easy due to difficulties in its p-type doping, there are a number of published reports which support a value of >50 cm²/Vsec at T≈150 K. Since

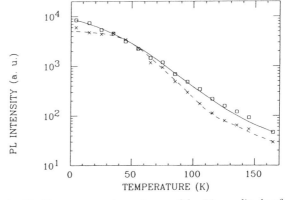

Fig.19. Temperature dependence of the PL amplitude of the etched heterostructure (crosses) compared with unetched sample (open squares). The lines are to guide the eye.

the electron-hole lifetime measured in our structures through time-resolved photoluminescence techniques is approximately 300 psec in the temperature range in question, one estimates an ambipolar diffusion length $L_A > 1000$ Å in the present circumstance (dominated by the hole). Because this value is comparable to or larger than the typical spatial size of the 'quantum wires' (or 'quantum dots' illustrated next), the higher temperature PL data directly supports the idea of low loss of photoexcited carriers through etch-induced surface states (such states would, of course be very likely also charged so that drift induced trapping would simply accelerate the nonradiative processes). This suggests application of such periodic structures to multielement arrays in future optoelectronic applications of II-VI heterostructures at short visible wavelengths.

6. Summary

We have reviewed recent work in which a particular QW system, that of (Zn,Cd)Se, has been shown to be a very useful type I heterostructure with a large excitonic oscillator strength being available for both absorptive and emissive devices. It has been particularly encouraging to witness the realization of diode laser action in such quantum confined system, aided by an appropriately designed p-n junction injector structure. While major challenges remain in improving the laser devices, especially with issues such as contacts (the principal problem at this writing), strain-induced defects such as dislocations, and residual absorption losses, profound advances have clearly occurred within the past twelve months or so. Consequently, efficient LEDs and practical blue and green diode lasers will probably ensue from upcoming research and development efforts in relativel short term (especially the LEDs). This progress should also impact other ZnSe-based heterostructures, especially those with higher sulfur content, in the extension towards shorter wavelength limits of II-VI optoelectronic devices into the near UV.

7. Acknowledgements:

The authors would like to acknowledge H. Luo, N. Samarth, and J. Furdyna at the University of Notre Dame for their contributions in MBE growth and development of the QW systems discussed in this paper. We especially thank M. Kobayashi and N. Otsuka at Purdue whose role has been critical in the success of the Brown/Purdue team. Key members of the team are the authors' students and associates who have made much of this work possible. These are: Jian Ding, Heonsu Jeon, Mary Hagerott, W. Walecki, and T. Ishihara at Brown, and Weize Xie and Don Grillo at Purdue. We wish to recognize the valuable contributions of G. Hua, S. Ogino, H. Iwata, and D. Lubelski at Purdue. We also very much appreciate the preprint information from our 3M colleagues about their pioneering work on diode lasers.

The work at Brown and Purdue Universities was supported by DARPA. Additional support was provided by NSF (Brown) and AFOSR (Purdue)

8. References:

Akasaki, I., 1992 e.g. in *Wide Bandgap Semiconductors*, Mat. Res. Soc. Symposium Proc. MRS (Pittsburgh)

Akimoto, O. and Hasegawa, H. 1966 *J. Phys. Soc. Japan* **22**, 181

Bylsma, R., Becker, W., Bonsett, T., Kolodziejski, L.A., Gunshor, R.L., Yamanishi, M. and Datta, S. 1985 *Appl. Phys. Lett.* **47**, 1039

Cammack, D.A., Dalby, R., Corneliassen, H., and Khurgin, J. 1987 *J. Appl. Phys.* **62**, 3071

Chadi, J. and Chang, K.J. 1989 *Appl. Phys. Letters* **55**, 575

Chang, S-K., Nurmikko, A.V., Wu, J.-W., Kolodziejski, L.A., and Gunshor, R.L., 1988 *Phys. Rev.* **B37**, 1191

Cheng, H., DePuydt, J., Potts, J. and Haase, M. 1988 *J. Cryst. Growth* **95**, 512

Clausen, E.M., H.G. Craighead, M.C. Tamargo, J.L. deMiguel, and L.M. Sciavone, *Appl. Phys. Lett.* **53**, 690 (1988) and J. Vac. Sci. and Technol. **B6**, 1889 (1988)

DePuydt, J., Haase M., Qiu J., and Cheng H. 1992, *Proc. Conf. on Blue-Green Lasers*, Albuquerque NM (Optical Society of America)

Ding, J., Pelekanos, N., Nurmikko, A.V., Luo, H. Samarth, N., and Furdyna, J.K. 1990(a) *Appl. Phys. Letters* **57**, 2885

Ding, J., Jeon, H., Nurmikko, A.V., Luo, H., Samarth, N. and Furdyna, J.K. 1990(b) *Appl.Phys. Lett.* **57**, 2756

Ding, J., Pelekanos, N., Fu, Q., Walecki, W., Nurmikko, A.V., Han, J., Kobayashi, M., and Gunshor, R., 1991 *Proceedings of the 20th International Conference on Semiconductor Physics*, Thessaloniki (Greece), World Publishing (Singapore), 1198

Ding, J., Jeon, H.,Ishihara, T., Nurmikko, A.V., Luo, H., Samarth, N. and Furdyna, J.K. 1992(a) *Surface Science* **267**, 616

Ding, J., Jeon, H., Ishihara, T., Nurmikko, A.V., Luo, H., Samarth, N. and Furdyna, J.K. 1992(b), submitted to Phys. Rev. Lett.

Durbin, S., Han, J., Sungki O, Kobayashi, M., Menke, D., Gunshor, R.L., Fu, Q., Pelekanos, N., Nurmikko, A.V., Li, D.,Gonsalves, J., and Otsuka, N., 1989 *Appl. Phys. Lett.* **55**, 2087

Farrow, R., Schetzina, J., and J. Cheung, J. eds. 1987 see e.g. *Materials for Infrared Detectors and Sources*, Mat. Res. Soc. Symposium Proc. **90**, MRS (Pittsburgh)

Fehrenbach, G.W., Schafer, W., Treusch, J., and Ulbrich, R.G. 1981 *Phys. Rev. Lett.* **49**, 1281

Fu, Q., Lee, D., Nurmikko, A.V., Gunshor, R.L. and Kolodziejski, L.A. 1989 *Phys. Rev.* **B39**, 3173

Fu, Q., Lee, D., Nurmikko, A.V., Kobayashi, M. and Gunshor, R.L. 1991 *Proceedings of the 20th International Conference on Semiconductor Physics*, Thessaloniki (Greece), World Publishing (Singapore), p.1353

Goede, O., Heimbrodt, W., Weinhold, V., Schnurer, E., and Eberle, H., 1987 *Phys. Stat. Sol.* (b)**143**, 511; 1988 ibid (b)**146**, K65

Gunshor, R., Kolozdiejski, L. and Nurmikko, A. 1990 in *Semiconductors and Semimetals*, T. Pearsall ed., Academic Press, New York, **33**

Gunshor, R.L., Kolodziejski, L.A., Melloch, M., Vaziri, M., Choi, C. and Otsuka, N. 1987 *Appl. Phys. Letters* **50**, 200

Haase, M., Qiu, J., DePuydt, J.M., and Cheng, H., 1991 *Appl. Phys. Letters* **59**, 1272

Hagerott, M., Jeon, H., Ding, J., Nurmikko, A.V., Xie, W., Kobayashi, M. and Gunshor,

R.L. 1992 *Appl. Phys. Letters* **60** (in press)

Hefetz, Y., Nakahara, J., Nurmikko, A.V., Kolodziejski, L.A., Gunshor, R.L. and S. Datta, S., 1985 *Appl. Phys. Letters* **47**, 989

Hefetz, Y., Lee, D., Nurmikko, A.V., Sivananthan, S., Chu, X., and Faurie, J.-P., 1986(a) *Phys. Rev.* **B34**, 4423

Hefetz, Y., Goltsos, W. C., Nurmikko, A. V., Kolodziejski, L. A., and Gunshor, R. L., 1986(b) *Appl. Phys. Lett.* **48**, 372

Hefetz, Y., Goltsos, W. C., Lee, D., Nurmikko, A. V., Kolodziejski, L. A., and Gunshor, R. L. 1986(c) *Superlattices and Microstructures* **2**, 455

Hvam, J.M. 1973, *Solid State Comm.* **12**, 95

Ishihara, T., Brunthaler, G., Walecki, W., Hagerott, M., Nurmikko, A.V., Luo, H., Samarth, N., and Furdyna, J.K., submitted to Appl. Phys. Lett.

Jeon, H., Ding, J., Nurmikko, A.V., Luo, H., Samarth, N., Furdyna, J.K., Bonner, W.A. and Nahory, R.E. 1990 *Appl. Phys. Lett.* **57**, 2413

Jeon, H., Ding, J., Patterson, W., Nurmikko, A., Xie, W., Grillo, D., Kobayashi, M., Gunshor, R.L., 1991(a) *Appl. Phys. Letters* **59**, 3619

Jeon, H., Ding, J., Nurmikko, A.V., Luo, H., Samarth, N. and Furdyna, J.K. 1991(b) *Appl. Phys. Letters* **59**, 1293

Jeon, H., Ding, J., Nurmikko, A., Xie, W., Grillo, D., Kobayashi, M., Gunshor, R.L. 1992(b) *Appl. Phys. Letters* **60**, April 27

Jeon, H., Ding, J., Nurmikko, A.V., Xie, W., Kobayashi, M., and Gunshor, R.L. 1992(b) Appl. Phys. Letters **60**, 892

Kawakami, Y., Yamaguchi, S., Wu, Y.-H., Ichino, K., Fujita, S., and Fujita, S. 1991 *Jap. Journal Appl. Phys.* **30**, L605

Klingshirn, C. and Haug, H. 1981 *Phys. Repts* **70**, 315; and references therein

Koch, S.W., Haug, H., Schmieder, G., Bohnert, W., and Klingshirn, C. 1978 *Phys. Stat. Sol.* **b89**, 431

Lee, D. et al, submitted for *Phys. Rev. Lett.*

Lee, D., Johnson, A.M., Zucker, J., Burrus, C., Feldman, R. and Austin, R. 1992 Appl. Phys. Lett. **60**, xxxx

Lee, D., Zucker, J., Divino, M., Austin, R., Feldman, R.D., Jones, K. and Johnson, A.M. 1991 *Appl. Phys. Lett.* **59**, 1867

Lee, D., Zucker, J., Johnson, A.M., Feldman, R.D. and Austin, R.F. 1990 Appl. Phys. Lett. **57**, 1132

Miles, R., Wu, G., Johnson, M., McGill, T., Faurie, J.-P. and Sivananthan, S., 1986 Appl. Phys. Letters **48**, 1383

Newbury, P.R., Shahzad, K. and Cammack, D. 1991 Appl. Phys. Lett. **58**

Ohkawa, K., Mitsuyu, T. and Yamazaki, O., 1987 J. Appl. Phys. **62**, 3216

Olego, D.J., Shahzad, K., Cammack, D.A., and Cornelissen, H. 1988 Phys. Rev. **B38**, 5554

Park,R.M., Trofer,M.B., Rouleau, C.M., DePuydt, J.M., and Haase, M.A., 1990 *Appl. Phys. Letters* **57**, 2127

Pelekanos, N., Ding, J., Fu, Q., Nurmikko, A.V., Durbin, S., Kobayashi, M. and Gunshor, R. 1991 *Phys. Rev.* **B43**, 9354

Pelekanos, N., Ding, J., Nurmikko, A.V., Luo, H., Samarth, N. and Furdyna, J.K., 1992 *Phys.Rev* **B45**, 6037 (1992)

Pelekanos, N., Fu, Q., Ding, J., Walecki, W., Nurmikko, A.V., Durbin, S., Kobayashi, M., and Gunshor, R., 1990 *Phys. Rev.* **B41**, 9966

Qiu, J., Qian, Q.-D., Gunshor, R.L., Kobayashi, M., Menke, D.R., Li, D., and Otsuka, N. 1990 *Appl. Phys. Letters* **56**, 1272

Proceedings of the International Conference on II-VI Semiconductors, Tamano (Japan) 1991 Special Issue of J. Crystal Growth, Elsevier (Amsterdam)

Samarth, N., Luo, H.,Furdyna, J.K, Alonso, R.G., Lee, Y.R., Ramdas, A.K., Qadri, S.B., and Otsuka, N. 1990 *Appl. Phys. Letters* **56**, 1163

Schmitt Rink, S., Chemla, D.S., and Miller, D.A.B., 1985 *Phys. Rev.* **B32**, 6601

Schmitt-Rink, S., Chemla, D.S., and Miller, D.A.B. 1989 *Adv. Phys.* **38**, 89

Suemune, I., Yamada, K., Masato, H., Kan, Y. and Yamanishi, M. 1989 *Appl. Phys. Lett.* **54**, 981

Sun, G., Shahzad, K., Khurgin, J. and Gaines, J. 1991 Conf. Lasers and Electro- Optics, Baltimore

Valster, A. and Acket, G.A., 1991 in *Quantum Optoelectronics 1991*, Technical Digest Series, (optical Society of America, Washington DC), **7**, 126

Van de Walle, C.G., 1989 *Phys. Rev.* **39**, 1871

Walecki, W., Nurmikko, A.V., Luo, H., Samarth, N., and Furdyna, J.K. 1991 *J. Opt. Soc. Am.* **B8**, 1799

Walecki, W.,Nurmikko, A.V., Samarth, N., Luo, H.,Furdyna, J. and Otsuka, N. 1990 *Appl. Phys. Letters*, **57**, 466

Xie, W., Grillo, D.C., Kobayashi, M., Gunshor, R.L., Otsuka, N. Hua, G.C., Jeon, H., Ding, J. and Nurmikko, A.V. 1992(a) *Appl. Phys. Letters* **60**, 463

Xie, W., Grillo, D.C., Kobayashi, M., Gunshor, R.L., Otsuka, N. Hua, G.C., Jeon, H., Ding, J. and Nurmikko, A.V. 1992(b) *Appl. Phys. Letters* **60**, April 20

Zimmerman R., Kilimann K., Kraeft,K., Kremp, D., and Roepke, G. 1978 *Phys. Stat. Sol.* **(b)90**, 175

Zimmerman, R. 1988, in Teubner-Texte zur Physik, v. **18**, VOB National (Berlin), p. 1-176

IV and IV–IV   Semiconductors

Chapter 9

Growth and doping of silicon by low-temperature molecular beam epitaxy

H-J Gossmann and D J Eaglesham

AT&T Bell Laboratories, Murray Hill, NJ 07974

ABSTRACT: Traditional concepts of epitaxial growth assume that there is a temperature, T_{epi}, separating the regime of epitaxial, single-crystalline growth of the film from the regime of amorphous growth. This concept is not applicable for Si MBE. Instead growth at a given temperature always proceeds epitaxially for a certain limiting *thickness*, h_{epi}, before the film becomes amorphous. Segregation and low incorporation can be avoided at growth temperatures below ≈ 400 °C. The growth of arbitrarily complex doping profiles by thermal, co-evaporative doping thus becomes possible. We address the characterization and control of defects in the low-temperature Si-MBE-films and provide a comparison to growth and doping by solid phase epitaxy. Finally we discuss some applications.

1. INTRODUCTION

Molecular beam epitaxy (MBE) has opened new frontiers in physics and materials science by its ability to synthesize, almost arbitrarily, new materials, which do not occur naturally (Esaki and Tsu 1970, Döhler 1972a 1972b, Ploog and Döhler 1972, Esaki 1986). It is a physical deposition method, based on vacuum evaporation of material. The beginnings of *silicon* molecular beam epitaxy (Si-MBE), i.e. employing elements of the fourth group of the periodic table (Si, Ge, Sn), date back to the early sixties, when the first homoepitaxial films were reported (Unvala 1962, Hale 1963, Tannenbaum Handelman and Povilonis 1964, Widmer 1964, Unvala and Booker 1964, Booker and Unvala 1965). At that time the typical growth temperatures for Si epitaxy, usually done by chemical vapor deposition (CVD), exceeded 1000 °C significantly, leading to substantial interdiffusion at interfaces and of dopants. Vacuum evaporation offered a dramatic reduction in growth temperature, greatly reducing the effect of bulk diffusion and making the growth of novel structures possible. *Compositional* superlattices, in which thin layers of different composition and with thicknesses ranging from a few angstrom to thousands of angstrom form a stack; or *doping* superlattices, in which thin doped regions in an otherwise homogeneous structure are employed. In either case the electronic properties are changed in predictable ways and "band-engineering" becomes feasible. It soon became apparent that this potential of vacuum evaporation could only be realized if growth was performed under ultra high vacuum (UHV) conditions (pressure $\ll 10^{-8}$ Torr) and meticulous attention was paid to cleanliness and substrate surface preparation. Indeed, UHV-practices, atomically clean substrates, and precision and control on a monolayer scale, discriminate what is now called Si-MBE from its origins as vacuum evaporation. In 1968 Thomas and Francombe were thus able to produce high quality material at a growth temperature of only 550 °C.

Thomas and Francombe (1969) were also the first to note difficulties with doping of films in Si-MBE that they could not attribute to solid state diffusion. As subsequently became apparent, this was due the large amount of surface segregation and the non-unity sticking coefficients of common dopants on Si at the growth temperature of 550 °C and has become known as the Si doping problem. We will discuss this further in section 3.

2. Si MOLECULAR BEAM EPITAXY

Si molecular beam epitaxy (Si-MBE) has been extensively reviewed in several articles and books (Joyce 1974, Bean 1981, Ota 1983, Allen 1985, Shiraki 1986, Kasper and Bean 1988). Here we will only briefly review the major components of the technique. A schematic of a typical Si-MBE system is shown in Figure 1.

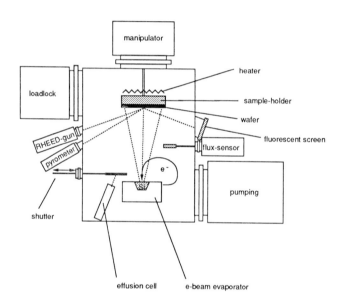

Figure 1. Si-MBE system (schematic).

Float-zone Si is evaporated thermally from an electron-beam heated hearth at the bottom of the ultra-high vacuum chamber (base pressure typically $< 10^{-10}$ Torr.) Oriented and polished Si substrates are cleaned chemically *ex-situ* and introduced into the growth chamber through a load-lock. An appropriate *in situ* cleaning step follows to achieve an atomically clean, smooth surface. The cleanliness of the substrate surface may be checked by Auger electron spectroscopy. Growth morphology and crystallinity are monitored by reflection high energy electron diffraction (RHEED) during growth. Evaporation rates are monitored using flux sensors and their output fed back to the electron-gun power-supply to achieve rate stabilization. Since Si and Ge, have negligible vapor pressures at room temperature, shutters are used to achieve abrupt transitions in composition. Doping is most easily achieved by thermal, coevaporative doping. Dopants are evaporated from an effusion source such that the ratio of dopant flux to Si evaporation rate equals the desired atomic fraction of dopant atoms in the film.

The accurate and precise determination of the wafer temperature during growth is of great importance. This is in particular true for low temperature MBE (LT-MBE) as we will discuss in section 7.2. Typically, temperature measurements in MBE are performed by infrared pyrometry or a thermocouple in close proximity of the wafer. These methods are always somewhat unreliable and even more so at temperatures close to room temperature.

A very accurate determination of temperature, even below room temperature is possible by laser-interferometry, described in detail by Donnelly and McCaulley (1990). Basically the thermal expansion of the wafer is measured during heating or cooling by counting fringes produced by interference from laser light reflected from the front- and back-sides of the wafer. In our experimental geometry, and using 1.52 µm laser light, one 2π phase shift corresponds to about 6 K, so that a precision of 1 K is easily achieved. The accuracy is essentially limited by the accuracy by which the wafer thickness is known, in the present case about ± 1 µm, which translates into $\pm 1K$ at 550 °C. The laser interferometric technique requires, however, double-sided polished wafers.

For the data presented in section 6-8 the following procedure was therefore adopted. The temperature vs time profile of the samples during growth was executed by setting the heater power to appropriate values. No feedback of the block-temperature (measured by a thermocouple) to the heater power-supply was employed. The heater power, P, was calibrated in terms of a temperature scale determined in user units (we will call these user units °U) with the aid of an infrared pyrometer above ≈ 480 °U. The pyrometer was sensitive for the wavelength range $0.8 - 1.1$ µm. Its emissivity setting had been chosen such that the user scale and the true temperature scale coincided at 577 °C = 577 °U, the eutectic point of Al-Si (Mizutani 1988). Lower temperatures were estimated from a linear extrapolation of the $P - °U$ relationship. Since the emissivity of the Si wafer is not necessarily constant below 577 °C, user and true temperature scales will in general differ in this range. Separate calibration runs were executed, on a 20 Ωcm p-doped Si-wafer, polished on both sides, to link the °U scale to the thermodynamical temperature scale as well as measure the response of the true wafer temperature to changes in heater power in the open-loop system. The backsides of the regular samples and the calibration wafer differ in emissivity, i.e. for a given temperature the heater power will be higher for the double-sided polished wafer. This effect was corrected for by determining the additional heater power required to bring the calibration wafer to the same temperature, as measured by the pyrometer, as the regular samples, for temperatures around 525 °U. With this correction, the power program used for the regular wafers was then executed on the calibration wafer and the temperature determined by laser interferometry.

That this procedure indeed accurately gives the actual and true temperature profile of the samples during growth rests on the assumptions that (1) The emissivity of a 20 Ωcm wafer, used for the calibration runs, agrees with that of a 1000 Ωcm wafer of the same dopant-type, used as sample substrate, to within $\pm 3\%$ (non-compliance would result in a temperature error of $\approx \pm 10K$). This assumption is needed to correct the heater power appropriately for the double-sided polished wafer. (2) The thermal conductivity through the wafer clamps and the block to the chamber environment is always the same, i.e for the same wafer the same heater power will give the same temperature for every run. While no data has been published for assumption (1), we have no reason to assume that it is wrong. Experience shows that (2) is true to better than ± 20 °C. We therefore arrive at a precision of ± 1 K and an accuracy of $\pm 20K$ for the temperatures quoted in sections 6-8.

3. THE Si DOPING PROBLEM

Doping in III-V-MBE is usually accomplished by incorporating the dopant atoms concurrently with the growth of the matrix, e.g. GaAs by coevaporating dopants and matrix atoms at appropriate rates. This approach is the easiest solution. However, in Si-MBE, the sticking coefficients of dopant atoms are well below unity at typical MBE growth temperatures, whereas Si and Ge atoms have a unity sticking coefficient (Thomas and Francombe 1969, Kuznetsov and Postnikov 1969, Tolomasov *et al* 1971, Kuznetsov and Postnikov 1974, Postnikov and Kuznetsov 1975, Bennet and Parish 1975, Postnikov and Perov 1977, Perov and Postnikov 1977, Becker and Bean 1977, Ota 1977, König *et al* 1979, Kuznetsov *et al* 1979). At growth temperatures above $\approx 450\ °C$ dopant incorporation is characterized by severe surface segregation. This is true for all common dopants in Si, such as the p-type dopants Ga (Becker and Bean 1977, Nakagawa *et al* 1988), and In (Knall *et al* 1984, Knall *et al* 1989), and the n-type dopant Sb (Ota 1977, Bean 1978, König *et al* 1979, König *et al* 1981, Tabe and Kajiyama 1983, Metzger

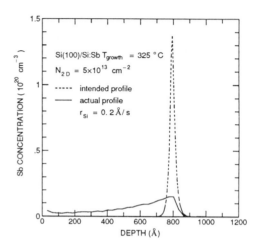

Figure 2. Intended (dashed line) and actual (solid line) concentration profile of Sb in a Si film grown by MBE at a growth temperature of $T_s = 325\ °C$ and a growth rate of $r = 0.2\ \text{Å/s}$ on Si(100).

and Allen 1984a 1984b, Barnett and Greene 1985, Nakagawa *et al* 1988, Jorke 1988). It is also true for p-type doping with B, although to a lesser extent (Kubiak *et al* 1984, Kubiak *et al* 1985, Ostrom and Allen 1986, Tatsumi *et al* 1987, Tatsumi *et al* 1988, Hirayama *et al* 1988, Andrieu *et al* 1988, de Frésart *et al* 1988, Lin *et al* 1989, Jackman *et al* 1988, Jackman *et al* 1989, Headrick *et al* 1989, Parry *et al* 1991). As a consequence, the dopant concentration at the beginning of growth will be significantly below the value given by the ratio of dopant and Si flux until the dopant surface coverage has built up to a sufficient value. After the dopant flux has been shut off, the dopant surface coverage slowly depletes leaving a trail of dopant in the nominally undoped part of the film. To illustrate the scope of the problem Figure 2 shows intended (dashed line) and actual (solid line) concentration profile of Sb in a Si film grown by MBE at a growth temperature of $T_{growth} = 325\ °C$ and a growth rate of $r = 0.2\ \text{Å/s}$ on Si(100). The width and height of the dopant concentration are off by more than an order of magnitude.

Doping is central to all applications of MBE grown material and many approaches have been proposed in the past to overcome the Si doping problem, such as low energy ion implantation or high temperature deposition (for a review see Bean 1981, Ota 1983, Allen 1985, Shiraki 1986, Kasper and Bean 1988). Mostly, these methods introduce difficulties of their own, such as defect formation, temperature-, growth-rate-, and species-dependent incorporation, complicated growth schedules and equipment, sensitivity to small variations in process parameters, or solid state diffusion. However, segregation is a thermally activated phenomenon and as such susceptible to kinetic suppression at sufficiently low temperatures. Jorke (1988) has described doping of Si(100) by Sb in a model from which it can be inferred that this regime is entered at growth temperatures significantly below 400 °C.

Traditional concepts of epitaxial growth are based on the assumption that for growth of a film with a finite thickness in finite time there is a temperature, T_{epi}, separating the regime of epitaxial, single-crystalline growth of the film from the regime of amorphous growth (see for example Venables 1975). Consider a substrate with N_0 adsorption sites per unit area. The distance between neighboring sites is then on the order $d \approx 1/\sqrt{N_0}$. The time it takes to build up one monolayer is $\tau = N_0/r$. Arriving atoms will stay exactly where they adsorb if the mean pathlength, $<x> \approx \sqrt{D\tau}$, traveled due to surface diffusion during time τ is smaller than d, where D is the surface diffusion coefficient. This then implies $r >> N_0^2 D$ and for atoms with directional bonds, such as Si, the film will be amorphous. Conversely, if diffusion is fast enough, $r << N_0^2 D$, atoms have time enough to find correct lattice sites before being locked in by neighbors and the film will be epitaxial (we are ignoring the possibility of strain, dislocations etc., which will not change the general conclusions of these order-of-magnitude considerations.) Since D is temperature dependent we can define the epitaxial temperature implicitly by setting $r = N_0^2 D(T_{epi})$. Surface diffusion is a thermally activated process and thus the dependence of T_{epi} on r will only be weak.

In order to minimize interdiffusion in heterostructures and of dopants, growth at the lowest possible temperature is desirable. Within the concept of the epitaxial temperature, this implies growth slightly above T_{epi} and consequently a large amount of effort has been spent in the determination of T_{epi}. The results for Si on Si(100) and Si(111) are summarized in Table I.

Table I. Epitaxial temperatures for MBE growth of Si on Si(100) and Si(111) (RT denotes room temperature.)

T_{epi} (°C)	Si(100)	Si(111)
Wiedmer 1964	-	550
Jona 1967	RT	420
Shiraki *et al* 1978	165	-
de Jong *et al* 1983	200	600
Gossmann and Feldman 1985	300	520
Allen and Kasper 1988	240	490

Although only data obtained from growth in UHV are listed, large variations in the values of T_{epi} are seen. Nevertheless, it is generally believed that epitaxial Si MBE films on Si(100) are obtained only for growth temperatures $T_{growth} \geq 400$ °C (Shiraki 1986). This is too high to suppress segregation kinetically, as shown in Figure 2.

4. GROWTH BY SOLID PHASE EPITAXY

An alternative approach [(solid phase epitaxy (SPE)] is to relax the requirement for epitaxial growth, and separate the growth of doped films into two steps. First, deposition takes place at very low temperatures where segregation is kinetically suppressed. For practical reasons, room-temperature is usually chosen, and the films are thus amorphous. In the second step, the amorphous material is regrown in an annealing cycle (Roth and Anderson 1977, Caber *et al* 1982, Streit *et al* 1984, Vescan *et al* 1985, Casel *et al* 1986, Casel *et al* 1990). The whole procedure takes place in UHV to preserve the clean substrate surface prior to growth of the amorphous film that is required for high quality regrowth and to prevent migration of impurities into the amorphous film before recrystallization that would otherwise interfere with regrowth (Bean and Poate 1980, Saitoh *et al* 1981, Christou *et al* 1981). Since the dopants are buried when the film is crystallized, surface segregation does not come into play and bulk diffusion is generally small at regrowth temperatures of the order of 600 °C, although for very sharp dopant transitions, as required for δ-doping, the possibility of transient diffusion during recrystallization needs to be considered. Si films doped by SPE exhibit reduced mobility and an activation of less than 100% (Vescan *et al* 1985, Casel *et al* 1986), requiring post-regrowth annealing at temperatures exceeding 900 °C (Casel *et al* 1990). This problem is illustrated in Figure 3. There the electrical activation is shown for Sb in Si(100) grown by SPE (open circles) as a function of concentration. Deposition took place at 150 °C, followed by annealing at 575 °C) (Casel *et al* 1990). The electrical activation is defined as the ratio between the electron concentration, $n_{ELECTRON}$ and the atomic concentration, n_{Sb}. As grown, the films exhibit an activation much smaller than unity; the activation improves upon annealing at 900 °C for 10 min (filled circles), although it is still significantly below one

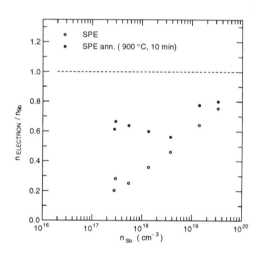

Figure 3. Electrical activation, defined as the ratio between electron concentration and atomic concentration for Sb in Si(100) at room temperature. Films grown by solid phase epitaxy (SPE) before (open circles) and after annealing at 900 °C for 10 min (filled circles) are shown (after Casel *et al* 1990).

5. THE EPITAXIAL THICKNESS

In the preceding section we have reviewed the traditional concept of the epitaxial temperature believed to separate the regime of crystalline and amorphous growth. This concept of an unique *temperature*, however, is not appropriate for Si MBE (Eaglesham *et al* 1990). Figure 4 shows a high resolution cross sectional transmission electron micrograph (XTEM) of Si grown at 50 °C on Si (100). The substrate had previously been cleaned chemically and a volatile oxide grown *ex-situ* and annealed at 850 °C *in-situ* (Ishizaka and Shiraki 1986), before a buffer layer was grown at 550 °C, followed by deposition of 2 monolayers of Ge. The Si can be seen to be initially epitaxial, with the amorphous phase appearing at a later stage. The question of whether 100 °C is above or below an epitaxial temperature is clearly irrelevant, and the important parameter becomes the epitaxial thickness h_{epi} which grows prior to the formation of the amorphous

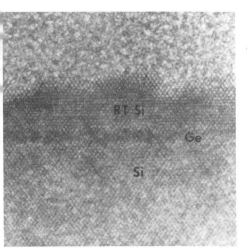

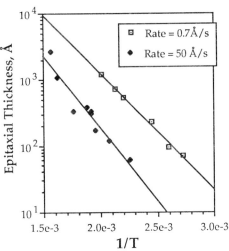

Figure 4. Thin layer of Si deposited at low temperature following growth of a high-temperature homoepitaxial buffer and 2 monolayers of Ge as a marker. The layer is initially epitaxial, but becomes amorphous after growth of an epitaxial thickness h_{epi}.

Figure 5. Temperature-dependence of h_{epi} for Si(100) at two different deposition rates. The "activation energy" implied by the fits to these "Arrhenius" plots is 0.4±0.1 eV.

phase. Several points about the phenomenon, which are readily apparent from Figure 4 and similar micrographs, should perhaps be emphasised. First, the crystalline quality of the Si does not degrade continuously from the Ge marker: the quality appears to remain high up to the point where the amorphous phase begins to form (although small stacking faults are sometimes seen very close to h_{epi}). Thus a continuous increase in extended defect densities below some threshold temperature for defect-free epitaxy is also implausible. (High resolution transmission electron microscopy is sensitive only to very high densities of extended defects, but plan view TEM, as well as the dopant activation and positron measurements described later, confirm zero extended defects and low point defect densities below h_{epi}). While some change in the growing layer must therefore be responsible for the ultimate breakdown of epitaxy, it does not appear to be build-up of extended defects. Second, the interface between the amorphous and crystalline phases is rough. Experiments where growth is interrupted at the point where the film is becoming amorphous show that this does not correspond to simultaneous formation of the amorphous Si (a-Si) everywhere on a very rough surface, but arises from local nucleation of a-Si at different times in different places on the surface. Studies of layers grown at different temperatures show that the roughness of the a-Si:c-Si interface always scales approximately with the epitaxial thickness, so that typically the "interface width" is $\approx h_{epi}/3$. Finally, the abruptness of the image of the Ge layer suggests that Ge segregation is not responsible for the phenomenon: this has been confirmed by control experiments without the marker layer.

Given that epitaxy is limited by h_{epi} as opposed to T_{epi}, this becomes the relevant parameter in controlling growth temperature in order to limit the thermally-activated segregation effects discussed above. We have measured the epitaxial thickness in Si(100) as a function of substrate temperature and deposition rate, as shown in Figure 5. The thickness appears to follow an "Arrhenius" relation with an "activation energy" of 0.4±0.1 eV at all rates, and an intercept h_0 that depends (rather weakly) on rate. [Note that strictly an Arrhenius relation applies only when a measured rate is exponentially dependent on temperature, whereas here we measure an exponential dependence of the time to nucleate the amorphous phase; plots of $1/h_{epi}$ vs $1/T$ (also a straight line) are thus a more rigorous Arrhenius plot of the "rate" of amorphous formation]. For practical purposes, Figure 5 defines a phase boundary for the straightforward use of low-temperature MBE to suppress segregation. Growth at low temperature must be restricted to temperatures such that h_{epi} for the desired deposition rate lies above the total thickness of the grown layer. This approach is followed for many of the abrupt doping profiles which will be described in Section 6.

The range of layers that can be grown at low temperature can, however, be considerably extended by more detailed investigation of the limited-thickness phenomenon. It is clear from Figure 4 that the low-temperature MBE has a "memory" effect, i.e. some change in the surface or bulk of the film during growth is responsible for the ultimate breakdown. The prime candidates here would be build-up of surface contamination, accumulation of point defects (which are not easily visible in Figure 4), or increases in surface roughness or step density. Since the starting surface was prepared by Si MBE at high temperature, the question then arises of what annealing/growth sequence is required to "reset" the surface to its starting condition. It turns out that deposition at high temperature is not necessary, and a simple anneal to ≈500 °C for >10s is sufficient. Annealing to lower temperatures, even for more extended periods, has no marked effect (also see section 7.2). It should be noted here that point defect accumulation would probably have to be a bulk-diffusion-limited process at the growth temperature. This observation that extended growth interrupts at, or even well above, the growth temperature are ineffective in increasing h_{epi} therefore makes point defects an unlikely culprit for causing breakdown of epitaxy. We therefore have a recipe for growing low defect density epitaxial Si to arbitrary thickness at arbitrary temperature by use of a periodic "flash" to ≥ 500 °C every time the layer thickness approaches ≈$2h_{epi}/3$. A typical thermal sequence involving this technique will be discussed later (see Figure 12a,b).

There is now considerable evidence that limited-thickness epitaxy is widespread or universal in semiconductor MBE. The phenomenon has been observed in Si(111) (structurally very different from (100)) and in GaAs homoepitaxy (III-V system), as well as in Ge/Si (heteroepitaxy) (Eaglesham *et al* 1991a, Eaglesham *et al* 1991b, Eaglesham and Cerullo 1991) so that a universal explanation for the effect should perhaps be sought. It should be noted, however, that both Si(111) and GaAs show high defect concentrations (extended and point, respectively), so that mechanisms which we have ruled out for Si(100) could possibly apply in these cases. The effect is broadly similar in all cases, with an interface width of about $h_{epi}/3$ and h_{epi} increasing along a thermally activated curve.

Some important differences are worth noting. GaAs shows two distinct temperature regimes, with an abrupt increase in h_{epi} near 200 °C which may be linked to As_4 dissociation. Low-temperature GaAs has been widely studied as a buffer layer and fast photodetector because of its short carrier lifetime induced by excess As antisite defects which precipitate on annealing (See e.g. Melloch *et al* 1991 and references therein). Our own studies, however, suggest that while the excess As concentration is extremely large, there appears to be no increase in this concentration during low-temperature growth, so that point defects are also unlikely to be responsible for the ultimate breakdown of epitaxy in GaAs.

Ge/Si and Si/Ge are the only heteroepitaxial systems we have studied to date. The epitaxial thickness of Ge/Si is seen to increase smoothly as a function of temperature (with an activation energy of 0.5 ± 0.1 eV), without any obvious discontinuity at the critical thickness h_c at which misfit dislocations are introduced into the layer. There is some conflict here, as studies of Si/Ge suggest a strong influence of strain on low-temperature growth of Si: if the thickness of the Ge "marker" layers is increased beyond the thickness where Ge relaxes to its bulk lattice parameter then the epitaxial thickness of low-temperature Si overlayers drops dramatically. When buffer layers of GeSi alloys are used (relaxed to their equilibrium lattice parameter with high- temperature anneals) h_{epi} for strained low-temperature Si layers on these buffers appears to decrease approximately linearly with increasing strain, as shown in Figure 6. High resolution TEM shows that the epitaxial crystalline Si does not have a significant density of misfit dislocations at the Si-Ge interface, so that at these temperatures Si is required to adopt the buffer lattice parameter. The strain energy associated with coherently strained Si on pure Ge at its bulk lattice parameter is ≈ 100 meV/atom: it is perhaps instructive to compare this number with the free energy difference between the amorphous and crystalline phases, which is ≈ 120 meV/atom near these temperatures. Strain could thus play a significant role in destabilizing the (coherent) crystalline Si with respect to a relaxed amorphous phase.

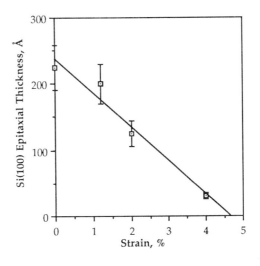

Figure 6. Strain-dependence of h_{epi} for Si(100) growth (at fixed temperature and rate) on thick GeSi alloy buffers of different Ge content. Buffer layers were well above the critical thickness for dislocation introduction, and were relaxed by annealing to high temperature prior to cooling for Si growth.

Most of our studies, have concentrated on the technologically important Si(100) system. The precise nature of the mechanism involved in the breakdown of epitaxy remains ambiguous at this stage. We have considered a broad range of possible processes for the breakdown of epitaxy, focussing on step build-up and impurity segregation because point defect densities appear to be low under optimum growth conditions and diffusion-limited defect accumulation is inconsistent with the observed annealing behavior. Impurity segregation also seems unlikely because SIMS of grown layers does not detect either a segregation profile or a residual peak at the amorphous-crystalline interface. It should be noted, however, that the sensitivity of SIMS to hydrogen is limited, hydrogen is probably the dominant background in the UHV chamber, and the annealing temperature to "reset" limited- thickness epitaxy is close to that for desorption of hydrogen (500 °C). While hydrogen is now known to inhibit epitaxy of Si(100) (Wolff *et al* 1989), there is also some evidence that at least for chemically-terminated Si(100):H subsequent Si growth is qualitatively different from that seen during h_{epi} (Eaglesham *et al* 1991c). Step build-up can be suggested as an alternative mechanism by plausibility arguments: step energies from STM measurements of thermal distributions (Swartzenruber *et al* 1990)) are 30 − 90 meV/atom, so that pairs of high-energy steps seen in STM images following low-temperature growth can be argued to have energies approaching 220 meV/atom, while the free energy difference between crystalline and amorphous Si is only 120 meV/atom: damped RHEED oscillations seen during low-temperature growth epitaxy show that step densities increase during low-temperature growth (see e.g. Aarts and Larsen 1988), and the threshold temperature for RHEED oscillation recovery (i.e. smoothing of an already rough surface) is close to 500 °C. However, hard evidence linking the breakdown to steps is not yet forthcoming, principally because of the difficulties involved in direct measurements of the behavior of single atoms at steps. Thus although the mechanism for breakdown of epitaxy has been the subject of intensive effort for more than a year now, a definitive answer can not yet be given.

6. DOPING OF Si

The procedures discussed above allow us to grow Si at arbitrary temperatures. In particular growth can be done at a temperature low enough to suppress segregation and to achieve a unity sticking and incorporation coefficient. Thus complete control of arbitrary doping profiles is possible with thermal coevaporative doping, without the need for high temperature postgrowth annealing (Gossmann *et al* 1990, 1991). Figure 7 illustrates the sharpness of doping profiles achievable in this way. The atomic concentration as determined by secondary ion mass spectrometry (SIMS) is shown as a function of depth for a δ-doped sheet of concentration 5.2×10^{14} cm^{-2} at a depth of 800 Å. Growth took place at 150 °C; one 3 min temperature flash to 450 °C was applied after 500 Å had been grown (corresponding to a depth of 300 Å). After growth was completed, an *in situ* anneal at 325 °C for 1 h followed. A full-width-at-half-maximum of 31 Å is achieved, corresponding to a slope of 30 Å/decade. Note that this is an apparent width, limited by the resolution of the SIMS technique. The true width is probably even smaller. Hall measurements give a carrier concentration of 2.7×10^{14} cm^{-2}, i.e. despite the extremely high concentration still more than half of the dopant atoms are electrically active.

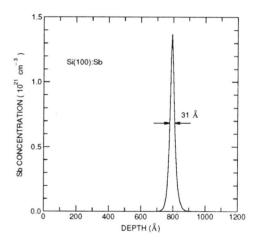

Figure 7. Secondary ion mass spectrometry (SIMS) profile of a δ-layer of 5.2×10^{14} cm^{-2} Sb at a depth of 800 Å. Growth took place at 150 °C; one 3 min temperature flash to 450 °C was applied after 500 Å had been grown (corresponding to a depth of 300 Å). After growth was completed, an *in situ* anneal at 325 °C for 1 h followed.

The activation of dopants was measured by the Hall-technique in the van der Pauw arrangement (van der Pauw 1958) at room-temperature. Figure 8 we show the results of these measurements for homogeneously doped films, in (a) for n-type carriers in Sb-doped films (squares), as a function of atomic concentration measured by SIMS and ion scattering; in (b) for p-type carriers in B-doped films. In both panels, the solid line corresponds to unity activation. In Figure 8(a) the arrow indicates the solid solubility limit of Sb in Si. A Hall-factor of unity was assumed for Sb, and of 0.75 for B (Lin *et al* 1981). All growth took place at 270 °C; the thickness of the films did not exceed h_{epi} at this temperature, so no high temperature flashes took place. Concentrations up to 2×10^{21} cm^{-3} are measured, which exceed significantly the solid solubility limit of Sb in Si. For both dopants 100% activation is achieved for concentrations up to 10^{21} cm^{-3}. The drift mobility for the samples in Figure 8a,b is shown in Figure 9a,b. The solid lines represent bulk values, based on As-doped Si in Figure 9a, and B-doped Si in Figure 9b (Masetti *et al* 1983).

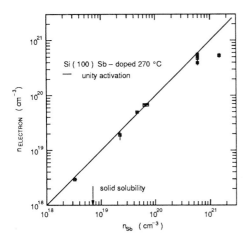
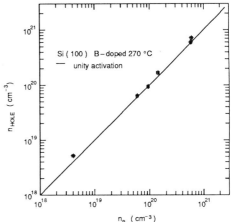

Figure 8a. Electron concentration in Sb-doped films (squares), measured by the Hall technique in the van der Pauw geometry at room temperature, as a function of atomic concentration measured by SIMS and ion scattering. The solid line corresponds to unity activation. The arrow indicates the solid solubility limit of Sb in Si.

Figure 8b. Hole concentration in B-doped films (squares), measured by the Hall technique in the van der Pauw geometry at room temperature, as a function of atomic concentration measured by SIMS and ion scattering. The solid line corresponds to unity activation.

7. DEFECTS

7.1 Electrical Measurements

While the low temperature films do not show extended defects, such as dislocations or voids, in transmission or cross-sectional electron microscopy (TEM), the existence of point defects can not be excluded from TEM observations. From the Hall measurements discussed in the preceding section we can place an upper limit of about 10^{18} cm^{-3} on the concentration of any electrically active, shallow defects. Figure 10 shows a capacitance-voltage (CV) profile of a Sb δ-doped Si layer with $N_{Sb}^{2D} = 4\times10^{12}$ cm^{-2} embedded in 2000 Å lightly (1×10^{16} cm^{-3}) n-doped Si. The film was grown at 250 °C with 90 s flashes to 450 °C every $\approx$700 Å. The measured carrier concentration of negatively charged carriers drops to less than 1×10^{17} cm^{-3} in the lightly doped background material after the δ-peak, placing this value as an upper limit on charged, shallow defects.

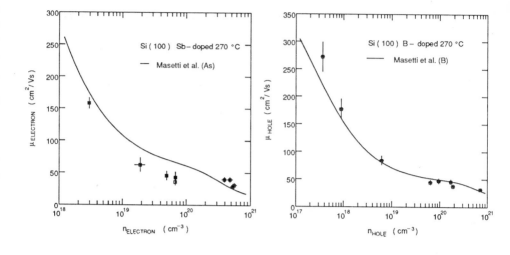

Figure 9a. Mobilities for the samples of Figure 8a as a function of electron concentration. The solid line represents the bulk mobility of n-type Si doped with As (Masetti *et al* 1983).

Figure 9b. Mobilities for the samples of Figure 8b as a function of hole concentration. The solid line represents the bulk mobility of p-type Si doped with B (Masetti *et al* 1983).

7.2 Positron Annihilation

The lineshape of the radiation emitted from annihilating positrons in the surface region of a solid gives quantitative, depth-resolved information on defect concentrations and types (Schultz and Lynn 1988). In particular the method is very sensitive to point-defects, which are difficult to observe in electron microscopy. It thus represents an unique way to survey film quality as a function of growth temperature, impurity concentration, or post-growth processing parameters in a defect concentration region inaccessible to TEM but important for the electrical characteristics of minority carrier devices.

In a typical positron annihilation experiment a beam of 0-20 keV positrons is directed towards the target. The positrons thermalize at a depth related to their primary energy and annihilate. The main quantity detected is the line shape of the emitted annihilation line at 511 keV. Usually it is quantified by the so-called S-parameter, the ratio of the counts in a narrow window around the center energy of 511 keV and the total counts in the annihilation line. If a positron thermalizes in a perfect crystal region and annihilates there, the overlap with core electrons is larger than if the positron gets trapped and annihilates in a open volume defect, for example a vacancy. Consequently, the line will be broader (the S-parameter smaller) in the former than in the latter case. Depth-sensitive information is obtained by recording the S-parameter, normalized to its bulk-value, S_b, as a function of positron energy. Defects are present in those regions where S/S_b exceeds a value of unity. Quantitative evaluations of defect distributions are possible but require the use of numerical procedures (Schultz and Lynn 1988).

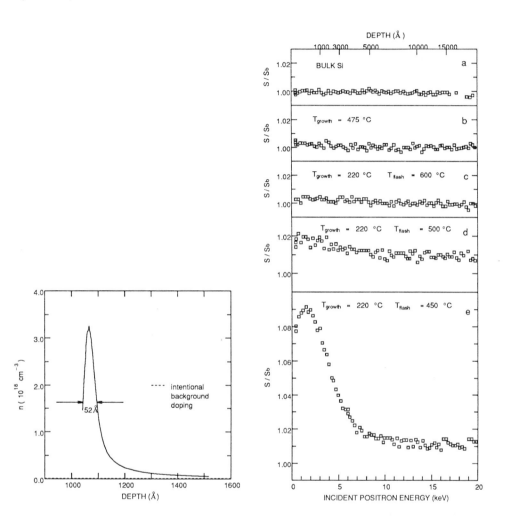

Figure 10. Capacitance-voltage profile of a Sb δ-doped Si layer with a sheet concentration of 4×10^{12} cm^{-2} embedded in 2000 Å lightly $(1\times10^{16}$ cm$^{-3})$ n-doped Si.

Figure 11. Normalized S-parameter for bulk Si (a) and 2000 Å thick Si films (b)-(e) grown by MBE at growth temperatures T_{growth} with temperature flashes to T_{flash}.

Positron annihilation spectroscopy was performed on Si films of at least 2000 Å thickness (Leung *et al* 1991). In most cases, this necessitated the use of high temperature flashes during growth to prevent the films from turning amorphous.

Figure 11 shows the normalized S-parameter as a function of positron energy for several MBE-grown samples (Leung *et al* 1991). A depth scale is indicated at the top. All films were 2000 Å thick. The growth temperature is indicated as T_{growth}. Panel (a) gives the result for a plain Si wafer for comparison. The value of $S/S_b = 1$ over the whole energy- (or depth-) range indicates the high quality of standard Si-wafers, as expected, and serves as a baseline for the MBE samples. For the data in panel (b) growth took place at 475 °C, which for the thickness grown does not require the use of temperature flashes. The positron data is basically indistinguishable from the bulk Si sample, placing an upper limit of $\approx 5 \times 10^{15}$ cm^{-3}, essentially the sensitivity of the present experiment, on the concentration of vacancy-like defects. Panels (c)-(e) show the result if the growth temperature is dropped to 220 °C. In this case high temperature flashes are required to prevent the film from becoming amorphous.

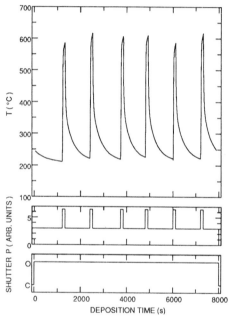

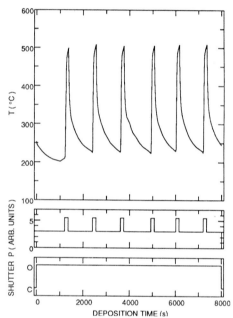

Figure 12a. Wafer-temperature as a function of time during MBE growth at a growth temperature $T_{growth} = 220$ °C with 2 min flashes to 600 °C. The heater power, P, is indicated. No growth interruption took place during the flashes, i.e. the shutter stayed open (O) all the time.

Figure 12b. Wafer-temperature as a function of time during MBE growth at a growth temperature $T_{growth} = 220$ °C with 2.5 min flashes to 500 °C. The heater power, P, is indicated. No growth interruption took place during the flashes, i.e. the shutter stayed open (O) all the time.

The time schedule for the power, P, to the sample heater and the resulting temperature profile is shown in Figure 12a (for the sample of Figure 11c) and Figure 12b (for the sample of Figure 11d). The temperature profile for the sample in Figure 11e is similar to the data shown in Figure 12a,b, except that the peak temperature during the flashes was only 450 °C. It is obvious that T_{flash} plays an important role for the quality of the films. A temperature of 600 °C during the flash results in films almost indistinguishable from the bulk sample, whereas $T_{flash} = 450$ gives rise to a significant amount of defects.

A quantitative analysis of the data in Figure 11e results in the defect profile of Figure 13 (solid line). Measuring from the interface, the first ≈ 1200 Å of the 2000 Å thick film grown at 220 °C with flashes to 450 °C are defect free, whereas the remainder has a defect concentration of $\approx 10^{18}$ cm^{-3}. Increasing T_{flash} to 600 °C results in a film with a defect concentration at the sensitivity limit of the experiment (5×10^{15} cm^{-3}). Note that in both cases the numerical analysis assumed that the film is composed of two regions characterized by their thickness and a constant defect concentration. The step-like jump of the solid line in Figure 13 is a consequence of this model. A one-level model, in which the whole film was assumed to be homogeneous, did not lead to satisfactory fits of the data in Figure 11.

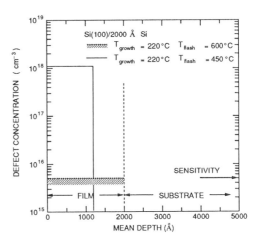

Figure 13. Defect concentration as a function of depth for a 2000 Å thick film on Si(100) grown at 220 °C. For the sample represented by the solid line high temperature flashes up to 450 °C were executed, for the one represented by the dashed line $T_{flash} = 600$ °C. In both cases a two-level model was assumed for the film. The sensitivity of the experiment for defects is indicated

8. APPLICATIONS

Low temperature MBE offers the capability to provide arbitrary complex doping profiles without the need for post-growth annealing. Thus a region in the doping of Si has become accessible that is characterized by ultra-high and ultra-sharp profiles. This makes possible the study of new phenomena that might occur at these high doping densities, such as the enhancement of dopant diffusion due to their Coulombic repulsion at very small distances. (Schubert *et al* 1992, Gossmann *et al* 1992). Using a sheet of dopants instead of a homogeneously doped layer ("δ-doping") is a very recent and exciting new field. Since the spatial extent of the dopants in this case is much smaller than the characteristic length-scales on which the carriers operate, such a dopant distribution can be represented by Dirac's delta function and the dopant concentration n of a dopant sheet at depth z_0 with a sheet concentration N is given by

$$n = N\delta(z - z_0). \tag{1}$$

Significant improvements of existing devices and completely new device concepts have been demonstrated or proposed (Gossmann and Schubert 1992). A review of these developments, however, is completely beyond the scope of this paper. Instead we will focus on one area where the capability of very high doping concentrations will be crucial, the contact resistance problem.

The continuing shrinkage of devices, forecast to reach design rules of 0.1 μm by the turn of this century, also leads to shrinkage of contact area, for example for the source and drain contact of a MOSFET. This is an ohmic contact, produced by a Schottky barrier to highly doped

Si; typical specific contact resistances achievable in VLSI are of the order of 10^{-7} Ωcm^2. For a $1\mu m \times 1\mu m$ contact pad this gives rise to a quite satisfactorily low contact resistance of 10 Ω. However, for a $0.1\mu m \times 0.1\mu m$ pad we would obtain 1000 Ω, significantly too high. Gossmann and Schubert (1992) have calculated the contact resistance for a Schottky contact to δ-doped Si. These calculations show that for a sheet concentration of 10^{14} cm^{-2} a very significant reduction in specific contact resistance over present values is achievable. The peak volume concentration of dopant in such a δ-spike would be enormous, reaching 10^{22} cm^{-3}. Nevertheless, recent work has demonstrated that these kinds of electrically active concentrations are achievable (van Gorkum *et al* 1989, Headrick *et al* 1990, Vink *et al* 1990, Roksnoer *et al* 1991, Gossmann and Unterwald, 1992). Figure 14 shows the electron concentration as a function of atomic concentration for Sb-δ-doped films. The electrical data

was obtained from Hall measurements, assuming a Hall factor of 1.00. The triangles are from films grown by SPE (van Gorkum *et al* 1989), crystallized at 550 °C and annealed at 750 °C, the squares from films grown by LT-MBE (Gossmann and Unterwald, 1992) at growth temperatures between 150 °C and 270 °C. Both growth techniques are capable of achieving an active concentration of 10^{14} cm^{-2}. However, the data from SPE-grown films levels off at this value despite a further increase in atomic concentration, whereas the LT-MBE samples show active concentrations up to 3×10^{14} cm^{-2}.

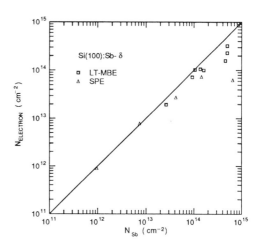

Figure 14. Electron concentration as a function of atomic concentration for Sb-δ-doped films. The electrical data was obtained from Hall measurements, assuming a Hall factor of 1.00. The triangles are for films grown by SPE (van Gorkum *et al.* 1989), the squares for films gorwn by LT-MBE (Gossmann and Unterwald, 1992).

That the active fraction of dopants in films grown by LT-MBE remains at unity up to higher concentrations than for other growth techniques is again demonstrated in Figure 15. There the hole concentration for B-δ-doped films is shown as a function of the atomic concentration. A Hall factor of 0.75 has been used to convert Hall concentration to hole concentration (Lin *et al* 1981). Squares denote LT-MBE films grown with $T_{growth} = 220$ °C and $T_{flash} = 600$ °C (Gossmann and Unterwald, 1992). Circles are for films grown by chemical vapor deposition (CVD) (Roksnoer *et al* 1991) and triangles for films grown by SPE (Headrick *et al* 1990) with an annealing temperature of 450 °C.

9. ACKNOWLEDGEMENTS

We appreciate the discussions and collaborations with L. C. Feldman, G. H. Gilmer,

G.Higashi, H. S. Luftman, and E. F. Schubert. We thank D. C. Jacobsen for ion scattering measurements and P. Asoka-Kumar, T. C. Leung, and K. G. Lynn for letting us use their data on positron annihilation in Si-MBE samples prior to publication. The expert technical assistance of F. C. Unterwald and the excellent substrate preparation by T. Boone and F. Schrey are greatly appreciated.

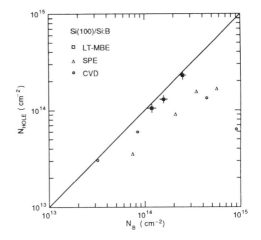

Figure 15. Hole concentration for B-δ-doped films as a function of the atomic concentration. A Hall factor of 0.75 has been used to convert Hall concentration to hole concentration. Squares denote LT-MBE films (Gossmann and Unterwald, 1992), circles chemical vapor deposition (CVD) films (Vink *et al* 1990) and triangles SPE films (Headrick *et al* 1990).

10. REFERENCES

Aarts J and Larsen P K 1988 in *RHEED and Reflection Electron Imaging of Surfaces* Ed P K Larsen and P J Dobson (Plenum, New York, 1988) p 449

Allen F G 1985 *Proc. Electrochem. Soc.* 85-7 3

Allen F and Kasper E 1988 *Silicon molecular beam epitaxy* Vol. I eds Kasper E and Bean J C (CRC, Boca Raton).

Andrieu S Chroboczek J A Campidelli Y André E and d'Avitaya F A 1988 *J. Vac. Sci. Technol. B,* 6 835

Barnett S A and Greene J E 1985 *Surf. Sci.* 151 67

Bean J C 1978 *Appl. Phys. Lett.* 33, 654

Bean J C and Poate J M 1980 *Appl. Phys. Lett.* 36 59

Bean J C 1981 *Impurity Doping Processes in Silicon* ed Wang F F Y (North-Holland, Amsterdam) ch. 4

Becker G E and Bean J C 1977 *J. Appl. Phys.* 48 3395

Bennett R J and Parish C 1975 *Solid State Electron.* 18 833

Booker G R and Unvala B A 1965 *Jpn. J. Appl. Phys.* 21 L712

Casel A, Jorke H, Kasper E and Kibbel H 1986 *Appl. Phys. Lett.* 48 922

Casel A, Kibbel H and Schäffler F 1990 *Thin Solid Films* 183 351

Christou A, Wilkins B R and Davey J E 1983 *Appl. Phys. Lett.* 42 1021

De Jong, T Douma, W A S Smit L, Korablev V V and Saris F W 1983 *J. Vac. Sci. Technol. B* 1 888

Döhler G H 1972a *phys. stat. sol. (b)* 52 79

Döhler G H 1972b *phys. stat. sol. (b)* 52 533

Donnelly V M and McCaulley J A 1990 *J. Vac. Sci. Technol. A* 8 84

Eaglesham D J, Gossmann H-J and Cerullo M 1990 *Phys. Rev. Lett.* 65 1227

Eaglesham D J and Cerullo M 1991 *Appl Phys Lett* 58 2276

Eaglesham D J, Gossmann H-J, Cerullo M, Pfeiffer L N and West K W 1991a *J Crystal Growth 111* 833

Eaglesham D J, Pfeiffer L N, West K W and Dykaar D R 1991b *Appl Phys Lett* 58 65

Eaglesham D J, Higashi G S and Cerullo M 1991c *Appl Phys Lett* 59 685

Esaki L and Tsu R 1970 *IBM J. Res. Develop.* 14 61

Esaki L 1986 *IEEE J. Quantum Electron.* QE-22 1611

Van Gorkum A A, Nakagawa K and Shiraki Y 1989 *J. Appl. Phys.,* 65, 2485

Gossmann H-J and Unterwald F C 1992 to be published

Gossmann H-J and Schubert E F 1992 to be published

Gossmann H-J, Gilmer G H, Karunasiri U and Unterwald F C 1992 to be published

Gossmann H-J and Feldman L C 1985 Initial stages of silicon *Phys. Rev. B* 32 6

Gossmann H-J, Schubert E F, Eaglesham D J and Cerullo M 1990 *Appl. Phys. Lett.* 57 2440

Gossmann H-J and Eaglesham D J 1991 *J. Materials* 43(10) 28

Hale A P 1963 *Vacuum* 13 93

Headrick R L, Weir B E, Levi A F J, Eaglesham D J and Feldman L C 1990 *Appl. Phys. Lett.,* 57, 2779

Headrick R L, Feldman L C and Robinson I K 1989 *Appl. Phys. Lett.* 55 442

Hirayama H, Tatsumi T and Aizaki N 1988 *Appl. Phys. Lett.* 52 1335

Ishizaka B and Shiraki Y 1986 *J. Electrochem.Soc.* 133 666

Jackman T E, Houghton D C, Jackman J A and Rockett A 1989 *J. Appl. Phys.* 66 1984

Jackman T E, Houghton D C, Denhoff M W, Kechang S, McCaffrey J, Jackman J A and Tuppen C G 1988 *Appl. Phys. Lett.* 53 877

Jona F 1967 *Proc. of the* 13th *Sagamore Army Materials Research Conf.* eds Burke J J Reed N L and Weiss V (Syracuse Univ., Syracuse) 399

Jorke H 1988 *Surf. Sci.* 193 569

Joyce B A 1974 *Rep. Prog. Phys.* 37 363

Kasper E and Bean J C 1988 *Silicon Molecular Beam Epitaxy* Vol. I and II (CRC, Boca Raton)

Knall J, Barnett S A, Sundgren J-E and Greene J E 1989 *Surf. Sci.* 209 314

Knall J, Sundgren J-E, Greene J E, Rockett A and Barnett S A *Appl. Phys. Lett.* 45 689

König U, Kibbel H, and Kasper E, 1979 *J. Vac. Sci. Technol.* 16 985

König U, Kasper E, and Herzog H-J, 1981 *J. Cryst. Growth* 52 151

Kubiak R A A, Leong W Y and Parker E H C 1984 *Appl. Phys. Lett.* 44 878

Kubiak R A A, Leong W Y and Parker E H C 1985 *Proc. Electrochem. Soc.* 85-7 169

Kuznetsov V P and Postnikov V V 1969 *Sov. Phys. Crystallogr.* 14 441

Kuznetsov V P and Postnikov V V 1974 *Sov. Phys. Crystallogr.* 19 211

Kuznetsov V P, Tolomasov V A and Tumanova A N 1979 *Sov. Phys. Crystallogr.* 24 588

Leung T C, Lynn K G, Nielsen B, Asoka-Kumar P, Gossmann H-J, Feldman L C and Unterwald F C 1991 to be published

Lin J F, Li S S, Linares L C and Teng K W 1981 *Solid St. Electron.* 24 827

Lin T L, Fathauer R W and Grunthaner P J 1989 *Appl. Phys. Lett.* 55 795

Masetti G, Severi M and Solmi S 1983 *IEEE Trans. Electron Dev.* ED-30 764

Melloch M R, Mahalingham K, Otsuka N, Woodall J M and Warren A C 1991 *J Cryst Growth 111* 39

Metzger R A and Allen F G 1984a *Surf. Sci.* 137 397

Metzger R A and Allen F G 1984b *J. Appl. Phys.* 55 931

Mizutani T 1988 *J. Vac. Sci. Technol. B* 6 1671

Nakagawa K, Miyao M and Shiraki Y 1988 *Jpn. J. Appl. Phys.* 27 L2013

Ostrom R M and Allen F G 1986 *Appl. Phys. Lett.* 48 221

Ota Y 1983 *Thin Solid Films* 106 3

Ota Y 1977 *J. Electrochem. Soc.* 124 1795

Parry C P, Newstead S M, Barlow R D, Augustus P, Kubiak R A A, Dowsett M G, Whall T E and Parker E H C 1991 *Appl. Phys. Lett.* 58 481

Perov A S and Postnikov V V 1977 *Sov. Phys. Crystallogr.* 22 475

Ploog K and Döhler G. H. 1983 *Adv. Phys.* 32 285

Postnikov V V and Kuznetsov V P 1975 *Sov. Phys. Crystallogr.* 20 71

Postnikov V V and Perov A S 1977 *Sov. Phys. Crystallogr.* 22 350

Roksnoer P J, Maes J W F M, Vink A T, Vriezema C J and Zalm P C 1991 *Appl. Phys. Lett.,* 58, 711

Roth J A and Anderson C L 1977 *Appl. Phys. Lett.* 31 689

Saitoh S, Sugii T, Ishiwara H and Furukawa, S 1981 *Jpn. J. Appl. Phys.* 20 L130

Schubert, E F, Gilmer, G H, Kopf, R F and Luftman H S 1992 to be published.

Schultz P J and Lynn K G 1988 *Rev. Mod. Phys.* 60 701.

Shiraki Y, Katayama Y, Kobayashi K L I and Komatsubara K F 1978 *J. Cryst. Growth* 45 287

Shiraki Y 1986 *Prog. Crystal Growth Charact.* 12 45

Streit D, Metzger R A and Allen F G 1984 *Appl. Phys. Lett.* 44 234

Swartzenruber B S, Mo Y-W, Kariotis R, Lagally M G and Webb M B 1990 *Phys Rev Lett* 65 1913

Tabe M and Kajiyama K 1983 *Jpn. J. Appl. Phys.* 22 423

Tannenbaum Handelman E and Povilonis E I 1964 *J. Electrochem. Soc.* 111 201

Tatsumi T, Hirayama H and Aizaki N 1987 *Appl. Phys. Lett.* 50 1234

Tatsumi T, Hirayama H and Aizaki N 1988 *Jpn. J. Appl. Phys.* 27 L954

Thomas R N and Francombe M H 1968 *Appl. Phys. Lett.* 13 270

Thomas R N and Francombe M H 1969 *Solid State Electron.* 12 799 1969

Tolomasov V A, Abrasimova L N and Gorshenin G N 1971 *Sov. Phys. Crystallogr.* 15 1076

Unvala B A 1962 *Nature* 194 966

Unvala B A and Booker G R 1964 *Phil. Mag.* 9 691

Van der Pauw L J 1958 *Philips Res. Rep.* 13 1

Venables J A and Price G L 1975 *Epitaxial Growth* ed Matthews J W (Academic, New York) Part B 381

Vescan L, Kasper E, Meyer O and Maier M 1985 *J. Cryst. Growth* 73 482

Vink A T, Roksnoer P J, Maes J W F M, Vriezema C J, van Ijzendoorn L J and Zalm P C 1990 *Jpn. J. Appl. Phys.,* 29, L2307

Widmer H 1964 *Appl. Phys. Lett.* 5 108

Wolff S H, Wagner S, Bean J C, Hull R, and Gibson J M 1989 *Appl Phys Lett* 55 2017

Chapter 10

Point defects and charge traps in the Si/SiO$_2$ system and related structures

Edward H. Poindexter

U. S. Army Electronics Technology and Devices Laboratory, Fort Monmouth, New Jersey
07703 U. S. A.

ABSTRACT: Recent studies have significantly advanced our understanding of atomic-scale
defects in the metal-oxide-silicon system. Several varieties of E' centers in the oxide have
been distinguished by their electrical behavior. The nature of the P$_{b1}$ center at the Si/SiO$_2$
interface remains a mystery, and there is evidence of a third center, beyond P$_{b1}$ and the
well-defined P$_{b0}$. The negative-bias-temperature instability has been modeled by a complex
and unique electrochemical mechanism. An equivalent model for radiation damage is not
yet developed. Most Si/SiO$_2$ defects, however, have been found to incorporate either
dangling $\cdot$Si$\equiv$ orbitals or hydrogenous species.

1. INTRODUCTION

During the past few years, new approaches and some fortunate experimental discoveries have
resulted in substantial improvement in defining the nature and behavior of atomic scale or point
defects in the metal-oxide-silicon (MOS) system. In this paper we will review a selection of
topics of notable current interest, and we will in some cases attempt to assess, extend, or generalize findings and interpretations. Although most of this paper is confined to elsewhere-pub-
lished studies, we have in a few places enriched the discussion with very recent, unpublished
findings which enable more definitive conclusions.

The several areas of MOS defects included here are unified by the common thread of electron
spin resonance (ESR) spectroscopy (Poindexter and Caplan 1983). ESR is the only technique
applied to the MOS system which offers a picture of the ultimate physical structure of the point
defects in question. Furthermore, the MOS structure may well be unique among technologically important systems in its susceptibility to ESR analysis. Although ESR is of very limited
usefulness as a general analytical tool, in no way rivaling optical spectroscopy or even nuclear
magnetic resonance (NMR), it is usefully applicable to many MOS defects. In every topic presented here, ESR was either the dominant contributor, or played a significant ancillary role in
the comprehension of the underlying phenomena.

We will review 4 areas of MOS defect study. The first comprises the point defects in uncon-
taminated oxide, where the very important E' center ($\cdot$Si$\equiv$O$_3$) has now been well characterized
in several variations by ESR and correlated electrical measurements. The E' center is the site of
positive oxide charge in radiation-damaged structures (Lenahan et al 1987). Second, we con-
sider the nature of the lesser centers, mainly P$_{b1}$ (possibly Si$_2$O$\equiv$Si$\cdot$), at the Si/SiO$_2$ interface.
The P$_b$ family of centers is the main source of interface traps in unannealed or stress-damaged
MOS structures (Poindexter et al 1984; Gerardi et al 1986). Next, we review a proposed new
electrochemical model for the negative-bias-temperature instability (NBTI) (Nicollian and
Brews 1982), since surprising new experimental results (Gerardi and Poindexter 1989; Blat et
al 1991) have suggested an interpretation quite unique in MOS chemistry (Gerardi et al 1991).

Last, some new ideas in radiation damage mechanisms will be presented (Lelis et al 1988; Saks and Brown 1990; Ma 1989); unhappily, despite an enormous amount of research, no unified model which resembles the NBTI model has been proposed.

2. MOS DEFECT OVERVIEW

In the past decade there has been discovered an impressive group of paramagnetic point defects in the MOS system, a number of which have been successfully identified with important electrical defects. Most of these defects are shown in Fig. 1. This paper will concern mainly E' centers and P_b centers. Two non-paramagnetic centers--water-related positive charge (Gerardi et al 1991), hydrogen-related radiation-induced charge--will also be discussed. The seemingly non-paramagnetic oxide trapped negative charge and/or water-related electron traps will not be discussed (Gale et al 1988; Young 1988). Likewise, P_b centers in buried implantation-oxides (Carlos 1987), and the nebulous silicon dangling bond center (g = 2.0055) in the bulk oxide (Dersch et al 1981), in silicon monoxide (Mizutani et al 1970), and in Si-implanted oxides (Kalnitsky et al 1990) will not be discussed.

The excess-oxygen center, called non-bridging-oxygen hole center (NBOHC), i.e., $O_3 \equiv Si$–$O \bullet$....H–O–$Si \equiv O_3$, and the peroxy radical ($O_3 \equiv Si$–O–$O \bullet$) are important centers, occurring frequently in bulk fused silica. They have been observed in thin thermal oxides, and seem to behave more or less like those in fused SiO_2 (Vikhrev et al 1977; Poindexter and Caplan 1987). However, they have not been shown to be electrically active and therefore will not be mentioned further.

3. E' CENTERS IN THE OXIDE

The essential element of the E' center is the moiety $O_3 \equiv Si \bullet$ (Weeks 1956). Altogether, more than 10 E'-like centers have been differentiated in crystalline quartz, in amorphous fused silica, and in thin film silica. A major subdivision is the presence or absence of a complementary adja-

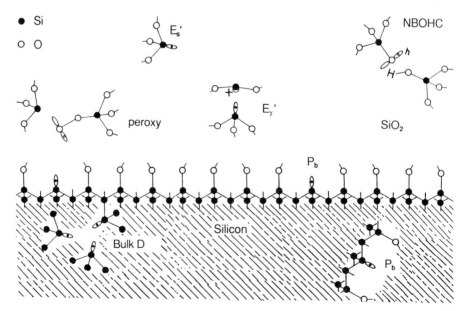

Fig. 1. Paramagnetic point defects which have been observed in Si/SiO$_2$ structures by electron spin resonance (ESR).

cent moiety O$_3$≡Si$^+$. Lesser variations include the presence of ancillary atoms, mainly O or H. All the thin film variations are summarized elsewhere in a very recent paper (Warren et al 1992). We will discuss at length two main subtypes, E$'_\gamma$ or O$_3$≡Si•....$^+$Si≡O$_3$, and E$'_s$ or O$_3$≡Si•, which have been clearly observed by ESR; and two others, E$'_1$ or O$_3$≡Si•....$^+$Si≡O$_3$, and E$'_\alpha$ or O$_3$≡Si•....$^+$Si≡O$_3$..O, which have been inferred.

It was some years ago demonstrated that E' centers were a major source--perhaps the dominant source--of positive oxide charge near the interface of radiation-damaged Si/SiO$_2$ structures (Lenahan et al 1987). The positive oxide charge developed after exposure to γ-rays was quantitatively 1:1 correlated with the radiation-induced positive charge, both temporally during post-radiation anneal, and spatially after successive removal of oxide by etching. The ESR visibility when the defect was in its positively charged state indicated that the E' center was of the E$'_\gamma$ or E$'_1$ variety. The main difference between the two is the ease of charge neutralization and rechargeability. This aspect was not specifically explored at the time; we will return to it later.

Since the E' center in any form represents a local oxygen deficiency or silicon excess, it might be expected to occur more readily in materials with such a non-stoichiomatic nature. The first silicon-rich samples to be explored were SiO$_2$ films sputter deposited on Si wafers (Zvanut et al 1988). The films were approximately 50 nm thick, atop a 1-2 nm native oxide. The electrical properties were tested by a pulsed capacitance-voltage (C-V) technique. It was observed through repeated cycles of (+) or (−) stress that there were reversible charge traps present in the oxide, preferentially detected near the Si/SiO$_2$ interface. The sign of the charge in the traps was positive or zero. The extreme values after positive or negative bias stress (+8,−12MV/cm) are plotted in Fig. 2. It was found that the charging rates as a function of bias could be quantitatively explained by tunneling of electrons to and from the silicon substrate under (−) or (+) bias, respectively. The nature of the charge trap is, of course, not disclosed by the electrical measurements; it was initially suspected that they were E$'_\gamma$ centers.

ESR was used to determine the nature of the trap (Zvanut et al 1989). Samples were bias-stressed to +7, −14 MV/cm by the corona method, to obviate metal gates (which are a problem for ESR). Because the traps were expected to be E$'_\gamma$, they would be charged (+) when ESR-active, and would be electrically neutral when not ESR-active and presumably rebonded to form O$_3$≡Si–Si≡O$_3$. Surprisingly, the opposite sense of charge was observed; the centers were ESR active in their neutral state. Furthermore, the successive variations in ESR spin concentration upon application of reversible bias stress were 1:1 quantitatively anti-correlated with the successive charges in positive charge, also shown in Fig. 2. The succession of ESR spectra after (+) or (−) bias stress is shown in Fig. 3.

The simplest explanation for this sense of charge versus ESR was that the E' centers were of the "hemi" or E$'_s$ variety, •Si≡O$_3$, which becomes •Si≡O$_3$ when it loses an

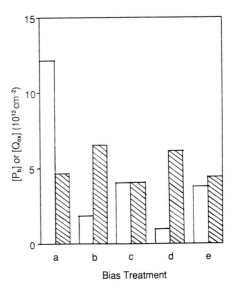

Fig. 2. Concentration of E$'_s$ centers (open bars) and positive oxide charge Q$_{ox}$ (hatched bars) in sputter-deposited SiO$_2$ on Si after successive bias stress: (a) unstressed, (b) (−) bias; (c) (+) bias; (d) (−) bias, (e) (+) bias.

electron. This observation was thin film evidence for the "hemi" or surface E'_s center, first proposed to explain the gaseous anneal of E' centers on SiO_2 surfaces in crushed SiO_2 (Hochstrasser and Antonini 1972).

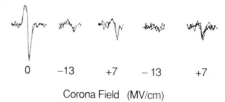

0   −13   +7   − 13   +7

Corona Field   (MV/cm)

Fig. 3. ESR signals from E'_s centers in ion-sputtered SiO_2 on Si after successive application of bias stress.

The next deliberately non-stoichiometric oxides to be probed for rechargeable E' centers were Si-rich films heavily doped with excess Si by ion implantation (Kalnitsky et al 1990). The excess Si was fused into the lattice by heat treatment at 1200 K for one hour. These films showed a repeatable charging and discharging in electrical tests, developing net positive charge near the positive electrode of the poly-Si gate MOS structure, and net negative charge near the negative electrode. The controlling charge transport mechanism in the system was found to be Fowler-Nordheim tunneling between electrode and the near-interface traps, with other transport processes filling in the gap between near-interface regions.

Testing of Si-implanted wafers by ESR and corona bias stress revealed that E' centers were present in the stressed wafers, and were preferentially located near the gate or substrate interfaces. The sequence of E' signals observed after repeated bias-stress cycles is shown in Fig. 4. Generally speaking, the data resemble those of the ion-sputtered oxides in Fig. 3. The interpretation, however, is different and not straightforward. The simple 1:1 correlation between oxide charge state and E' visibility was obscured by the occurrence of both (+) and (−) charges and the spatial symmetry and reversibility of trap charging. Moreover, the concentration of E' centers was only about 20 percent of the total (+) charge concentration. Hence, some careful deconvolution was necessary to establish the E' versus charge correlation.

Fig. 5 shows the sign of the effective oxide charge developed by opposite gate voltage stress. The samples contain more than one kind of charge trap , with about as many (−) as (+) species. The (−) trap in itself is not a problem, as the existence of any negatively charged E' species was rather strongly ruled out by the small binding energy for the extra electron in any feasible E'-like site. However, the reversible spatial symmetry of sample charging was a problem. The electrical measurements revealed the location and sign of the sample charges by the C-V measurements of Fig. 5, coupled with a modeling of the current flow over 7 orders of magnitude, which resulted in a good spatial map of the charges. The modeling was found to be strongly constrained by the nature of the transport mechanisms, and little leeway as to adjustment of mechanism or resultant charge disposition was found possible.

The ESR, on the other hand, gives equivalent signals regardless of the location of defects in the oxide. Simple numerical correlation with the oxide charges, reversible in location and far outnumbering the E', was not possible. To obtain depth profiles, an etchback experiment was performed on a sample preferentially positively charged near the outer oxide surface by a (+) corona stress. The E' signal was at first very large in the positively-charged outermost oxide layer. It dropped off rapidly as oxide was removed; in fact, it could not be rejuvenated in the middle of the oxide by reapplied (+) stress. Only a feeble signal could be recreated near the substrate. It was shown by calculation of trap-to-trap tunneling rates that the very low resistivity of the heavily

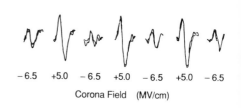

−6.5   +5.0   − 6.5   +5.0   − 6.5   +5.0   − 6.5

Corona Field   (MV/cm)

Fig. 4. ESR signals from E'_γ centers in silicon-implanted SiO_2 on Si after successive applications of bias stress.

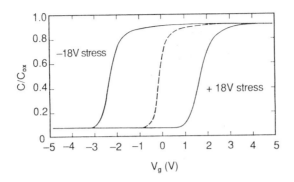

Fig. 5. Capacitance-voltage (C-V) analysis of MOS devices with Si-implanted (2×10^{15} cm^{-2}, 25 keV) oxides.

implanted mid-oxide region precluded the field gradients required to ionize any E' sites situated in this region. However, near the substrate the lower implant density permitted ionization.

Another sample was implanted differently to yield a relatively lower concentration of Si ions near the substrate and higher near the outer surface, as compared to the oxide above. In this sample, the opposite E'-signal versus sign of corona stress was observed, as expected from the logic developed in the first sample. This finding was coupled with the idea that a change in local sample conditions by a field gradient cannot simultaneously cause one species of trap to charge positively, and an intermingled different species to charge negatively; such a sample could not exist in an equilibrium state with all defects neutral. Thus it was established that the E' centers in these samples are positively charged when they are visible to ESR and are neutral when invisible. They are of the composite, dual-moiety variety, E'$_\gamma$ or $O_3 \equiv Si \cdot .... ^+Si \equiv O_3$. This determination was in good accord with the simple idea that a silicon-rich oxide, with excess Si fused into the lattice, must *a priori* harbor microscopic non-stoichiometric structures such as $O_3 \equiv Si-Si \equiv O_3$, the likely precursor for the E'$_\gamma$ center. Further, the high temperature anneal would probably allow the Si–Si moiety to adjust to its most stable bonded distance, 0.25nm.

The ESR results and the C-V determinations are, however, only part of the story in the E' centers in sputter-deposited and Si-implanted SiO₂ films. The charging dynamics and nature of any current flow give additional new information on the E' centers. In the case of the sputtered oxides, the charging and discharging of the traps, by tunneling between trap and substrate, showed a strong hysteresis, Fig. 6. This hysteresis was ascribed to a Franck-Condon defect relaxation (Zvanut et al 1988; Gurney 1931; Silsbee 1967). It was early calculated for the E'$_\gamma$ center that the lowest-energy configuration comprises the paramagnetic moiety $O_3 \equiv Si \cdot$ in a protuberant configuration , i.e., with the Si atom positioned such that the tetrahedron was more acute than in the normal SiO₂ lattice (Feigl et al 1974). The Si of the complementary (+) charged $O_3 \equiv Si^+$ moiety is retracted well back, actually, slightly behind the O_3 plane, so that the moiety is approximately flat. The first proposal of this asymmetric-relaxation model of the E' center has led to the informal name "Feigl-Fowler-Yip" (FFY) center. The same general configuration-versus-energy situation was expected to prevail for the isolated fragment representing the hemi or surface E'$_s$ centers. Then a negative bias promotes tunneling of an electron from a stable distended and neutral moiety $O_3 \equiv Si \cdot$ to the silicon valence band, yielding immediately an unstable distended positively-charged moiety $O_3 \equiv Si^+$. This process was experimentally found to require an energy corresponding to a defect level about 5 eV below the SiO₂ conduction band (Zvanut et al 1988). Immediately after losing its electron, the defect relaxes to its preferred nearly flat configuration, releasing a small amount of energy to the lattice.

Discharging the center by electron tunneling to the flat, charged center required a reverse gradient sufficient to align Si valence band with the receiving level of the defect, which is at about 4.5 eV below the SiO_2 conduction band. Subsequent defect relaxation into the protuberant shape, also releasing energy to the lattice. The configuration relaxation makes it necessary to supply more energy to ionize or charge the E_s' defect, and prevents electrical recovery of the original input energy upon discharge or neutralization. The charge states and structural configuration of the E_s' center are shown in Fig. 7, which portrays the sequence in events after successive application of bias. The energy levels of the various states of the $\cdot Si \equiv O_3$ fragment have been computed theoretically by a molecular orbital approach (Fowler et al 1990). The results are shown in Fig. 8.

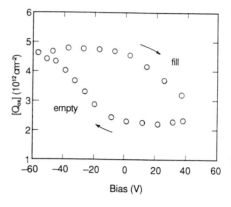

Fig. 6. Variation of (+) charge trapped in E_s' centers near the interface with sputter-deposited SiO_2 during bias stress cycling.

Theory and experiment are in generally good agreement, differing by only about 1 eV.

Analysis of charging and discharging of the classic 2-part E_γ' center found in the Si-implanted oxides is less straightforward, as noted earlier. No simple hysteretic tunneling plot was possible. Rather, the observed charging and discharging rates under the applied field gradients were compared with predictions derived from several feasible transport mechanisms. For example, electron hopping from E' site to E' site via field ionization and retrapping is dominated by the first step. We have (Terry et al 1985)

$$\gamma_{fi} = 10^{15} \, exp\left(-6.45 \times 10^7 \, \phi^{3/2} \, \mathcal{E}^{-1}\right), \tag{1}$$

where γ_{fi} is the ionization probability in s^{-1}, ϕ is energy barrier in eV and $\mathcal{E}$ electric field gradient in V/cm. The rate calculated for the observed field gradients with $\phi = 3$ eV was $\sim 10^{-7} \, s^{-1}$, much, much slower than the required $10^{-2} \, s^{-1}$. With a more realistic $\phi \geq 4.5$ eV, the rate would be vanishingly small.

The first layer of E' trap sites near the electrode interface must then clearly discharge by direct tunneling, and more distant traps by successive trap-to-trap tunneling. The appropriate expressions are similar, but the process would be expected to be controlled by the region just one monolayer more distant from the electrode, where the density of E' sites is lowest. The applied

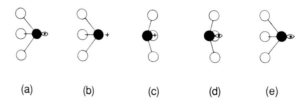

Fig. 7. E_s' center disposition in sputtered oxides at successive stages of bias-stress cycle: (a) initial, unstressed center; (b) immediately after loss of electron during (−) bias; (c) subsequent relaxation to stable near-planar form; (d) upon gain of electron during (+) bias; (e) relaxation to stable distended form, as in (a).

field must raise the relaxed, occupied level of $O_3\equiv Si-Si\equiv O_3$ to align with electron-receiving, unrelaxed $O_3\equiv Si\cdot....^+Si\equiv O_3$ level. We have (Löwdin 1965)

$$\gamma_{TT} = 4 \times 10^{14} \, exp\left(-8d\,\phi^{1/2}\right), \quad (2)$$

where γ_{TT} is trap-to-trap tunneling probability in s^{-1} and d is distance between traps in nm. For a barrier of 5 eV, an alignment energy difference of 1 eV, a gradient of 5 eV, and a trap spacing of 2 nm. The tunneling rate is about $10^{-2}\,s^{-1}$, consistent with observations.

Because of the lack of 1:1 correspondence between positive charge and [E'] in the Si-implanted samples, numerous other transport scenarios for both charging and discharging, and involving the trap parameters for the more numerous additional and unidentified trap species were examined. However, no other mechanism came even close to explaining the observed phenomena and parameters. It was thereby felt reasonably well established that the E' centers present were of the E_γ or composite dual-moiety FFY species, and that a Franck-Condon relaxation phenomenon similar to the E'_s center was operative. The sequence of events upon trapping and detrapping under successive bias application is shown in Fig. 9.

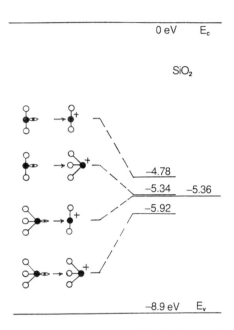

Fig. 8. Calculated energy levels for transitions involving relaxed/unrelaxed charged/paramagnetic variants of the E'_s site in amorphous SiO_2.

The energy level scheme for the E'_γ center has also been calculated by the some means as for the E'_s center (Rudra and Fowler 1987). The energy levels are not very different from those of the hemi E'_s center. This may seem surprising in view of the fact that the lowest state of the E'_s center has a dangling orbital, whereas the E_γ center is rebonded as Si–Si across the space representing a missing oxygen.

This raises the question of a distinction between the two archetypes of composite or 2-unit E' centers, the E'_γ and E'_1. Published literature on the subject often glibly refers to the E'_γ or E'_1 centers as oxygen-vacancy centers, without careful specification of just what that means. Clearly, in an SiO_2 lattice, any Si......Si structure without an intervening oxygen can be called an oxygen vacancy. However, there is much more to be said. If the vacancy is produced by a knock-on or ionization event in crystalline SiO_2, the initial Si......Si spacing will be 0.31 nm. If the "vacancy" is the result of oxygen deficiency in a synthesis process, or excess Si from an annealed implantation, the initial Si......Si distance will more likely be 0.25 nm, as in bulk silicon itself. The strength of any bonds formed across the longer distance will be much less than that formed across the shorter distance. It has been calculated, for instance, that the long bond in an unrelaxed lattice is 5.9 eV below the SiO_2 conduction band, but for the shorter distance, 7.7 eV (Robertson 1984; O'Reilly and Robertson 1983). The calculations by the authors who also studied the E'_s center (Rudra and Fowler 1987) as noted above in essence confirm this view, though their energy values differ from the earlier ones. The bond-length criteria would provide a good basis, then, for distinguishing between the E'_1 center of α-quartz and the E'_γ center of amorphous SiO_2 thin films.

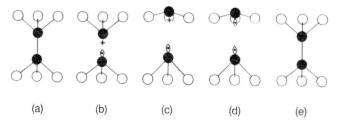

Fig. 9. E'_γ center disposition in Si-implanted thermal oxides, near the gate, at successive stages of bias-stress cycle: (a) initial, non-paramagnetic precursor moiety; (b) immediately after loss of an electron during (+) bias; (c) subsequent relaxation to stable configuration; (d) after recapture of an electron during (–) bias; (e) relaxation and rebonding to non-paramagnetic state.

The charging and discharging of E' centers are readily observed in thin-film SiO_2 on Si, because of the relative ease of applying gradients in the MV/cm range. Not only is this of direct interest in study of traps, but it also allows the thin film research to shed additional light on E' centers in bulk crystalline or fused amorphous SiO_2. Two other E' centers besides the E'_s and E'_γ were evident in the studies on sputtered and implanted oxides. When initially implanted, the oxide contained predominantly one E' center, which responded only once to the applied field; it disappeared after one ± sequence and never recovered, leaving a non-dischargeable residual E' center. A center which is so easily annealed at 300K by electric field alone, and which remains unrechargeable, suggests a displaced oxygen which remains very near and is readily nudged back into place. The healed E' center, of course, cannot be regenerated by electric field. This center may be the easily annealed E'_α center of fused SiO_2 (Griscom 1985) After 1200 K anneal in N_2, the recyclable E'_γ centers appeared, accompanied by the same reduced quantity of non-dischargeable E' centers. The sputtered oxides also showed about 30 percent non-dischargeable E' centers.

It is tempting to explain these non-dischargeable centers as composite E'_1 centers, where the Si–Si distance is long (0.31 nm), and where the activation energy for rebonding prevents healing of the center by simple electron trapping at 300 K. While this might be the case, it cannot be settled without good information on the distance between centers. If much greater than 2.5 nm, the easy-to-trap tunneling will disappear, to be supplanted by the difficult field ionization. Furthermore, the affair may be complicated by the presence of other tunnelable species which can perform at high concentration. Thus, all-in-all, the best that can be said is that there do indeed appear to be two or more sub-species of the composite or FFY E' center. However, some of the non-dischargeable E' centers might also be hemi or E'_γ centers which were too far from other like centers to allow easy electron transport.

The important test of chargeability of E'_1 centers in α-quartz, needless to say, has not been attempted. Such tests would clarify experimentally the relative Si–Si bond energies in α-quartz and amorphous SiO_2. Note that the non-dischargeable center cannot be comprised of paired hemi-centers with very large distance between Si atoms. Such a pair, though not able to rebond would show dipolar broadening of the ESR signal, which is not observed.

4. THE P_{b1} MYSTERY

The great success of ESR in identification of the P_b centers over 12 years has led to quite a number of studies of interface traps benefitting from this solid foundation. The general acceptance of P_b as the major source (some say the only source; we will not discuss that issue here.) of interface traps in unannealed MOS structures has tended to obscure the remaining mystery concerning the P_{b1} center, a variety of P_b occurring on the (100) Si surface. The overview summarized in Fig. 1 presents P_b centers as a single species, specifically, P_{b0}, the only variety occurring on the (111) surface. This was the first species of P_b discovered (Nishi 1971;

Poindexter and Caplan 1978). It is also the simplest, both physically and chemically, having only one microstructural embodiment, and a single well-substantiated chemical structure, viz., $\equiv Si_3–Si\bullet$. The dangling orbital is aligned perpendicular to the plane of the interface; the other (Caplan et al 1979), tetrahedral orbitals are never observed by normal ESR spectroscopy to be in a paramagnetic state.

The situation on (110) is a little more complicated structurally. There are two EPR-visible P_b centers whose orientations are in accord with the canted intercept of the (111) tetrahedral axes at the interface plane. The other possible dangling orbital dispositions lying on the interface plane are not detected by ESR. This surface has attracted no further ESR attention (and very few derivative electrical studies) since the discovery of P_b (110) more than 10 years ago. Presumably, the center is chemically the same as P_b on (111), though surprises are not unprecedented in this sort of system.

The device-pertinent (100) Si surface, of greatest interest in electronic applications, is the most complicated in regard to P_b centers. Again, there are two P_{b0} of different orientation, in accord with the canted intercepts of the tetrahedral bond directions with the interface plane. The two allowed dangling orbitals directions comprise the entire suite of such defect structures on (100) Si; there are no missing or "forbidden" dangling orbitals. The most important feature of the (100) face, however, is the presence of a second chemically-distinct P_{b1} defect, which also occurs in two orientations. The symmetry and orientation of the g-value ellipsoid from ESR studies are readily consistent with the defect moiety $Si_2O\equiv Si\bullet$, which also occurs in two possible interface orientations. This structural assignment, however, was never rigorously confirmed. The ESR determined disposition of P_b centers on the 3 principal wafer planes, (111), (110), and (100) is shown in Fig. 10 (Poindexter and Caplan 1983).

We will present here the various bits of evidence on the non-P_{b0} type of P_b center, including P_{b1}, in roughly historical order, without any attempt at conclusive interpretation. The first peculiarity noted for P_{b1} was during the process of its definition as a new and distinct defect (Caplan et al 1979). The complex g-value plot observed for (100), as compared to the very simple (111), needed some chemical help for deconvolution. Quite a large number of variously oxidized and annealed (100) samples had been prepared in advance, and these were searched for a possibly different balance of $[P_{b0}]$ to the other species, not then separately characterized. While no systematic pattern of process history was evident, two or three samples showed a much reduced P_{b0} signal, allowing P_{b1} to be observed clearly. This was the basis for the initial assignment as $Si_2O\equiv Si\bullet$ (Poindexter et al 1981). Despite the exceptions described there, a seeming chemical contradiction to the above findings arises in the case of native oxide growth. A (100) wafer, oxidized in dry O_2 and finished with fast pull in O_2, shows about equal amounts of P_{b0} and P_{b1} (Poindexter et al 1981). With 1300 K oxidation, the concentration of either is about 5×10^{11} cm^{-2}. If a (100)

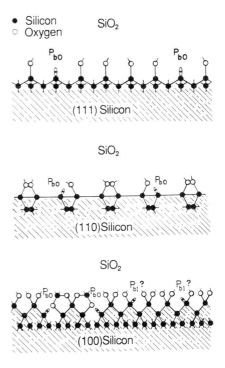

Fig. 10. Structure of P_b centers on Si wafers of 3 different orientations.

wafer is stripped of oxide in HF and then allowed to grow a native oxide, only P_{b0} reappears (Gerardi et al 1986). The rapid regrowth of P_{b0} on etched (100) Si is very similar to the equally rapid regrowth of P_b on etched (111) Si. The P_{b1} center on (100) is thus conspicuously different from these other two, which would seem to be chemically very similar from etching tests.

The preferential regrowth during native oxidation of P_{b0} on (100) enhances the bias-controlled variation in P_{b1} occupancy in an MOS test structure. To maximize the apparent change in $[P_b]$, the back side of the wafer is etched to remove furnace-grown oxide. The P_{b0} recovers partially, giving an invariant signal which adds to the bias controlled front surface signal, reducing the apparent contrast between zero bias and maximum bias $[P_{b0}]$. Since P_{b1} does not regrow, the full contrast is displayed for its variation, which is very fortunate, as its signals are quite weak and needed somewhat indirect methods to supplement the direct measurements.

A major physical peculiarity occurs in the case of P_b signals in electrically detected magnetic resonance (EDMR) (Henderson 1984). EDMR arises from spin-dependent recombination (SDR), by which a non-equilibrium conduction band electron recombines with a valence band hole via an intermediate paramagnetic level in the bandgap. Though somewhat complicated and controversial in detail (Lepine 1972; Kaplan et al 1978; Vranch et al 1988), the basic principle is simple. At some stage in the recombination, two bodies with spin must come together--either a carrier electron and a paramagnetic center, or a carrier electron and a hole. Pairing is very much more likely if the two bodies have opposed spins. Hence the recombination process tends to remove momentary spin-opposed pairs from the system, leaving a surplus of momentary spin-aligned pairs. If one of the bodies has a readily observed ESR, then saturation of its resonance will flip its spin rapidly and create momentary opportunities of opposed-spin pairs to be generated and thus allow recombination to proceed. The EDMR phenomenon may be observed electrically by measurement of current, appearing much like a normal ESR signal, or it may be observed in the usual ESR method after creation of a large non-equilibrium concentration of electrons by irradiation with above-band gap light. The latter has been called the photoconductive resonance or PCR (Shiota et al 1977). The inherent sensitivity of EDMR or PCR can be orders of magnitude greater than the normal ESR (Kaplan et al 1978).

When PCR was extensively applied to Si wafers of orientation (111), (110), and (111), unexpected and unexplained results were obtained (Poindexter et al 1988). On (111) Si, not only was the P_{b0} signal observed, with attendant dangling orbital orientation perpendicular to the Si/SiO_2 interface, but centers of the 3 other tetrahedral bond directions were observed. On (110) Si, the usual canted P_{b0} were observed; and in addition, similar centers with the in-plane orientation. The major surprise, however was on (100). Again, the usual P_{b0} orientations were observed, with no additional non-standard orientations, there being no such orientations on (100). Very strangely, P_{b1} was not observed at all. Instead, a different center emerged. This strange result on (100) was confirmed by nearly concurrent experiments in which SDR was detected electrically on (100) MOS capacitors (Henderson et al 1989), with signals detected only from P_{b0} centers. The PCR g-maps for the 3 surfaces are shown in Fig. 11, together with the normal ESR g-maps for comparison.

Another important and unexpected finding was made during the SDR/PCR studies. As noted above, no P_{b1} signal was seen on (100)Si, but a second signal was observed, in addition to that of P_{b0}. Its g-value matrix and orientation in the lattice were different from either P_{b0} or P_{b1}. Unlike the case of the P_{b0} and P_{b1} centers, there was no evidence of this new center by normal ESR. This may mean that its concentration is too low; SDR can be up to 10^6 times as sensitive as ESR. It might also mean that a new center which is visible only by its SDR interaction is present. It is difficult to imagine just what the physical character of such a center might be.

A decisive explanation for the SDR-visibility of all the possible orientations of the P_{b0} centers on (111) and (110) silicon is not possible at this time. Why, indeed do the normally invisible (to ESR) bond directions suddenly appear in SDR? Initially, it might be ascribed to the much greater potential sensitivity of SDR. However, the PCR signals from both the usual ESR-

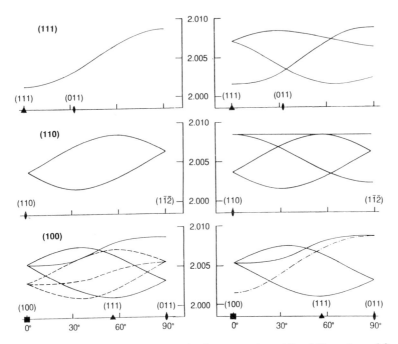

Fig. 11. Comparison of g-value anisotropy for P_b centers in oxidized Si wafers of 3 orientations. Left: observed by normal ESR. Right: observed by photoconductive ESR or PCR. Solid lines: P_b (111, 110) and P_{b0} (100) centers. Dashed lines: P_{b1} (100). Dot-dashed line: a new center with unassigned structure.

detected bond directions were no stronger than those from the non-standard bond directions. The question then arises as to whether the new signals represent recombination through some transient paramagnetic condition of the normally electron-paired bonds, or rather from simply very small numbers of broken bond moieties of these non-standard orientations.

A very surprising chemical peculiarity was next discovered. In an exhaustive study, of passivation of interface traps by hydrogen, and subsequent depassivation (Reed and Plummer 1988), excellent self-consistent kinetic results were obtained for P_b on (111) Si and for the assumed P_{b0} on (100) Si. The passivation of P_{b0} was found to be modeled by the reactions:

$$P_{b0} + H \overset{k_1}{\rightleftharpoons} HP_{b0} \tag{3}$$

$$2H \overset{k_2}{\rightarrow} H_2 \tag{4}$$

$$HP_{b0} \overset{k_r}{\rightarrow} H + P_{b0} \tag{5}$$

The appropriate kinetic expression is

$$\frac{d}{dt}[P_{b0}] = -k_1[P_{b0}][H] + k_r([P_{b0}]_o + [HP_{b0}]_o - [P_{b0}]) \tag{6}$$

$$\frac{d}{dt}[H] = -\frac{k_1}{T_{ox}}[P_{b0}][H] + \frac{k_r}{T_{ox}}([P_{b0}]_o + [HP_{b0}]_o - [P_{b0}]) - 2K_2[H]^2 \tag{7}$$

Everything was well fit by [H] and [P_{b0}] values which were consistent with the carefully weighed evidence from other types of study. However, on (100) Si, the passivation of the P_{b1} component in the total Q_{it} was found not to consume any hydrogen. We emphasize that the data in this study were taken over a very wide range with many data points, which made possible detailed extraction of parameters. Among the latter, the values obtained for E_a (k_1, P_{b0}), E_a (k_2, P_{b0}), and E_a (k_r, P_{b0}), were 0.75 eV, 0.75 eV, and 1.6 eV, respectively, were all very reasonable. Several possible explanations of the P_{b1} behavior were considered, but no definite conclusion was possible.

Another important piece of evidence is the search for nuclear hyperfine structure (hfs) from nearby O^{17} nuclei (Brower 1989). Clearly, if the $Si_2O\equiv Si\cdot$ model were correct, the oxygen atoms directly bonded to the $\equiv Si\cdot$ would be naturally expected to generated a hfs. To examine this matter, samples were specifically grown in dry O_2 enriched to > 50 percent in O^{17}. Both (111) and (100) wafers were tested. The results were, unfortunately, in a gray zone. The only effect was a broadening ΔH of the ESR signal in the O^{17}-enriched samples as compared to normally oxidized wafers; no hoped-for hfs was produced. Furthermore, the broadening results were not very illuminating in themselves. The maximum ΔH was observed for P_b (111) and was 4.8 G. After that, ΔH for P_{b1} (100) was 3.5 G, and for P_{b0} (100), 2.2 G. These results suggest that more oxygens are actually located near P_b (111) than near either of the (100) centers, and most surprisingly, more O than near P_{b1}--supposedly, the oxygen-bonded defect.

Molecular orbital theory has recently been brought to bear on the P_{b1} versus P_{b0} distinction (Edwards 1988). Such calculations have been most effective and in good accord with experiments on the P_b (100) center. The ESR Si^{29} hyperfine structure, the geometrical configuration, and the (+/0) and (0/–) energy level positions in the Si bandgap are all found to be in agreement with experimental findings (Cook and White 1987; Edwards 1988; Carrico et al 1986). Especially informative are the energy level positions. On (111), these are separated by an electron correlation energy U of about 0.55 eV, and are located at E_v + 0.3 eV and E_v + 0.85 eV, respectively. Less accurate measurements on (100) Si find P_{b0} to be very nearly the same as P_b (111), with P_{b1} levels definitely bracketed between P_{b0}, at roughly E_v +0.45 eV and E_v + 0.8 eV. The MO calculations (Edwards 1988), however, find a center with the composition originally assigned by ESR to P_{b1}, i.e., $Si_2O\equiv Si\cdot$, to have very narrowly spaced--even negative-U-levels sited at the Si conduction band. It should be ESR-invisible, and is in severe disagreement with the simplistic early model (Edwards 1988).

Several other models consistent with the ESR g-value symmetry were also studied theoretically (Edwards 1988). A moiety sited among strained bonds which otherwise resembles a P_{b0} center was found to be a better candidate for the P_{b1} resonance than is the $Si_2O\equiv Si\cdot$ structure. This conclusion is in conflict with the SDR findings and the very similar chemistry of the P_b (111) and P_{b0} (100) centers; noted above. The suggested models for the strained-bond centers are shown in Fig. 12.

Another very pertinent finding is the recent observation by ESR of P_b-type centers generated on (100) wafers by radiation (Kim and Lenahan 1988). The wafers, initially prepared to have very few interface traps, showed the expected radiation-induced P_b centers. Only P_{b0} centers were produced. The same thing occurred in 3 variously prepared oxides, indicating that the phenomenon was characteristic of the interface crystallography, rather than some peculiarity in the oxide or oxidation process. The early radiation experiments (Lenahan et al 1981) on (111) Si, of course, showed only P_{b0} centers, i.e., the moiety $\cdot Si\equiv Si_3$, which evidently occurs on

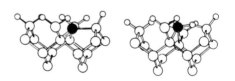

Fig. 12. Molecular clusters used for theoretical modeling of P_b centers on (100) Si . Left: strained site with dangling-orbital center adjacent to a reconstructed surface Si–Si bond. Right: dangling orbital center near a strain-relief site.

Si wafers of all orientations. For our purpose here, this latest finding reinforces the idea of the chemically identical and structurally near-identical nature of P_b (111) and P_{b0} (100) centers. Thus, although this has no direct bearing on the identity of P_{b1}, it does substantiate the close similarity of the P_{b0}-like centers--of significance to the situation below.

In contrast to much of the above, the invalidity of the original P_{b1} assignment has received recent experimental support. Wafers of (111) and (100) Si oxidized in dry O_2, then passivated by exposure to (moist) room air upon cooling, could be depassivated by anneal in Ar, and recycled as desired by sequential exposures to room air or Ar while hot (Stathis and Dori 1991). In these tests, P_b (111) and P_{b1} behaved very similarly, i.e., both were undetectable after passivation, and reappeared strongly after inert anneal. In contrast, P_{b0} hardly varied at all. It emerged from the oxidation cool-down with a strong signal which actually increased slowly with subsequent recycling. The experiments seem straightforward. Unhappily, these experiments on P_b chemistry and the several early ones above are in striking discord, and a simple and believable rationalization is not immediately apparent.

In addition to the numerous approaches to the P_{b1} mystery discussed above, there is another quite different line of study which involves interface traps generated by radiation or electron-injection (Ma 1989). Equally suitable for this present section, it will instead be included in Sec. 6, Radiation Damage.

5. NEGATIVE-BIAS-TEMPERATURE INSTABILITY

The negative-bias-temperature instability (NBTI) is a long-standing unexplained problem in MOS integrated circuit devices (Nicollian and Brews 1982). The NBTI is manifest as a temporal increase in both interface traps (Q_{it}) and oxide fixed charge (Q_f), under the action of negative bias stress in the MV/cm range, and/or thermal stress, up to 700 K, Fig.13. Each factor individually can produce the effect; but a much stronger or at least faster effect is produced by their combined action. The effect is not generally reversible. The potentially devastating effect on device operation has been successfully minimized in manufacturing, essentially by empirical approaches. There is no extant chemical explanation of the NBTI; nor indeed, has any serious attempt been made to conceive such an explanation, despite quite a number of studies (Deal et al 1967; Hofstein 1967; Goetzberger et al 1973; Nakagiri 1974; Breed 1975; Jeppson and Svensson 1977; Sinha and Smith 1978; Shiono and Yashiro 1979; Risch 1981). The lack of a model now seems especially unfortunate, as the nature of results to be described here suggest that the NBTI comprises several very key aspects of the entire MOS chemical regime.

The initial clues to the nature of the NBTI came from an ESR study of perhaps fortuitous samples. As noted in previous sections, the P_b center is the main source of interface traps in unannealed, thermally stressed, and radiation-damaged MOS structures. Thus it is natural to apply ESR spectroscopy to the identification of NBTI-generated Q_{it}. Inspection of the above-cited NBTI studies suggested that hydrogenous species (hyd) in the oxide played some role, although there was insufficient control

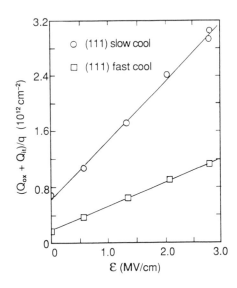

Fig. 13. The electrically-observed negative-bias-temperature instability in MOS structures. Data of Deal et al (1967).

or report of the deliberate or incidental [hyd] in the tested structures to provide the basis of any particular model. An extensive ESR survey of numerous high quality industrial and lower quality research-grade oxidized SiO_2 wafers showed great variation in generation of P_b centers by electric field or heat, applied separately (Gerardi and Caplan 1986), with little hint of pattern in various nominally dry- or wet-grown oxides. Among the industrial grade wet or dry oxides, essentially no generation of P_b by heat or electric field was observed, whether oxidized by the usual device-processing recipes in dry O_2 or wet O_2 (which varies from ~10% H_2O to pure steam, but which produces the same fast oxide formation). Laboratory-grown oxides with highly variable [hyd] showed equally variable behavior, from essentially no bias-stress effect, to samples so unstable that their depassivation was too fast for feasible measurement.

An especially interesting result was that stable dry-oxidized samples were not destabilized by anneal in molecular hydrogen at 700 K (Gerardi and Poindexter 1989), but that exposure to atomic H rendered the samples very susceptible to thermal or electric field stress This chemical difference strongly implicated hydrogenous species in the P_b generation. Thus a large suite of wafers of controlled [hyd] was assembled, with most samples specially oxidized in variably damp ambient. Some critical thermal and electrical depassivation findings on samples of extreme [hyd] content were adopted from the seemingly best-defined published results. (Deal et al 1967; Reed and Plummer 1988; Brower 1990). The wide unexplored region of intermediate [hyd] content was filled in by the specially-made samples. The surprising results observed in this middle range makes it seem peculiar that they were never noticed before.

The location of hyd and its concentration [hyd] cannot be experimentally determined in thermal oxides with much certainty. An estimation by some of the relevant and oft-cited studies (Deal and Grove 1965; Beckmann and Harrick 1971; Shelby 1977; Hartstein and Young 1981; Gale et al 1988) leads to discordant results an essential aspects of [hyd]--especially [H_2O], and the behavior of H_2O in different oxides. Fortunately the dependence of the NBTI on [hyd] is so large and conspicuous that a crude semi-quantitative ordering by [hyd] is sufficient to develop a chemical depassivation model. Hence [hyd] was established by a semi-empirical procedure which was a compromise between the conflicts in the various published approaches, yielding mainly a monotonic scale of [hyd] with only weak meaning in absolute terms (Gerardi et al 1992). The [hyd] was reduced to an equivalent [H_2O], as later suggested by the observed results on the P_b depassivation. Observations of P_b were by ESR, and electric field stress was applied to (unmetallized) oxidized samples by the corona method.

Experiments were confined to the P_{b0} center on (111) silicon, in view of the complexities of the (100) face and the P_{b1} center noted earlier. The initial state of the eventual P_b centers was assumed to be hydrogen passivated, as shown in Fig. 14. The low temperature (700 K) of the thermal depassivation, and even more, the field-alone depassivation at 300 K, made it highly unlikely that active P_b centers were being generated by breaking of Si–O bonds at the interface. Typical examples of field-induced P_b generation at 300K are shown in the ESR traces of Fig. 15, where wet, damp, and dry oxides are compared.

The field-induced depassivation time in the sensitive midrange of [hyd] was typically about 10s. At the extreme dry and wet ends of the range, exposure to field for up to 10^4 sec produced no visible P_b signals. The same mid-range samples which showed easy field-induced P_b generation also showed easy field-less thermal depassivation at ~600-700 K, with time constants slightly longer (~20 s); again no effect was seen for heating times up to 10^5 sec at the dry or wet extremes. The ultimate [P_{be}] generated in any sample

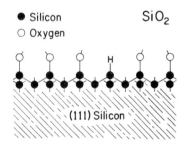

Fig. 14. The hydrogen-passivated P_b center on (111) Si.

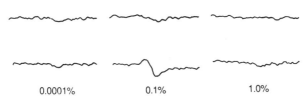

0.0001%   0.1%   1.0%

Fig. 15. ESR signals showing effect of electric field (–13 MV/cm) on passivated P_b centers as a function of [H_2O] in the oxidizing ambient. Upper traces: initial condition. Lower traces: after field stress, 60 s.

never exceeded ~$2.5 \times 10^{12}/cm^2$, in accord with many other observations in samples prepared on treated in diverse ways (Poindexter and Caplan 1983). The limiting values of [P_b] obtained from electric field stress are presented in Fig. 16. The conspicuous hump in the midrange of [hyd] is clearly evident.

The variation in depassivation time, especially thermal depassivation, suggested an activation energy variation. The E_a was determined by very simple 4-point anneals, again, judged adequate for display of expected large trends. Indeed, a drastic variation in E_a was observed, also plotted in Fig. 16. The correlation with field-induced [P_b] is shown. These two aspects of stress-induced [P_b] generation comprise a very rich signature for a depassivation reaction; somewhat unusual in itself, the signature is perhaps unique among reported studies of Si/SiO₂ chemistry.

Together with the generation of P_b centers, it was crudely observed with capacitance-voltage test on a mercury probe that positive oxide charge Q_f was approximately concurrently developed. Within the limited precision attempted, [P_b] = Q_f. This and the above-demonstrated need for [hyd] suggested a simple electrochemical reaction for the field-induced depassivation. We have

$$\equiv Si - H + A + hole^+ \xrightarrow{\Delta G} \equiv Si \cdot + AH^+, \tag{8}$$

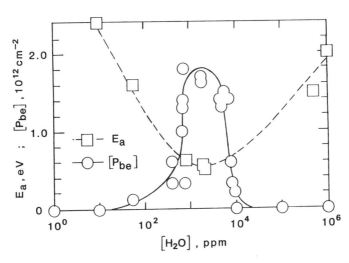

Fig. 16. The limiting value [P_{be}] of electric-field-induced P_b centers on (111) Si and the activation energy E_a for thermal depassivation as a function of [H_2O] in oxidizing ambient.

where A is some as yet unspecified hydrogenous species. The holes are attracted to the interface by the negative electric field $\mathcal{E}$. That some A is needed seems reasonable, but the snuffing effect of too much A is not instantly explainable. A survey of electrochemical studies reported in numerous books and papers reveals a phenomenon known as the "volcano" effect (Balandin 1955), in which the rate of a reaction is maximized at some intermediate value of bond energies in reactant or product, e.g., in Si–H or AH^+. Even more to the point is the recent experimental demonstration (Closs and Miller 1988) of the so-called Marcus inversion regime (Marcus 1960), which are broader in scope and which avoid the limitations of metallic electrochemical electrodes.

The essence of the Marcus reaction model is illustrated in Fig. 17. In the figure, the left hand parabola of each pair represents the manifold of vibration energy levels of an H atom bonded to Si in the reactant moiety $Si_3\equiv Si–H$ at the passivated Si/SiO_2 interface. The right-hand parabola is the energy level manifold for the H atom in its post-stress site, here assumed as AH^+ from Eq.8. For our purposes here, the energy change in the reaction is ΔG, and the energy of activation for the reaction is E_a. A certain percentage of the passivated silicon orbital sites, in the presence of attacking reactant A, will be thermally agitated by an amount E_a, and the reaction will proceed, creating $Si\equiv Si\cdot$ at the interface and positively charged moieties AH^+. Though not shown by the left-hand parabolas, should the reaction be slightly endothermic, it might still proceed if product were removed promptly to inhibit any reverse or passivating reaction.

If now it were possible to increase the exothermicity of the reaction, i.e., increase ΔG, then the middle situation might occur--where E_a is zero and the reaction proceeds instantly. Finally, if the ΔG were still further increased, the activation energy increases; and the reaction is retarded. This latter aspect--the slowing of a reaction with increase in exothermicity--was very controversial for about 30 years. The reaction coordinate model presented by Marcus may seem familiar and reasonable to those versed in electron trapping phenomena in semiconductors, but it was not intuitive to chemical researchers; and it required carefully controlled experiments to isolate it from numerous competing and obscuring influences (Closs and Miller 1988).

The proposed Marcus model for depassivation will be laid aside for the moment, that we may examine nearly concurrent experiments on direct electrical measurement of the NBTI in metallized MOS capacitors (Blat et al 1991). These experiments were tailored in light of the reaction in Eq. 8 above. Samples were prepared by both dry and wet oxidation at 1300 K to oxide thickness 57 nm. One dry sample was not annealed after aluminum deposition; the other was heated at 750K in $H_2 + N_2$ for 30 m. Samples were stressed at 2,3,4,5 MV/cm at 620K for increasing times. Interface trap density Q_{it} and near interface charge Q_f were measured by the charge-capacitance (Q-C) technique (Brews and Nicollian 1984; Miller et al 1989), supplemented by the conductance measurement of the lower values of Q_{it}. The individual Q_{it} and Q_f were extracted on the basis that the two levels of P_b, which dominates Q_{it}, straddle the midgap (Blat et al 1991).

It was immediately evident that the stress-generated Q_{it} was equal to Q_f, strongly confirming the crude measurement on the ESR samples. The second important observation was that the NBTI occurred only in the wet samples--steam grown oxides, or dry oxides subject to post-metallization anneal. This again supported the observations by ESR; though at first, it may seem a contradiction that the very wet samples show the NBTI when wet ESR samples did not. This is essentially a matter of chemical kinetics, and will be briefly discussed later.

The NBTI experiments were planned to obtain quantitative kinetic data, on a few samples, to complement the broad range of qualitative data from the many ESR samples. The build-up of Q_{it} and Q_f was monitored by following the easier quantity (Q_f) above, since they were shown to be equal in a number of measurements. The experiments were analyzed via the kinetic equation for the reaction of Eq. 8. We have for the rate of creation of product $\equiv Si\cdot$ or AH^+ :

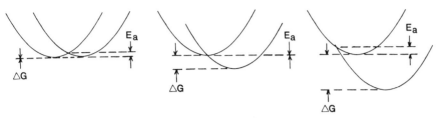

Fig. 17. The activation energy E_a for reactions where the free energy may be varied, showing the decline to zero and ultimate increase as ΔG increases.

$$\frac{d[AH^+]}{dt} = \frac{d[\equiv Si \cdot\ ]}{dt} = k_f[\equiv Si \cdot\ -\ H][A][hole^+] - k_r[\equiv Si \cdot\ ][A\ H^+], \tag{9}$$

where the forward and reverse reaction rate constants are k_f and k_r respectively. They are assumed to be of Arrhenius form, e.g.,

$$k_f = k_{f0}exp(-E_a/kT) \tag{10}$$

The observed temporal growth of Q_f (i.e., $[AH^+]$) is shown in Fig. 18 for several values of electric stress field $\mathcal{E}$. Inspection shows that the time constants are essentially equal and independent of $\mathcal{E}$. This may seem surprising, as it might be expected that holes attracted to the interface would weaken the Si–H bond. The limiting values of Q_f were of the form

$$Q_f = Q_{obs} - Q_{f0} \tag{11}$$

where Q_{obs} is the total observed charge, and Q_{f0} is the initial charge. Both include contributions from charge located at the interface, which had to be deconvolved. It was found that the final Q_f introduced by the NBTI was directly proportional to $\mathcal{E}$.

$$Q_f \sim \mathcal{E} \tag{12}$$

From MOS electrostatics, we have

$$[hole^+] \sim \mathcal{E} \tag{13}$$

The final value of the Q_f introduced by bias stress was thus determined to be linearly proportional to the initial $[hole^+]$.

The single-logarithmic time-growth of product also means that there is only one reactant species which influences the rate; as noted above from different reasoning, this single species is the hole. The reactants A and $\equiv$Si–H are present in concentration substantially greater than the limiting product concentration, which here was $\sim 6 \times 10^{12}\,cm^{-2}$ at 5 MV/cm. And finally, the single-logarithmic dependence also means that there is no back reaction which would also introduce a quadratic term. Stated another way, $k_r = 0$. The NBTI reaction is first order in $[hole^+]$, and the observed phenomena are indeed well-modeled by Eq. 8.

The invariance of the NBTI rate constant k_r, and the general nature of the Marcus-type argument on E_a, suggested theoretical scrutiny of the reacting system. Molecular orbital (MO) calculations were applied. The MINDO/3 (modified intermediate neglect of differential overlap) approach was used, developed in the open-shell technique (MOPN). The reactant A was taken to be H_2O, in line with an especially straightforward interpretation of the P_b results. A cluster comprising 22 Si atoms, 6 O atoms, and 2 H atoms, plus non-reacting H atom terminators for all peripheral dangling orbitals was used, as shown in Fig. 19.

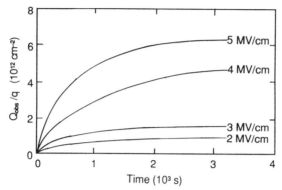

Fig. 18. Generation of oxide charge by the negative bias temperature instability (NBTI) at 620 K as a function of electric field.

An approximate reaction path was calculated by restricting motion to the attacking H_2O molecule and the passivating H atom. Several very pertinent results ensued. First, the strength of the Si–H bond is essentially independent of the presence of holes attracted to the silicon surface by the negative field. This may be very surprising, as it is perhaps intuitive that a hole which permeates a bonding orbital array should weaken the bond. The MO calculation shows that the hole prefers the highest energy valence band state, and does not affect the bonding orbitals or electrons. More important, this surprising result was confirmed by the NBTI observations; the forward rate constant is unaffected by negative bias; i.e., the energy to break the Si–H bond is invariant with bias. Second, the bonding energy and electron dispositions of the reactant-product system do indeed indicate that the Si–H bond eventually breaks as the H_2O is brought closer, resulting in H_3O^+ formation. Third, although the depassivation reaction can take place, it is slightly endothermic, about 1.4 eV for the calculated case of an isolated single H_2O reactant molecule. Finally, the activation energy for the forward (depassivation) reaction is about 2 eV--quite consistent with the value measured for dry-oxide depassivation.

Additional pertinent MO calculations (Grahn 1962) and verifying measurements (Conway 1981) have been done by others for reactions involving the subreaction

$$H^+ + nH_2O \xrightarrow{\Delta G} H_{2n+1} O_n^+ \qquad (14)$$

It has been calculated that $\Delta G = 8$ eV for n = 1, but increases to 13 eV for n = 4. Thus this partial reaction becomes very much more exothermic by increase in availability of H_2O. The increase in the partial ΔG is more than enough to overcome the endothermicity of the single-H_2O depassivation calculation described above, and would likely drive the reaction parabolas through the minimum E_a configuration.

At this juncture, attention should be paid to the identity of attacking reactant A. Numerous inferences have pointed to H_2O;

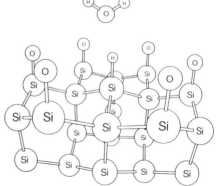

Fig. 19. Si/SiO$_2$ interface fragment used for molecular orbital calculations on depassivation of P_b centers by attack of H_2O.

indeed, it is a very favorable candidate. In addition to being party to a variable energy-versus-concentration phenomenon, it is favored by several other factors. The molecule would be favorably oriented by the negative field gradient, attacking with its exposed O end toward the $\equiv$Si–H moiety. The (+) charged H_3O^+ product would be promptly removed by the field, greatly hindering any possible back reaction. Reactions in which H_2O attacks surface-bonded H are well modeled in electrochemistry. The presence of field-attracted holes readily assists the reaction. The presence of a large quantity of H_2O near the interface would also greatly increases the local dielectric constant, which in itself can modify ancillary bond energies. Last, but not least, there are reasonable sources of H_2O in wet SiO_2 at low temperature. Physisorbed H_2O is readily dislodged with a activation barrier of about 0.5 eV (perhaps the source of the minimum E_a observed); chemisorbed H_2O is available above 500 K.

Nonetheless, despite the promising features of the H_2O - H_3O^+ scheme, there is no spectroscopic demonstration of the presence and disposition of H_2O in the SiO_2. Other feasible reactant species should be considered. One such species is atomic H, which has been reasonably asserted to be a depassivating reactant in both thermal (Reed and Plummer 1988) and radiation-stress (Saks 1990). Atomic H is also reasonable from on MO-theoretical view (Edwards 1990). Weighing against H in the NBTI, however, are several issues. First, there is no ready source of H in any form of SiO_2 at 300 K; hydroxyl or silanol groups in SiO_2 do not decompose until ~1200 K (Iler 1979). There is no known product complex with an [H⁺]-dependent reaction energy. Any feasible depassivation reaction would not be aided by negative field gradient. All-in-all, H seems less likely than H_2O as attacking reactant.

Last, in any case where the partial reaction

$$H_2O + H^+ \rightarrow H_3O^+ \tag{15}$$

is tacitly incorporated, it is generally feasible to write analogous electrochemical equations which involve the partial

$$H^+ + OH^- \rightarrow H_2O . \tag{16}$$

Again, possible sources of OH in SiO_2 do not behave usefully; silanols are decomposed at above 1200 K; they emit H atoms, not free OH⁻ (Iler 1979). There is no product complex with variable bonding energy. Finally, OH⁻ radicals have not been implicated in countless studies of hydrogen species in SiO_2 (Iler 1979; Boehm and Knözinger 1983; Kiselev and Lygin 1975). All-in-all, H_2O seems the most likely--if unproven-- depassivating reactant in the case of the NBTI. The H_2O model is portrayed in Fig. 20, with the successive stages in the reaction. In (a), the prospective reactant H_2O is located near a passivated P_b center. In (b), the electric field has reoriented it into attack position. In (c), the proton has been dislodged from the $\equiv$Si–H site and combined with the H_2O to yield H_3O^+. Finally, in (d), the now-positively charged H_3O^+ is pulled away by the electric field, thus preventing any reverse reaction.

Many of the arguments involving the P_b generation or the NBTI have relied on the bias dependence of the effects. Yet, purely thermal depassivation proceeds most readily in the very some samples which favor field-induced effect. The thermal depassivation will not benefit from bias-attracted holes, molecular reorientation, etc. But the extremely low E_a is expected to drive the reaction, with the great deceleration wrought by paucity of holes offset by the equally great acceleration from the Arrhenius expression, Eq. 10. Indeed, these factors are both or order $10^{\pm5}$, and nearly cancel (Gerardi et al 1992). This effect of heat also explains why the NBTI is strong in very wet samples, while the P_b depassivation was not detectable. The combined effect of field and heat discriminate between the dry end, where indeed the required H_2O may not always be available, and where E_a seems to be ~2.5 eV, higher than at the 1.6 - 2.0 eV at the wet end. An additional temperature-related aspect is the apparently finite supply of reactant A at 300 K, as compared to 620 K. This is reasonably ascribed to much slower diffusion of A at 300 K.

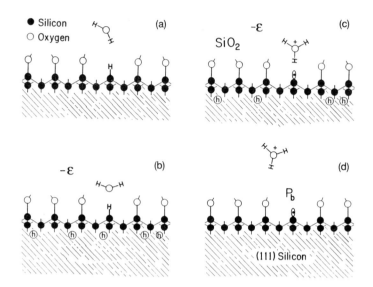

Fig. 20. Electric-field depassivation of P_b sites by attack of H_2O molecule: (a) passivated, un-stressed condition; (b) orienting of H_2O and attraction of holes to interface by electric field; (c) H_2O captures a hole and the H atom from the P_b site; (d) H_3O^+ has been removed from the P_b vicinity by the field, leaving an active P_b site.

We note in conclusion two other areas where consideration of the role of H_2O might be fruitful. The most obvious is in radiation hardness, where a reaction complementary to Eq. 8 is suggested by the positive-bias enhancement of damage:

$$\equiv Si - H + H_3O^+ + electron^{(-)} \rightarrow \equiv Si\cdot + H_2O + H_2\uparrow \qquad (17)$$

Without offering detail, the features of this reaction do suggest a careful study with variable H_2O content. And second, the fact that there seems to be an excess of SiH and A in the NBTI kinetics suggests that the so-called oxide fixed charge inevitable after oxidation (which is re-duced by inert anneal) and the so-called oxidation triangle (Deal et al 1967) may reflect the bal-ance of buffered H_2O-related reactions near the interface, rather than other factors, such as Si–SiO_2 lattice mismatch.

6. RADIATION DAMAGE

Radiation damage has been perhaps the most heavily researched aspect of MOS technology. Recent findings and interpretations are summarized in a comprehensive overview (Oldham et al 1989). An impressive amount of experimental data is available in the literature, which is rea-sonably well categorized and correlated from an empirical, physically-oriented viewpoint. Physically plausible working models have been suggested over the years on the basis of certain especially appealing experiments; but the chemical aspect has been markedly weaker, and for the most part, casually treated or neglected altogether. The phenomena of interest here are the time-dependent build-up of interface traps (Q_{it}) and near-interface oxide charge (Q_{ox}) following an initial damage event. The latter may be either impingement of radiation such as γ-rays, x-rays, neutrons; or it may be associated with injection of hot electrons or holes into the oxide. The course of events is represented in Fig. 21.

Following an initial particle collision or ionization event in the oxide, secondary entities travel away from the collision site and migrate to the gate or Si/SiO₂ interface, with a transport rate determined by diffusion constant and, for charged particles, the presence of an electric field. The relevant experiments and conditions, of course, are at 300 K. It has long been known that by far the greater part of the ultimate damage at or near the Si/SiO₂ interface is produced by secondary events, not by the initial interception. Even ion implantation focussed on the interface produces little direct damage there. These secondary mechanisms are the subject of the great bulk of MOS radiation research. Two main types of models for the subsequent Q_{it} generation have developed, viz., those based on hole transport and those based on hydrogen transport. Both seem to be factors, in the case of Q_{ox}, but hydrogen is coming to be viewed as the dominant player in regard to Q_{it}. None of the research to date however, has enabled any rich reaction signature like that defined in the NBTI effect, nor have experiments been planned for careful control of [hyd] over a wide range. We will consider first the H-related portion of the radiation dam-

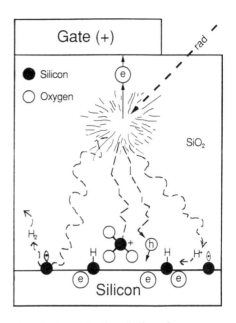

Fig. 21. Model of radiation damage sequence in an MOS structure.

age process, since that aspect is connected more closely to the NBTI model discussed in Sec. 5.

It was early proposed that radiation-generated holes in the oxide could promote dislodgement of hydrogen, leaving positively charged centers (perhaps E_s' centers) in the oxide (Revesz, 1977). The hydrogen then migrates to the interface and reacts to produce interface traps (Sah 1976; Svensson 1978):

$$\equiv Si - H + H \rightarrow \equiv Si\cdot + H_2\uparrow \tag{18}$$

These proposals were made before any experimental demonstration of the existence of P_b centers, but their existence was reasonable and soon verified by ESR (Lenahan et al 1981) for radiation-damaged Si/SiO₂ structures. Substantial reinforcement of the general hole-hydrogen concept was provided by critical experiments in which the ultimate Q_{it} was found to vary as the square root of applied field (McLean 1980). The data are shown in Fig. 22. Such a functional dependence was ascribed to a Schottky-like phenomenon (electron emission over a barrier) incorporated into a chemical reaction which released H⁺ ions and left a neutral center of no device import behind. (Note that any electrons released in the radiation encounter or subsequent reactions are quickly transported to gate or substrate without meaningful effect.) It is also significant that the wet oxide shows more damage than the dry one.

The argument was further developed by consideration of hydrogen transport through the oxide, as influenced by electric field. The idea was generally offered as a good explanation for the time dependence of Q_{it} build-up as a function of electric field, which would drive H⁺ to the gate or substrate for (−) or (+) fields, respectively (McGarrity et al 1978). Some recent critical experiments have added more substance to the H⁺-transport model (Saks and Brown 1989). The application of bias during and after irradiation was divided into two regions selected to display the two stages of the damage scenario. In the first stage, biases ranging over ± 2 MV/cm were applied for 10^{-2} s; then the bias was switched to $+2$ MV/cm. The final Q_{it} as a function of initial bias is shown in Fig. 23. Remarkably, the final Q_{it} is roughly symmetrical, with maximum at zero first-stage bias. The symmetry supports a scalar chemical process, i.e., without directional diffusion or boundary-electrode electrochemical reactions; the peak was, however, initially unexpected for a supposedly field-aided release of H^+.

Later experiments, however, clarify the situation (Saks and Brown 1990). The temporal generation of Q_{it} during the second stage, always (+)biased, shows an informative peak in Q_{ox}, Fig. 24. After the initial release of H^+ ions during the first bias interval, the H^+ travels to the interface, driven there by the positive field. For a while after arrival near the interface, the H^+ is a source of positive charge Q_{ox}. However, it is eventually neutralized by capture of an electron attracted to the interface under the action of positive bias.

$$H^+ + e^- \rightarrow H \tag{19}$$

Soon thereafter, the neutral H reacts with passivated P_b sites:

$$H + \equiv Si - H \rightarrow Si\cdot + H_2\uparrow \tag{20}$$

As has been noted in Sec. 5, such a reaction in theory is energetically feasible. The peak in Q_{it} occurs roughly at the time where the rate of growth of Q_{it} is the greatest. Up to this time, H^+ has been transported rapidly to the interface, yielding an increase in Q_{ox} there. (The bulk of radiation-induced interface Q_{ox}, however, is due to E'_γ centers, created earlier by trapping of a hole (Lenahan and Dressendorfer 1984). The reactions of Eq.(1) and Eq.(2) get under way more slowly, being essentially bottlenecked by the reaction of Eq.(3). When this depassivation step picks up speed, H^+ ions are consumed rapidly, resulting in a decline of Q_{ox}. The model is developed considerably by quantitative calculation on diffusion of H^+ through SiO_2, and is found to be generally self-consistent.

Several inconsistencies with H-based models have been discussed elsewhere (Reed 1989). Notable are the disappearance of ESR-observed atomic H at temperatures below 130 K (Griscom 1984), contrasted with the apparent effectiveness of post-aluminization anneal of P_b centers apparently by neutral atomic H at 750 K. There is also the consideration of the very high binding energy of H^+ in H_3O^+, which would suggest that free protons in wet SiO_2 are likely to scavenge bound H_2O or even vicinal silanols, in accord with Sec. 5. Nonetheless, the experiments embodied in the works cited here--and others uncited--offer the most intriguing and self-consistent basis for attempts at a more chemically-oriented model of radiation-generated interface traps.

The main source of interface Q_{ox} in radiation-damage Si/SiO_2 structures is the E'_γ center (Lenahan and Dressendorfer 1984), $O_3\equiv Si\cdot....^+Si\equiv O_3$. Some of these centers have been shown to be rechargeable under strong electric field, which can cause either direct tunneling or Fowler-Nordheim tunneling between substrate and defect center, as discussed in Sec. 3. The E'_γ centers discussed earlier were produced by implantation of excess Si into the oxide. A

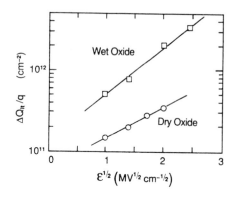

Fig. 22. Generation of radiation-induced interface traps as a function of electric field.

similar charging and discharging phenomenon has recently been observed in electron-irradiated MOS structures (Lelis et al 1988). Moderately hard (proprietary) oxides 97 nm thick were grown on p-type Si. They were electron-irradiated to 100 krad by 4 ms LINAC pulses at a bias of + 1MV/cm, 300 K. The resultant Q_{ox} and Q_{it} were analyzed in a special fast charge-separation technique. The resultant Q_{ox} was found to be recyclable by repeated application of bias at ± 1MV/cm. The findings have been replotted in Fig. 25. The result was interpreted as arising from the recharging of E'_γ centers. The recyclability of E'_γ centers was shortly thereafter demonstrated (though on a different type of sample, as noted in Sec. 3). The similarity between the data of Fig. 25 and the E' signal patterns in Sec. 3 is very suggestive.

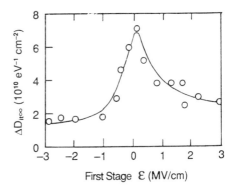

Fig. 23. Interface traps ΔD_{it} generated by radiation as a function of bias applied during the first stage after radiation of an MOS structure.

The time development of interface states in radiation - or injection-damaged MOS structures has recently been examined from a point of view (Ma 1989) different from the damage transport mechanism studies above. Silicon wafers of several orientations (100), (111), (110), and (311) were variously oxidized in dry O_2, wet O_2, HCl, and trichloroethane (TCA), and variously metallized with Al, Mo, TiS_2, and poly-Si gates. Damage was generated by Fowler-Nordheim electron injection or x-rays up to 24 Mrad. The particular focus of the study was the bandgap distribution of Q_{it} as the samples aged after the damage event, up to 4000 h, at temperatures from 300 K to 450 K, with and without bias. The essential features of the observed effects were only slightly affected by all these variations in treatment, and will be ignored here.

It has been shown that traps at the interface of oxidized (111) Si are of the single P_b variety, as shown in Fig. 10. these centers in as-oxidized, undamaged wafers are amphoteric, with possible charge states (+), (0), and (–), and corresponding levels in the Si bandgap at $E_v + 0.3$ eV

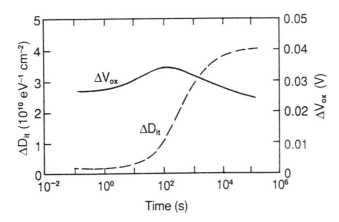

Fig. 24. Increase and then decrease of (+) charge, as measured by ΔV_{ox} developed near the Si/SiO₂ interface by radiation.

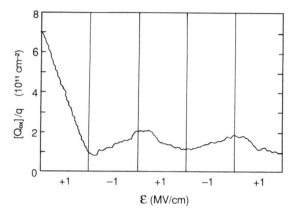

Fig. 25. Bias-induced cycling of radiation-induced (+) charge Q_{ox} near the Si/SiO$_2$ interface.

and $E_v + 0.85$ eV (Poindexter et al 1984). The peak heights of these levels as displayed by low-frequency capacitance-voltage (C-V) analysis are approximately equal; the areas under the peaks, of course, are rigorously equal in principle. The (111) interface trap distribution D_{it} is shown in Fig. 26.

On (100) Si, there are two different P_b-like centers, shown in Fig. 10. The P_{b0} center has nearly the same peak structure as P_b (111), as might be expected from the very similar ESR g-maps. The P_{b1} levels are bracketed by the P_{b0} peaks. When present in approximately equal amounts, the overall D_{it} distribution will show a prominent peak at about $E_v + 0.8$ eV, where the P_{b0} and P_{b1} levels are nearly coincident; and a less distinct broadened peak near $E_v + 0.4$ eV, where the respective levels are separated by about 0.15 eV. The (100) trap distribution is also shown in Fig. 26.

The D_{it} distribution after radiation or injection damage showed an initially promising , but later confounding behavior. First tried, damaged (100) wafers initially showed a prominent D_{it} asymmetry, with a peak at about $E_v + 0.8$ eV and no real hint of any peak in the bandgap below that energy, down to about $E_v + 0.35$ eV. This is shown in Fig. 27. After 3.5 h at 375 K, a

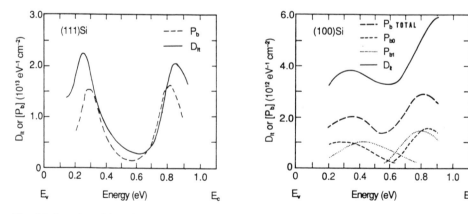

Fig. 26. Density of interface traps D_{it} and P_b centers in oxidized silicon wafers. Left: (111) Si. Right: (100) Si.

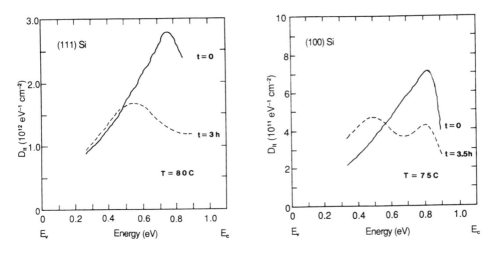

Fig. 27. Time development of interface trap spectrum after irradiation of MOS structures. Left: (111) Si. Right: (100) Si.

two-humped distribution develops, with peaks at E_v + 0.5 eV and E_v + 0.8 eV, also in Fig. 27. Rationally, these two distributions might be fairly well modeled by an initial generation of P_{b0} and P_{b1} centers in equal amounts, followed by a transformation of P_{b1} into P_{b0} centers-- with some decay of the overall total Q_{it}. The final trap identity partially agrees with the result cited earlier on the radiation-induced generation of P_{b0} exclusively on (100) Si (Kim and Lenahan 1988). The issue of time or temperature dependence, however, is left unsettled by lack of data.

It was then deemed suitable to try the same experiments on (111) wafers (Ma 1989), in hopes of avoiding the problem of two centers, with P_{b1} particularly intractable. Results for (111), treated about like the (100) above, are also shown in Fig. 27. They are surprising, and perhaps disconcerting. The initial D_{it} distribution is much like that on damaged (100) silicon, with a peak at E_v + 0.75 eV. This cannot be the normal P_b (111) center, as that would show two peaks. Continuing, after aging for 3 h at 380 K, the distribution changes to yield a new peak at E_v + 0.5 eV--again, like the (100) Si situation. However, there is no peak at E_v + 0.8 eV. If indeed there is a peak in the upper half of the Si bandgap, it must lie above that energy. The measurements do not allow any extension above E_v + 0.9 eV, and there is no clear suggestion of a trend in D_{it} above this cutoff. The hoped-for simplification in (111) Si as compared to (100) did not materialize; and further, the pattern of D_{it} (111) is in itself not readily identifiable with P_b (111), P_{b0}, or P_{b1}.

An above-midgap peak in radiation- or injection-damaged MOS structures has long been a commonly-observed feature; and so far, no good model has been offered. ESR has shown nothing. In an attempt at explanation of the discordant (111) results, a shift in defect energy levels has been tentatively ascribed to strain (Ma 1989). The defect may be initially distorted from its relaxed configuration, i.e., more or less tetrahedral than normal. The theory of a strained P_b center has been presented previously (Edwards 1987). It is calculated that a small flattening (< 5 percent of lattice constant) of the $Si_3 \equiv Si \cdot$ moiety could produce a raising of the energy levels by several tenths of an eV. The strain dependence of P_b, at least, has been reasonably well substantiated by compression of samples at high pressures (Johnson et al 1989). As to whether this is indeed the explanation of the radiation-damage energy levels is unsettled at this time.

The time-dependent transformation of these uncharacteristic interface traps adds unwanted richness to the nature of interface states in damaged structures, rather than clarification. It further appears to be somehow relevant to the unsettled nature of the P_{b1} center. Indeed, taken together with the findings discussed in Sec. 4. there is strong suggestion of at least one and maybe two as yet undefined defects lurking behind the scenes. Undamaged (100) interfaces may harbor a "P_{b2}" defect, perhaps to explain the SDR results. Radiation-damaged structures may contain still another defect, the upper-bandgap peak, which may or may not be the same as "P_{b2}"; and which we leave unchristened here.

ACKNOWLEDGMENTS

I am grateful to C. R. Helms for technical and editorial suggestions, and to B. T. Poindexter for computer graphics, word-processing concepts, and manuscript preparation. Research colleagues too numerous to name here are cited in co-authored references; I thank them all.

REFERENCES

Balandin A A 1955 *Voprosy Khimicheskoy Kinetiki, Kataliza, i Reaktionnoy Sposobnosti* ed V Kondratiev and N M Emanuel (Moscow: Acad. Sci. U.S.S.R.) p 461
Beckmann K H and Harrick N J 1971 *J. Electrochem. Soc.* **118** 614
Biegler T, Gonzalez E R and Parsons R 1971 *Collection Czech Chem. Commun.* **36** 414
Blat C E, Nicollian E H and Poindexter E H 1991 *J. Appl. Phys.* **69** 1712
Boehm H P and Knözinger 1983 *Catalysis Science and Technology* ed J R Anderson and M Boudart (Berlin: Springer) pp 40-142
Breed D J 1975 *Appl. Phys. Lett.* **26** 116
Brews J R and Nicollian E H 1984 *Solid State Electron.* **27** 963
Brower K L *Semicond. Sci. Technol.* **4** 970
Brower K L *Phys. Rev. B* **42(II)** 3444
Caplan P J, Poindexter E H, Deal B E and Razouk R R 1979 *J. Appl. Phys.* **50** 5847
Carlos W E 1987 *Appl. Phys. Lett.* **50** 1450
Carrico A, Elliott R and Barrio R 1986 *Phys. Rev. B* **34** 1872
Closs G L and Miller J R 1988 *Science* **240** 440
Conway B E 1981 *Ionic Hydration in Chemistry and Biochemistry* (Amsterdam: Elsevier) pp 394-404
Cook M and White C T 1987 *Phys. Rev. Lett.* **59** 1741
Deal B E and Grove A S 1965 *J. Appl. Phys.* **36** 3770
Deal B E, Sklar M, Grove A S and Snow E H 1967 *J. Electrochem. Soc.* **114** 266
Dersch H, Stuke J and Beichler J 1981 *Phys. Stat. Sol.(b)* **105** 265
Edwards A H 1987 *Phys. Rev. B* **36(II)** 9638
Edwards A H 1988 *The Physics and Chemistry of SiO$_2$ and the Si-SiO$_2$ Interface* ed C R Helms and B E Deal (New York: Plenum) pp 271-283
Edwards A H 1990 unpublished results
Feigl F J, Fowler W B and Yip K L 1974 *Solid State Commun.* **14** 225
Fowler W B, Rudra J K, Zvanut M E and Feigl F J *Phys. Rev. B* **41** 8313
Gale R, Chew H, Feigl F J and Magee C W 1988 *The Physics and Chemistry of SiO$_2$ and the Si-SiO$_2$ Interface* ed C R Helms and B E Deal (New York: Plenum) pp 177-186
Gerardi G J and Caplan P J 1986 unpublished results
Gerardi G J and Poindexter E H 1989 *J. Electrochem. Soc.* **136** 588
Gerardi G J, Poindexter E H, Caplan P J and Johnson N M 1986 *Appl. Phys. Lett.* **49** 348
Gerardi G J, Poindexter E H, Harmatz M, Warren W L, Nicollian E H and Edwards A H 1991 *J. Electrochem. Soc.* **138** 3765
Goetzberger A, Lopez A D and Strain R J 1973 *J. Electrochem. Soc.* **120** 90
Grahn M 1962 *Arkiv Fysik* **21** 13
Griscom D L 1984 *J. Non-Cryst. Solids* **68** 301
Griscom D L 1985 *J. Non-Cryst. Solids* **73** 51
Gurney R W 1931 *Proc. Roy. Soc. London A* **134** 137
Hartstein A and Young D R 1981 *Appl. Phys. Lett.* **38** 631

Henderson B 1984 *J. Appl. Phys.* **44** 228
Henderson B, Pepper M and Vranch R L 1989 *Semicond. Sci. Technol.* **4** 1045
Hochstrasser G and Antonini J F 1972 *Surf. Sci.* **32** 644
Hofstein S R 1967 *Solid State Electron.* **10** 657
Iler R K 1979 *The Chemistry of Silica* (New York: Wiley) pp 622-654
Jeppson K O and Svensson C M 1977 *J. Appl. Phys.* **48** 2004
Johnson N M, Shaw W and Yu P Y 1989 *Semicond. Sci. Technol.* **4** 1036
Kalnitsky A, Ellul J P, Poindexter E H, Caplan P J, Lux R A and Boothroyd A R 1990 *J. Appl. Phys.* **67** 7359
Kaplan D, Solomon I and Mott N F 1978 *J. Physique Lett.* **39** L-51
Kim Y Y and Lenahan P M 1988 *J. Appl. Phys.* **64** 3551
Kiselev A V and Lygin V I 1975 *Infrared Spectra of Surface Compounds* (New York: Wiley) pp 75-120
Lelis A J, Boesch H E Jr, Oldham T R and McLean F B 1988 *IEEE Trans. Nucl. Sci.* **35** 1186
Lenahan P M and Dressendorfer P V 1984 *J. Appl. Phys.* **55** 3495
Lenahan P M, Brower K L, Dressendorfer P V and Johnson W C 1981 *IEEE Trans. Nucl. Sci.* **NS-28** 4105
Lenahan P M, Warren W L, Dressendorfer P V and Mikawa R E 1987 *Z. Phys. Chem. N.F.* **151** 235
Lepine D J 1972 *Phys. Rev. B* **6** 436
Löwdin P-O 1965 *Advances in Quantum Chemistry* ed P-O Löwdin (New York: Academic) pp 248-286
Ma T P 1989 *Semicond. Sci. Technol.* **4** 1061
Marcus R A 1960 *Discuss. Faraday Soc.* **29** 21
McGarrity J M, Winokur P S, Boesch H E Jr and McLean F B 1978 *The Physics of SiO$_2$ and its Interfaces* ed S T Pantelides (New York: Pergamon) pp 428-432
McLean F B 1980 *IEEE Trans. Nucl. Sci.* **NS-27** 1651
Miller J A, Blat C and Nicollian E H 1989 *J. Appl. Phys.* **66** 716
Mizutani T, Ozawa O, Wada T and Arizumi T 1970 *Japan. J. Appl. Phys.* **9** 446
Nakagiri M 1974 *Japan. J. Appl. Phys.* **13** 161
Nicollian E H and Brews J R 1982 *MOS (Metal-Oxide-Semiconductor) Physics and Technology* (New York: Wiley-Interscience) pp 794-798
Nishi Y 1971 *Japan. J. Appl. Phys.* **10** 52
Oldham T R, McLean F B, Boesch H E Jr and McGarrity J M 1989 *Semicond. Sci. Technol.* **4** 986
O'Reilly E P and Robertson J 1983 *Phys. Rev. B* **27** 3780
Parsons R 1958 *Trans. Faraday Soc.* **54** 1053
Poindexter E H and Caplan P J 1978 *The Physics of SiO$_2$ and its Interfaces* ed S T Pantelides (New York: Pergamon) pp 227-231
Poindexter E H and Caplan P J 1983 *Prog. Surf. Sci.* **14** 201
Poindexter E H, Caplan P J and Gerardi G J 1988 *The Physics and Chemistry of SiO$_2$ and the Si-SiO$_2$ Interface* ed C R Helms and B E Deal (New York: Plenum) pp 299-308
Poindexter E H, Caplan P J, Deal B E and Razouk R R 1981 *J. Appl. Phys.* **52** 879
Poindexter E H, Gerardi G J, Rueckel M-E, Caplan P J, Johnson N M and Biegelsen D K 1984 *J. Appl. Phys.* **56** 2844
Reed M L 1989 *Semicond. Sci. Technol.* **4** 980
Reed M L and Plummer J D 1988 *J. Appl. Phys.* **63** 5776
Revesz A G 1977 *IEEE Trans. Nucl. Sci.* **NS-24** 2102
Risch L 1981 *Insulating Films on Semiconductors* ed M Schulz and G Pensl (Berlin: Springer) pp 39-42
Robertson J 1984 *J. Phys. C* **17** L221
Rudra J K and Fowler W B 1987 *Phys. Rev. B* **35** 8223
Sah C T 1976 *IEEE Trans. Nucl. Sci.* **NS-23** 1563
Saks N S and Brown D B 1989 *IEEE Trans. Nucl. Sci.* **36** 1848
Saks N S and Brown D B 1990 *IEEE Trans. Nucl. Sci.* **37** 1624
Shelby J E 1977 *J. Appl. Phys.* **48** 3387

Shiono N and Yashiro T 1979 *Japan. J. Appl. Phys.* **18** 1087

Shiota I, Miyamoto N and Nishizawa J-I 1977 *J. Appl. Phys.* **48** 2556

Silsbee R H 1967 *J. Phys. Chem. Solids* **28** 2525

Sinha A K and Smith T E 1978 *J. Electrochem. Soc.* **125** 743

Stathis J H and Dori L 1991 *Appl. Phys. Lett.* **58** 1641

Svensson C M 1978 *The Physics of SiO₂ and its Interfaces* ed S T Pantelides (New York:
    Pergamon) pp 328-332

Terry F L Jr, Wyatt P W, Naiman M L, Mathur B P and Kirk C T 1985 *J. Appl. Phys.* **57**
    2036

Vranch R L, Henderson B and Pepper M 1988 *Appl. Phys. Lett.* **53** 1299

Warren W L, Poindexter E H, Offenberg M and Müller-Warmuth W 1992 *J. Electrochem.
    Soc.* in press

Weeks R A 1956 *J. Appl. Phys.* **27** 1376

Young D R 1988 *The Physics and Chemistry of SiO₂ and the Si-SiO₂ Interface* (New York:
    Plenum) pp 487-496

Zvanut M E , Feigl F J and Zook J D 1988 *J. Appl. Phys.* **64** 2221

Zvanut M E, Feigl F J, Fowler W B, Rudra J K, Caplan P J, Poindexter E H and Zook J D
    1989 *Appl. Phys. Lett.* **54** 2118

Chapter 11

Growth and characterization of silicon carbide polytypes for electronic applications

J A Powell[*], P Pirouz[**], and W J Choyke[***]

[*]NASA Lewis Research Center, Cleveland, OH 44135
[**]Department of Materials Science and Engineering, Case Western Reserve University, Cleveland, OH 44106
[***]Department of Physics, University of Pittsburgh, Pittsburgh, PA 15260

ABSTRACT: Recent advances in boule and thin film crystal growth have accelerated the development of silicon carbide as a useful semiconductor. As a result, polished wafers and simple devices (e.g. LEDs, diodes) are now commercially available. Of particular importance has been the recent success in controlling the polytype in the growth of bulk crystals and thin films. Much has been learned regarding the role of defects in various SiC crystal growth processes. New models have been proposed to explain observed polytype growth and transformation processes. There have been significant recent optical and electrical measurements relating to donor and acceptor states in SiC polytypes. Discussions of the above plus some background information on SiC polytypes will be presented in this review.

1. INTRODUCTION

Silicon carbide (SiC) has attracted attention for more than thirty years as a potentially useful semiconductor (O'Connor and Smiltens 1960). While silicon was still in an early stage of development, SiC had already exhibited characteristics that held out promise for high temperature and high frequency devices, and for blue light-emitting diodes (LEDs). Properties of SiC that have attracted attention are its wide bandgap (2.2 to 3.3 eV), high electric breakdown (4 x 10^6 V/cm), high thermal conductivity (5 W/cm-°C), and excellent thermal and chemical stability. The successful development of any material for semiconductor applications depends to a large degree on the capability of producing high-quality single-crystals of that material with controlled electronic properties. This has been true for both Si and SiC. Whereas Si crystal growth technology has advanced to a high degree, the development of SiC has been hindered by a lack of a suitable crystal growth technology. Two significant factors in SiC's crystal growth problems are its ability to grow in many different crystal structures, called polytypes, and its inability to be melted at a reasonable pressure.

Lely (1955) developed a laboratory crystal growth process for producing small single crystals of SiC. These crystals, although rather pure, often consisted of a mixture of various polytypes. However, with proper doping, n- and p-type crystals were produced. During this early period, blue LEDs and high temperature devices were demonstrated (Henisch and Roy 1969, Marshall et al 1974). Because of the outstanding progress in Si technology and a lack of progress in SiC crystal growth, interest (and funding) in SiC had decreased significantly by the 1970s.

Developments in boule and thin film SiC crystal growth that began in the late 1970s and early 1980s has created a resurgence of interest in SiC. This renewed interest has also been fueled by an increasing need for semiconductor sensors and devices capable of operation in high temperature and high radiation environments, and increased performance in high frequency and high power applications. Progress over the last several years has been such that commercial production of SiC wafers and devices has begun.

An excellent and classic review of polytypic structures is the book by Verma and Krishna (1966). Three review papers on SiC polytypes of particular interest (Jepps and Page 1983, Pandey and Krishna 1983, and Tairov and Tsvetkov 1983) are contained in Krishna (1983). Recent proceedings of international SiC conferences (Harris and Yang 1989, Rahmann et al 1989) and several review papers (Davis and Glass 1991, Davis et al 1991) have described various aspects of the development of SiC semiconductor technology during the last decade. In this paper we review (1) processes for producing boule and thin film single crystal SiC, (2) some theories and models used to explain various aspects of these processes, (3) various types of defects (planar, linear, and point defects) that have a significant effect on crystal growth, transformation, and/or properties of SiC polytypes, and (4) some recent optical and electrical measurements relating to donor and acceptor states in SiC polytypes.

2. POLYTYPIC STRUCTURES

The development of SiC crystal growth processes has been made much more challenging by its ability to grow in many different crystal structures, known as polytypes. More than 100 SiC polytypes have been reported. In this section, the structure of these polytypes and various notations for designating the structures will be described.

2.1. Polytypes of Close-Packing Spheres

The occurrence of a given compound in more than one crystal structure is known as *Polymorphism*. The compound is then said to be *polymorphic* and each stable phase of the compound is known as a *polymorph*. In general, each polymorph is thermodynamically stable in a given regime of temperature and pressure. In a polymorphic compound, the structure of each polymorph may be completely different and no structural relation may exist between the different polymorphs. *Polytypism* is a special type of polymorphism where the different polytypes differ from each other in a specific way. Unlike, polymorphs, it is still not clear whether all the polytypes of a material exhibiting polytypism are thermodynamically stable phases; it is possible that at least some polytypes may form on kinetic grounds in a metastable form.

A polytypic compound generally has a structure which can be considered as the stacking of similar sheets of atoms or sheets of symmetrical variants. These sheets of atoms are identical in one polytype and also in the different polytypes of the same material, or they are related to each other by some symmetry operator. Thus in two dimensions within the sheet, there is no difference between the different polytypes. It is only in the stacking of

these sheets in the third direction (the c-axis), perpendicular to the
sheets, that the difference between the polytypes arises.

The simplest example would be a close-packed sheet of spheres called a
c-plane. Such a sheet would have 6-fold symmetry. Denoting the center of
each sphere by a point defines a layer of points (two-dimensional lattice).
The points in this layer are denoted as the A sites, and related to each
other by translations through vectors a_1, a_2, a_3 where a_{hex} = a = $|a_1| = |a_2| = |a_3|$
and the three vectors make an angle of $120°$ to each other. Note that these
three axes taken together with the reverse directions $-a_1$, $-a_2$, $-a_3$ form a set
of axes at $60°$ with respect to each other. See Fig. 1.

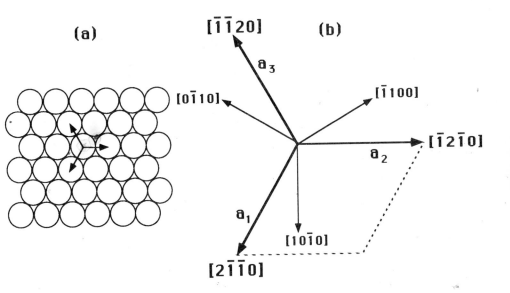

Fig. 1. (a) Close packing of spheres on the (0001) plane (also known as the
c-plane, or the basal plane), (b) a_1, a_2, a_3 axes in the Miller-Bravais
notation. The rhombus indicates the two-dimensional unit cell.

These three directions define the three crystallographic axes $[2\bar{1}\bar{1}0]$,
$[\bar{1}2\bar{1}0]$, and $[\bar{1}\bar{1}20]$, respectively, in the Miller-Bravais notation The
vector, say, a_1 is often defined in the form $\frac{a}{3}[2\bar{1}\bar{1}0]$ where its magnitude
is $a_{hex} = \sqrt{2}a_{cub}/2$ (a_{cub} is the cubic lattice constant) and its direction is along
$[2\bar{1}\bar{1}0]$. We can also define a two-dimensional unit cell by taking one of the
points as the origin and drawing a rhombus with sides defined by the vectors
a_1 and a_2 from the origin to two of the closest points in the layer (Fig. 1b).
A third axis, called the c-axis, can be defined which passes through the
origin and is pointed along a direction perpendicular to the sheet (i.e. the
c-plane).

Let us now consider the stacking of such sheets of spheres. The most stable
situation is to place the second sheet such that its spheres fit into the

valleys of the first sheet. There are two different ways of doing this. The second sheet can be placed vertically on top of the first sheet (unstably) and then displaced through a vector $\frac{a}{3}[01\bar{1}0]$ (or equivalently through $\frac{a}{3}[\bar{1}010]$, or $\frac{a}{3}[1\bar{1}00]$). Thus, the centers of the spheres in the second sheet have been translated with respect to those of the first sheet and define a set of new sites denoted by, say, the symbol B. An alternative way is to place the second sheet vertically on top of the first sheet (unstably) and then displace it through a vector $\frac{a}{3}[0\bar{1}10]$ (or equivalently $\frac{a}{3}[10\bar{1}0]$, or $\frac{a}{3}[\bar{1}100]$). In this latter case also, the spheres of the second sheet will be in stable positions in the valleys of the first sheet. Since, in this case, the centers of the spheres in the second sheet have been translated with respect to those of the first sheet by different vectors, they define a set of new sites different from A and B; these will be denoted by the symbol C. Note that the B and the C sites define two different sets of valleys in the first sheet of spheres. Placing spheres in the B sites does not leave any room for atoms in the C sites, and vice versa. Thus, three different sheets of spheres can be considered that are identical to each other but are displaced by a vector of the type $\frac{a}{3}<01\bar{1}0>$ with respect to each other. Note that a simple translation of a sheet of spheres located at the A sites through a vector $\frac{a}{3}<01\bar{1}0>$ transforms it to a sheet of spheres located at the B (or C) sites. This will be important in our later discussion of a dislocation mechanism for polytypic transformation.

In this simple example, all the polytypes can be considered as different stacking of A, B, or C sheets of spheres along the c-axis with the proviso that two sheets of spheres with the same letter cannot be placed next to each other. Thus, ...ABCABC... defines one polytype, ...ABCACB... another, ...ABAB... still a third, etc. ...ABAB... and ...ACAC... are the same polytypes. However, ...ABCB/BCABC... is not allowed. Note that the stacking of these two-dimensional close-packed sheets of spheres with the above requirement results in a three-dimensional structure which is also close-packed. If we assume an infinite stack of layers, there is always a periodicity in the stacking order, i.e. starting from, say, a layer A, after a certain integer number, n, of layers, we get to a next equivalent A layer. By equivalent layers here we mean that identical stacking sequences along the c-axis are produced. It follows that n is the order of periodicity along the c-axis in the stacking sequence.

2.2. Polytypes of SiC

The above example was for a close-packing of spheres. The discussion can be generalized to the case of SiC by replacing each sphere with a SiC tetrahedron such that the tetrahedra are joined to each other at their corners (i.e. two neighboring tetrahedra share a corner atom). A SiC tetrahedron arises out of the tetrahedral bonding between silicon and carbon atoms and is shown in Fig. 2(a). Note that a carbon atom is at the centroid of four silicon atoms (or vice versa) and is bonded to each of these atoms by a predominantly covalent bond.

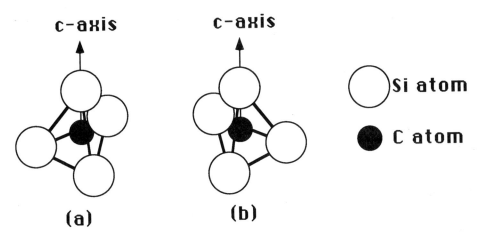

Fig. 2. (a) A SiC tetrahedron; (b) the twinned variant of the SiC tetrahedron. The two variants are related to each other by a 180° rotation around the c-axis.

Any SiC structure can be considered by replacing each sphere in the spherical structure discussed above with one of the corner-sharing tetrahedra shown in Fig. 2. In fact, the structure of all tetrahedrally-bonded materials, such as Si, Ge, diamond, GaAs, ZnS, etc. can be considered in terms of networks of corner-sharing tetrahedra.

One of the triangular bases of each tetrahedron is normal to the c-axis, i.e. a c-layer now consists of tetrahedra joined to each other at their corners and the interlayer spacing, d_{0001}, is equal to the height of the tetrahedron. Because the carbon atom is at the centroid of an equilateral tetrahedron, its center divides the height in a ratio of 3:1, i.e. its distance from the top silicon atom is three times larger than its distance from the base triangle. As a result, a basal plane may be considered to consist of two narrowly spaced sheets: a sheet of C atoms above a sheet of Si atoms. Thus, the SiC crystal may also be envisaged as narrowly-spaced double sheets of atoms stacked along the c-axis.

A tetrahedron shown in Fig. 2(a) has 3-fold symmetry about the c-axis. Hence, rotating the tetrahedron by 180° around the c-axis produces a different variant (a "twin" variant) as shown in Fig. 2(b). Hence there are A, B, and C layers of (untwinned) tetrahedra, as well as layers of the twinned variants A', B', and C' where the primes refer to layers consisting of twinned tetrahedra (Smith et al 1978). The complication with SiC polytypes arises because most of the polytypes are stacking of all the six types of layers A, B, C as well as A', B', and C'.

The edges of the base triangle in a SiC tetrahedron are along the $\langle 11\bar{2}0 \rangle$ directions. Hence, the projection of the SiC tetrahedron (Fig. 3a) and its twinned variant (Fig. 3b) on a $\{11\bar{2}0\}$ plane will appear in the form of acentric triangles. Note that in this and other figures of this section, the trace of basal planes is shown as horizontal dashed lines. The two triangles are related to each other by mirror symmetry in the {1$\bar{1}$00} plane.

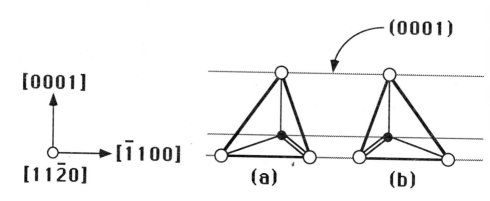

Fig. 3. The projection of the SiC tetrahedra in a {11$\bar{2}$0} plane. The two variants are related to each other by a 180° rotation about the [0001] direction, or by mirror symmetry in a {1$\bar{1}$00} plane.

In Fig. 3, the lines within the triangles show the Si-C bonds; the double lines indicate the two bonds from the C atom to two Si atoms (which overlap in the projection) at the base of the tetrahedron. We shall use such a projection of the SiC tetrahedron in our further discussion. The relation of the Si-C bonds, and the narrowly spaced double basal planes (e.g. *Cα*), to a tetrahedron in a particular SiC polytype (the cubic one) is shown in Fig. 4.

A little attention shows that corner sharing of tetrahedra is <u>not</u> possible between layers of the same letter, e.g. *AA* (Fig. 5a) or *AA'* (Fig. 5b), i.e. two neighboring layers of tetrahedra must have different letters. However, corner sharing of tetrahedra <u>is</u> possible if one variant of a tetrahedron is stacked upon by the same variant of the following letter, e.g. *AB* (Fig. 5c) or by the twinned variant of the preceding letter, e.g. *AC'* (Fig. 5d). These rules can be generalized to stacking upon a twinned variant. In this case, corner sharing is possible by the same (i.e. twinned) variant of the preceding letter, e.g. *A'C'* (Fig. 5e), or by the untwinned variant of the following letter, e.g. *A'B* (Fig. 5f).

In principle, an infinite variety of stacking sequences of sheets of tetrahedra can be considered which satisfy the above conditions. Each sequence then defines a different polytype. The simplest example would be the continuation of the *AB* stacking which, according to the above rules, is only possible if one variant of one letter and the twinned variant of the preceding letter are stacked along the c-axis, i.e. ...*AC'AC'AC'*... This is in fact the wurtzite structure which is common to II-VI compound semiconductors. Such a structure has 6-fold symmetry about the c-axis, i.e. it possesses hexagonal symmetry. [To be more precise, the c-axis is a 6_3 screw axis.] In fact, Jagodzinski (1949) suggested that a layer *C'* which is stacked in an *AC'A* type sequence has a local hexagonal environment and he used the letter h to denote this. Hence, one way to show the wurtzite structure is ...hhhh... On the other hand, in the sequence *ABC*, which if continued produces a cubic crystal, the sheet *B* has local cubic environment

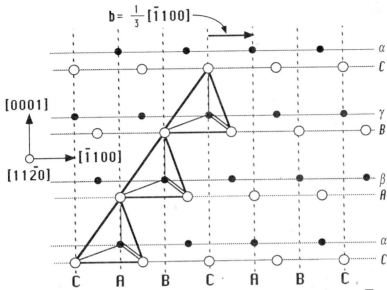

Fig. 4. The projection of the cubic polytype of SiC in the $\{11\bar{2}0\}$ plane. The three triangles are projections of three SiC tetrahedra at *A*, *B*, and *C* sites. The Si-C tetrahedral bonds within the tetrahedra are shown as lines within the triangles. The $\{0001\}$ planes are shown as dashed horizontal lines and the arrow indicates a $\frac{1}{3}[\bar{1}100]$ vector. Note that this structure is made from only one variant of the SiC tetrahedra.

and Jagodzinski used the letter k to denote this. Thus the cubic ...*ABC*... structure, which is often known as the zincblende (or the sphalerite) structure, may also be represented as ...kkk... A parameter, known as hexagonality, is defined for SiC polytypes as the fraction of letters h in a sequence of h's and k's within a period that represents a given polytype.

Another common notation for denoting the stacking sequence in a polytype is by Ramsdell (1947) which makes no distinction between layers consisting of untwinned tetrahedra and those consisting of twinned tetrahedra. Rather, in the Ramsdell notation, a polytype is denoted by giving the order of periodicity, n, together with the Bravais lattice of the crystal. Thus the wurtzite structure, i.e. the ...*AC'*... sequence, is shown simply as ...*AC*.... In this case, the order of periodicity, n, is 2 and the periodicity is $c = 2d_{0001}$. where d_{0001} is the interlayer spacing. Since the Bravais lattice of this structure is hexagonal (H), it is represented as 2H. On the other hand, the sphalerite structure, with the sequence ...*ABC*..., has an order of 3 and the periodicity is $c = 3d_{0001}$. As mentioned before, such a structure shows cubic (C) symmetry and in the Ramsdell notation it is represented as 3C. In cubic crystals the three number Miller indices are used instead of the four number Miller-Bravais indices and, because of a change of the coordinate axes, the (0001) planes become (111) planes; thus the periodicity of the sphalerite structure is $c = 3d_{111}$. In the Ramsdell notation, while the variant type in a stacking sequence is immaterial, the order of the letters is of course very important.

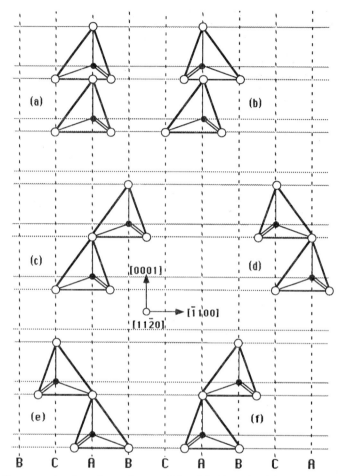

Fig. 5. Corner sharing <u>is not</u> possible if the stacking sequence is between tetrahedra of (a) same variants of the same letter, e.g. *AA*, or (b) different variants of the same letter, e.g. AA′. On the other hand, corner sharing <u>is</u> possible if an untwinned variant of the tetrahedron is stacked upon by (c) the same variant of the following letter, e.g. *AB*, or by (d) the twinned variant of the preceding letter, e.g. *AC′*. Similarly a twinned variant of a tetrahedron can be stacked upon by (e) the twinned variant of the preceding letter, e.g. *A′C′*, or by (f) the untwinned variant of the following letter, e.g. *A′B*.

Out of the infinite number of stacking variations, 3C is the only sequence which shows cubic symmetry. For this reason it is given a special name and is known as ß-SiC. All the other polytypes of SiC, which show non-cubic symmetry, are collectively known (for historical reasons) as α-SiC. It can be easily shown that if the stacking order in the Ramsdell notation is 2m, with m as an integer, then the crystal shows hexagonal (H) symmetry, while if the sequence is 2m+1, the crystal shows rhombohedral (R) symmetry (except for 3C which additionally shows cubic symmetry). The most common polytypes observed in SiC are 3C, 4H, 6H, and 15R. 2H, which is probably the simplest stacking sequence, is rarely observed.

A different notation is due to Zhdanov (1945). In this case, the number of consecutive layers of each variant within a period are given. In Zdanov notation, the order of the letters is immaterial, but the sequence of variants is important. Thus, the 6H-polytype with an ...*ABCB'A'C'*... stacking is shown as 33 (See Fig. 6) and the 15R polytype, with a stacking sequence ...*ABCB'A'BCAC'B'CABA'C'*... is shown as 323232 or, in short, as (32)$_3$. A list of common polytypes are listed in Table 1 together with stacking sequences and various notations for designating the polytypes.

Table 1. Notations for Selected SiC Polytypes

Ramsdell Notation	Stacking Sequence	Zhdanov Notation	Hexagonality
3C	...*ABC*...	∞	0.0
2H	...*AB'*...	11	1.0
4H	...*ABA'C'*...	22	0.5
6H	...*ABCB'A'C'*...	33	0.33
15R	...*ABCB'A'BCAC'B'CABA'C'*...	(32)$_3$	0.4

The tetrahedral arrangement of bonding in SiC gives rise to a very polar nature for the basal plane of SiC crystals. For a SiC surface oriented near the basal plane (i.e. a vicinal (0001) surface), the surface will be terminated with either Si or C atoms. Hence, the (0001) SiC surface is either a "Si face" or a "C face". In discussions that follow, the tilt angle of a vicinal (0001) SiC surface is the angle between the actual surface and the (0001) plane.

3. ORIGIN OF POLYTYPIC STRUCTURES

Silicon carbide was discovered early in the 19th century (Berzelius 1824) and polytypism in SiC was identified nearly 100 years ago by Frazier (1893). Yet, polytypism in SiC has still not been satisfactorily explained. A number of attempts have been made to explain the occurrence of polytypism and the mechanism of transformation from one polytype to another (Frank 1951, Jagodzinski 1954, Mitchell 1955, Pandey and Krishna, 1975, Daniels 1966, Madrix et al, 1968). However, none of the mechanisms proposed so far can explain all the experimental observations, and they are often at variance with them. Several review papers that contain extensive discussion of these mechanisms are Pandey and Krishna (1983) on the origin of polytypes and Jepps and Page (1983) on transformations in SiC. This paper will present recent results of investigations of the controlled growth and transformations of SiC polytypes.

Polytypism can be considered in two different ways. Firstly, from a thermodynamic point of view, the stability of each polytype over a certain

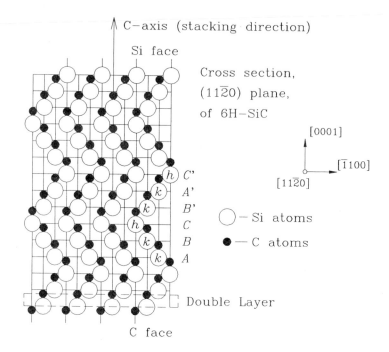

Fig. 6. Schematic cross-section [(11$\overline{2}$0) plane] of the 6H-SiC polytype.

range of temperature and pressure can be considered. In this respect, great progress has been made recently by means of ab initio calculations of the total energy of different polytypes (Heine et al 1992). Calculations of the relative stability of polytypes 2H, 4H, 6H, and 3C indicated that 4H and 6H are the most stable, followed by 3C, with 2H being much less stable than the others. Secondly, polytypism can be considered from a kinetic or growth viewpoint. That is, dominant factors controlling the formation and/or transformation of polytypes are factors such as growth rate, dislocations, contamination (impurities), etc. Results reviewed in this paper demonstrate that defects (i.e. dislocations and possibly impurities) play a significant role in the growth and transformation of SiC polytypes. As is well known, intentional defects (e.g. point defects in the form of substitutional donor and acceptor atoms) control many optical and electrical properties.

4. DEFECTS AND GROWTH MECHANISMS

In general when a molecule from the vapor phase deposits on a substrate surface at high temperatures, it will migrate on the surface until one of two things happens to it. During migration, it may either attach itself to the substrate or it may leave the surface and evaporate back into the vapor phase. At any instant of time there is a net number of molecules, N_d, depositing on the substrate surface and a net number, N_v, leaving the surface. The growth of the film depends on the competition between these two numbers. If $N_d < N_v$ the film thickness, t_f, decreases (i.e. etching takes place); if $N_d = N_v$

then t_f remains constant; finally, if $N_d > N_v$ film growth takes place and t_f increases. We are primarily interested in the latter condition. The attachment of the molecule to the surface may take place in a number of ways. The migrating molecule may attach itself to an already existing group of depositing atoms (i.e. to a stable nucleus). Alternatively, it may attach itself directly to the atoms of the substrate surface at a preferential site, such as a surface step or a defect. Herein, a defect is defined to be some structural imperfection (e.g. a dislocation), or some extraneous material present at the surface (i.e contamination). A third possibility is that a group of migrating atoms may find themselves at the same location on the substrate surface at the same time and form an embryo which has a supracritical size. In such a case, a stable nucleus forms to which other migrating atoms may attach themselves.

In general, surface steps and defects are very favorable sites for the attachment of migrating atoms. The reason is that at a step (or defect), there are many surface atoms with unsatisfied bonds which can trap the migrating atoms in order to satisfy their bonding. On a flat surface, on the other hand, there are only surface atoms directly underneath the migrating atom which have unsatisfied bonds.

When the substrate surface is parallel to a low index plane and no nucleation sites are present on the substrate surface, then two-dimensional nucleation on the substrate surface is required. Once a nucleus has formed, its periphery form steps to which migrating atoms can attach themselves. In general, the formation of a stable nucleus requires a high supersaturation. The process, however, can be treated with the usual concepts of a critical size for the embryo upon which a stable nucleus forms. Due to the high supersaturation, the _lateral_ growth of the nucleus, i.e. the migration of its associated steps on the substrate surface, is very rapid since the probability of a molecule from the vapor phase, or from those migrating on the substrate surface, sticking at a step edge is very high. If many nuclei have formed, then of course coalescence of the growing islands takes place resulting in a net increase in the thickness of the deposited film, i.e. resulting in _vertical_ growth.

When growth takes place on "off-axis" substrates, the nucleation sites of surface steps and perhaps of some types of defects may not regenerate as growth proceeds. As mentioned above, a pre-requisite for two-dimensional nucleation on the substrate surface is a high supersaturation. When the supersaturation is below a critical value necessary for the formation of stable nuclei, then growth will soon stop. This is because any existing surface step will rapidly migrate laterally to the end of the specimen and, literally, will grow itself out of existence. In this case there will soon be no preferential sites left on the substrate surface for trapping of the migrating vapor molecules and they will be free to roam around on the substrate surface until they evaporate back to the vapor phase.

In 1949 Frank suggested a very ingenious mechanism by which growth from a vapor phase (or, indeed, from a liquid phase) can occur when the supersaturation is below the critical value for the formation of two-dimensional nuclei (Frank 1949). In Frank's mechanism, steps will be generated continuously on the substrate surface, i.e. there is a persistent source of steps which does not disappear as lateral or vertical growth continues. The source of persistent surface steps is a dislocation which intersects the surface. By definition, a dislocation is the boundary between a part of the crystal which is displaced and a part which is not displaced. The relative amount of displacement is given by the magnitude of the dislocation Burgers vector, **b**, which is usually a lattice vector. The direction of **b** gives the direction in which the displacement of the crystal has taken place. If the dislocation is of the screw type, i.e. if its Burgers vector is parallel to its line direction, then the displacement field around the dislocation line is like a helicoid. Hence, when a screw dislocation intersects the crystal surface it forms a spiral step on the surface with a height |**b**|. See Fig. 7.

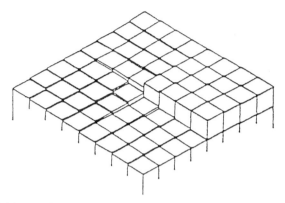

Fig. 7. Frank's mechanism of crystal growth (from Frank 1949).

As mentioned above, **b** is usually a lattice vector, e.g. |**b**|=d_{hkl} where d_{hkl} is the interplanar spacing for the set of planes (hkl) normal to the screw dislocation (i.e. normal to **b**). As the vapor molecules attach themselves to this step, the step moves laterally and <u>lateral</u> growth occurs. As the step migrates across the surface, <u>vertical</u> growth by a monolayer, i.e. an amount |**b**|=d_{hkl}, takes place. Simultaneously, however, the dislocation has lengthened by a monolayer. Since the nature of dislocation remains the same (i.e. its Burgers vector does not change), it will have associated with it the same displacement field, and the same spiral step will exist on the new surface. Thus, the screw dislocation will be a regenerative source of steps on the substrate surface and the crystal face grows up a spiral staircase. Shortly after Frank's proposal, it was suggested that for a dislocation to act as a growth center, it does not have to be a pure screw dislocation but it suffices for it to have a screw component in order to produce a step on the surface (Burton et al 1951).

Soon after Frank's suggestion in 1949, spiral growth was observed on a number of growing surfaces by decorating the step edges and observing them by Nomarski optical microscopy. One of the first surfaces on which such growth was observed was in fact SiC (Verma 1951, Amelinckx 1951a,1951b).

From the above discussion, it is clear that the requirements for a dislocation to act as a self-regenerating source of steps is that it should (i) intersect the surface, and (ii) it must have a Burgers vector with a component that produces a step on the surface. Thus, if a dislocation is normal to the surface and has a Burgers vector perpendicular to its line direction (i.e. its displacement field is parallel to the surface), then no surface steps will be produced. Such a dislocation with the Burgers vector normal to its line direction is called an edge dislocation.

It had been concluded from the above considerations that the screw type of dislocation can act as a regenerative source of nucleation steps, but an edge type of dislocation cannot (because there are no surface steps associated with an edge dislocation). It was thus very surprising that, during growth of alkali halides, Bethge noted that some of the dislocations which acted as growth sources were edge dislocations (Bethge 1962,1964)! These results were later confirmed by Keller (1968) in the same laboratory. Since then, there have been various explanations on how an edge dislocation can act as a growth source. These include various effects at the point of dislocation intersection with the surface, e.g. electrostatic forces Bethge (1962), impurity segregation (Chernov 1955), and surface relaxation (Frank 1981, Giling and Dam 1984). The most plausible explanation was given by Bauser and Strunk (1981,1982) who suggested that, near the intersection with the surface, the dislocation dissociates into two partials which bound a ribbon of stacking fault. Since the partial dislocations are no longer of the pure edge type, they will have a screw component. As a result, a step forms between the two partial dislocations which can act as a regenerative step source (see Fig. 8).

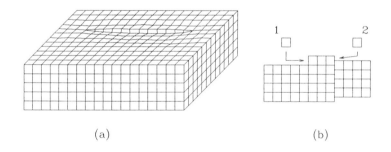

(a)   (b)

Fig. 8. The model of Bauser and Strunk for a Bethge dislocation (after Bauser and Strunk 1982).

4.1. Some Experimental Results and Discussion on Growth Dislocations

An investigation of dislocations in SiC films indicate that the few growth dislocations that were observed were of the Bethge type rather than the Frank type. A number of samples consisting of 6H-SiC films grown on vicinal (0001) 6H-SiC substrates (Powell et al 1990b) were studied. Films grown on substrates with very low tilt angles often exhibited hillock-like features as shown in Fig. 9.

Some hillocks similar to those seen in Fig. 9 exhibited faint spiral growth patterns emanating from the center of the hillock. These hillocks suggested growth around the points of intersection of dislocations in the substrate with the substrate surface. In order to check this, plan-view TEM specimens were prepared from a sample with a high density of hillocks and investigated

Fig. 9. Nomarski optical micrograph of a 6H-SiC film that has grown on a low-tilt-angle (0.2°) vicinal (0001) 6H-SiC substrate. Note hillock features.

by conventional electron microscopy. At the center of each growth feature, a dislocation was indeed observed which was clearly related to the growth. An example is shown in Fig. 10 which is a plan-view TEM micrograph with the foil plane parallel to the (0001) plane.

The TEM foil in Fig. 10 has been thinned from the back, i.e. from the side opposite to the growth surface. The roughly circular patch in the micrograph is the growth structure on the film surface (in the form of a pyramid). The two small hole-like regions with dark contrast at the center of the circular patch are the points at which one of the cross-slipped dislocations has intersected the specimen surface. The partials of the dissociated dislocation on the left have constricted on the basal plane and the resulting perfect screw dislocation has cross-slipped onto the prism plane. The dissociated dislocations on the right have likewise constricted and cross-slipped. In this case, the perfect screw dislocation on the cross-slip plane appears to be connected to a small segment of dislocation which runs from one hole to the other. The specimen is slightly tilted.

The two widely dissociated dislocations in Fig. 10 are of the screw type with a Burgers vector $\frac{1}{3}$ [11$\bar{2}$0] and both lie on the (0001) slip plane. They have dissociated on the basal plane according to the following reaction:

$$\frac{1}{3} [11\bar{2}0] = \frac{1}{3} [10\bar{1}0] + \frac{1}{3} [01\bar{1}0] \tag{1}$$

Near the center of growth, the partial dislocations have both constricted and the resulting perfect screw dislocations appear to have cross-slipped to the prism planes, i.e., the dislocation changed planes from the basal plane to a prism plane (i.e. a {10$\bar{1}$0} plane). As a result of cross-slip, the line direction of both screw dislocations has changed so that they are nearly

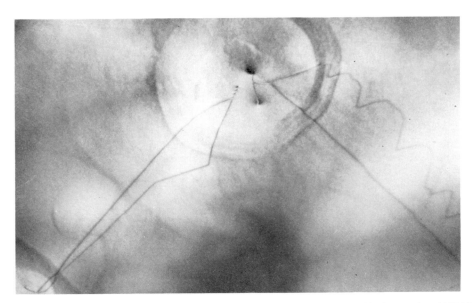

Fig. 10. Plan-view TEM micrograph with the foil plane parallel to the (0001) plane.

parallel to the c-axis, i.e. they intersect the substrate surface at some point. However, the Burgers vector of each dislocation is invariant and is still $\frac{1}{3}[11\bar{2}0]$. As a result, although the first condition (i.e. intersection of the dislocation with the substrate surface) has been satisfied, the second condition (i.e. a component of the Burgers vector normal to the surface) is not satisfied, and the dislocations would be of the Bethge type. Although, the few observed growth dislocations in 6H-SiC appear to be of the Bethge type, the validity of the Bauser-Strunk explanation has not as yet been clarified.

More discussion on the effect of defects will be presented in the section on growth processes.

5. DEFECTS AND TRANSFORMATIONS

Many types of polytypic transformations have been observed in SiC. Examples include: 2H→(3C/6H) (Krishna et al 1971, Powell and Will 1972), 3C→6H (Baumann 1952, Yoo et al 1990), 6H→3C (Kieffer et al 1969, Yang et al 1992). These are just a few examples; we have included several of the latest examples. A detailed discussion of this subject can be found in Jepps and Page (1983). Although various models have been proposed to explain transformations in SiC polytypes, none has satisfactorily explained the mechanism of polytypic transformations. In this section, we describe a recently proposed dislocation model that provides a mechanism for both twinning and polytypic transformations (Pirouz 1989a).

The actual mechanism of polytypic transformation is a kinetic problem and has been considered both as a diffusive process and also as a dislocation process. The dislocation model proposed by Pirouz (1989a) involves the glide of partial dislocations and thus implies that the phase transformation depends both on temperature as well as on stress. After a brief description of this model in Section 5.1, we will present some experimental evidence supporting its validity in Section 5.2.

5.1 A brief discussion of the dislocation model for transformations

Because of the lack of space we shall only consider the polytypic transformation 6H→3C which is the one that is investigated in our experiments (Pirouz et al 1991, Yang et al 1992). First, however, dislocations in 6H-SiC need to be discussed further.

Dislocations in 6H-SiC lie on the basal (0001) planes and have a Burgers vectors $\frac{1}{3}[11\bar{2}0]$. SiC is known to have a very low stacking fault energy [≈ 2.5 mJ/m^2 in 6H-SiC (Maeda et al 1988)], i.e. the dislocations in this material are dissociated into widely separated partials bounding a band of stacking fault. According to equation (1), a perfect screw dislocation, with a $\frac{1}{3}[11\bar{2}0]$ vector, dissociates into two 30° partials with $\frac{1}{3}[10\bar{1}0]$ and $\frac{1}{3}[01\bar{1}0]$ Burgers vectors. Note that these are the same vectors as the ones discussed in Section 2.1. The core of these two 30° partials consists of either all Si atoms or all C atoms. In this discussion, we assume that dissociated dislocations in SiC are of the "glide" type (rather than the "shuffle" type, see Hirth and Lothe 1982). Then, the partial with a core of Si atoms is designated as Si(g) while the dislocation with a core of C atoms is designated as C(g). The core structure of these two 30° glide partials is shown in Fig. 11a. In Fig. 11b, it has been assumed that the broken bonds in the core of the dislocations have reconstructed. Core reconstruction in SiC would presumably be relatively easy because both C and Si are tetravalent.

The accepted mechanism for dislocation motion in semiconductors, which have a high Peierls barrier, is kink nucleation and kink migration (Hirth and Lothe 1982). Both of these processes involve the breakage of bonds along the dislocation line and re-formation of bonds in the next Peierls valley (primary Peierls valley for kink pair nucleation and secondary Peierls valley for kink migration). Note that in Fig. 11, the motion of one of the partials involves the sequential breakage (and re-formation) of Si-Si bonds, while the motion of the other partial involves the breakage (and re-formation) of C-C bonds. Using a crude analogy of the comparative bond strength of bulk diamond versus bulk silicon, it may be reasonably expected that the formation and migration of kinks along the Si(g) partial is much easier than that along the C(g) partial. It follows that the thermal energy required for the motion of Si(g) partial is much less than that required for the motion of the C(g) partial. In other words, the activation energy, E, for motion of these two partial dislocations is very different and $E_{Si} \ll E_C$. Consequently, it may be expected that the Si(g) partial would become mobile at much lower temperatures than the C(g) partial.

The mechanism of the actual 6H→3C transformation can now be discussed. The mechanism depends on a pinned segment of a screw dislocation which is dissociated on the (0001) plane into a Si(g) and a C(g) 30° partial dislocation. With the more mobile Si(g) as the leading partial, under an

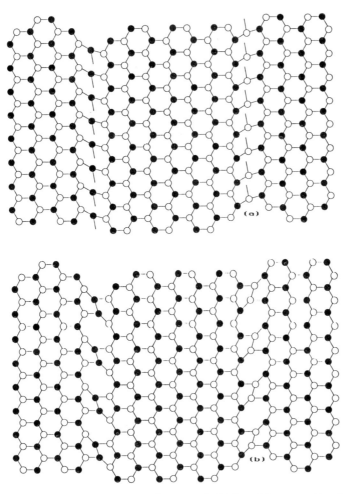

Fig. 11. The structure of the 30° partial dislocations on the basal plane of 6H-SiC. (a) Unreconstructed core, (b) Reconstructed core resulting in the elimination of all the dangling bonds. Note that the core of one of the partials consists of all Si atoms while the core of the other partial is all C atoms.

appropriately resolved shear stress, the leading partial expand and form a loop on the (0001) plane in a manner similar to the Frank-Read mechanism. See Fig. 12. On the other hand, the C(g) partial, which has a very low mobility at low temperatures (say below 1500°C), lags behind and cannot form a loop. Hence, the loop formed by the Si(g) partial is faulted. After the formation of a faulted loop, the leading Si(g) partial approaches the trailing C(g) partial from behind and the screw dislocation cross-slips onto the (1010) prism plane according to the Friedel-Escaig mechanism (Pirouz and Hazzledine 1991). Subsequent to this, there is a tendency for the screw dislocation to cross-slip back to the basal plane. This can only happen if the screw dislocation can dissociate on the basal plane without violating the stacking sequence, i.e. without forming a high energy *AA* stacking sequence.

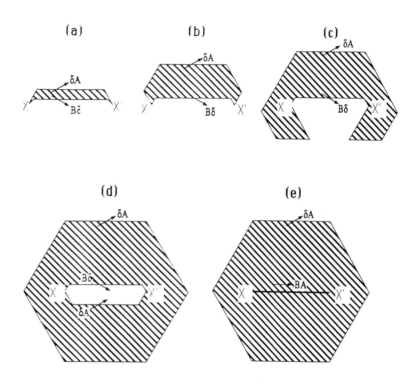

Fig. 12.   The expansion of a pinned screw dislocation on the basal plane of SiC at temperature where the mobility of the partials is very different (from Pirouz and Hazzledine 1991).   **δA** is the Si(g) partial and **Bδ** is the C(g) partial dislocation. **X** and **X'** are pinning points for the dislocation.

For the 6H→3C transformation, the sequence of stages is illustrated in Fig. 13.   In this figure, each column represents the stacking sequence of a 6H crystal at a particular stage of transformation to 3C.   The progression of stages is from left to right.   Consider the stacking of double layers parallel to the basal plane in 6H as shown in the first column of Fig. 13. In the second column, a faulted loop has formed on a basal plane by the mechanism of Fig. 12, thus shearing the atoms above the slip plane from *A'* sites to *B* sites.   In Fig. 13, this is shown by a horizontal arrow.   As mentioned above, after the formation of the faulted loop, because of compressive stresses on the two partials, the screw dislocation cross-slips into the cross slip plane [the {1$\bar{1}$00} prism plane].   Cross-slip is shown by vertical arrows in Fig. 13.   Because of a reversal in the nature of the screw dislocation from C(g)/Si(g) to Si(g)/C(g) (Pirouz and Hazzledine 1991), there is a driving force on the screw dislocation to immediately cross-slip back on the basal plane, dissociate, and form another faulted loop.   The next faulted loop shears the atoms from the *C'* sites to the *A* sites.   The remaining stages are illustrated in Fig. 13.

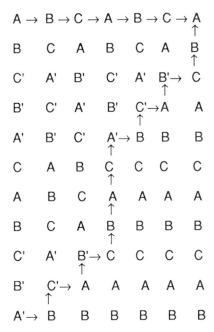

Fig. 13. Stages in the change of the stacking sequence from 6H to 3C during the 6H→3C polytypic transformation by the sequential formation of a faulted loop on a basal plane (horizontal arrows) and cross-slip (vertical arrows) of the screw dislocation to the next basal plane.

The sequence of faulted loop formation on the (0001) plane followed by cross-slip on the (10$\bar{1}$0) plane is also shown schematically in Fig. 14. On the right hand side of this figure, the stacking sequence of 6H-SiC is shown. Within each faulted loop, the crystal has sheared and the sites changed according to the sequence described in the previous paragraph. Thus, as shown in Fig. 14, the final stacking sequence within the loops will be ...BAC... and a 6H→3C transformation has taken place.

5.2. Observation of a 6H→3C Transformation.

In this section, we summarize experiments that demonstrate the 6H→3C transformation in SiC samples subjected to stress and high temperature (Yang et al 1992). The specimens to be transformed were cut from an Acheson crystal with a diamond saw into the shape of 1x1x3 mm^3 parallelepipeds and in the orientation shown in Fig. 15. The Acheson crystal was checked by TEM and was found to be a quite perfect 6H single crystal (Fig. 16) with only a very low density of defects in the form of stacking faults. Figure 16 illustrates that the preparation of the cross-sectional TEM foil does not cause transformation.

The primary glide plane is (0001) and the cross-slip plane is (10$\bar{1}$0). The

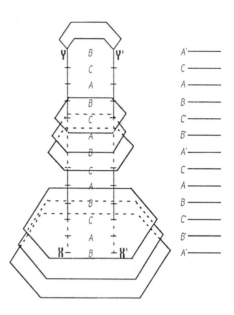

Fig. 14.   Schematic of the change in the stacking sequence from 6H to 3C during the 6H→3C polytypic transformation. The scheme follows the model due to Pirouz (1989).

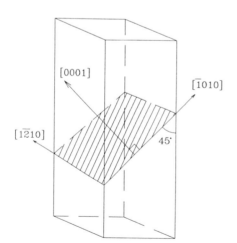

Fig. 15.   The orientation of the deformation specimen.

resolved shear stress on two of the $\frac{1}{3}<1\bar{2}10>$ dislocations is equal on the basal plane as well as on the corresponding $(10\bar{1}0)$ cross-slip plane. This particular orientation was chosen in order to have a large resolved shear stress both on the basal (0001) plane and also on the $(10\bar{1}0)$ cross-slip

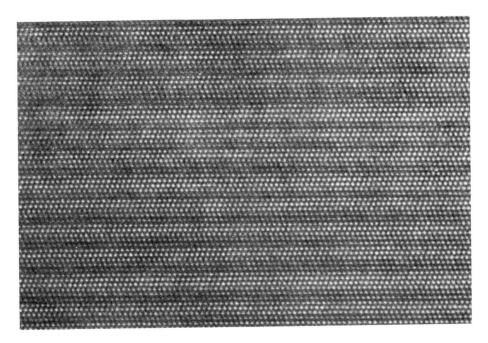

Fig. 16. A cross-sectional HREM micrograph from the as-received 6H-SiC
Acheson crystal (from Yang et al 1992).

plane.

Each specimen was pre-loaded by 20 MPa and then ramped up to the
deformation temperature. Subsequently the applied stress was increased to
the desired value and the specimen kept under load for the desired times.
After deformation, TEM specimens were prepared with foil normals parallel to
[0001] and [11$\bar{2}$0] directions and examined in JEOL 200CX and JEOL 4000EX
electron microscopes. The latter microscope, with a point-to-point
resolution of 0.18 nm, was used for HREM.

A specimen that was heated to 1400°C during the applied stress underwent
significant transformation from 6H to 3C. Optical and TEM results are shown
in Figs. 17 and 18. A specimen heated to 1100°C exhibited less
transformation.

According to the dislocation model (Pirouz 1989a), there are two requirements
for polytypic transformation in SiC: (i) a large difference in the mobility
of the 30° partial dislocations constituting a perfect screw dislocation, and
(ii) a sufficiently large resolved shear stress on the basal plane to form a
faulted loop from a pinned segment of the highly mobile 30° Si(g) partial
dislocation. These two requirements were satisfied in the deformation of two
specimens heated to 1100°C and 1400°C. In both specimens, there was partial
transformation of the 6H polytype to the cubic 3C polytype. The results are
consistent with the dislocation model.

Fig. 17. A Nomarski optical micrograph of the (11$\bar{2}$0) surface of a sample after deformation at 1400°C. Note the slip traces which are parallel to the traces of the basal plane (from Yang et al 1992).

1 mm

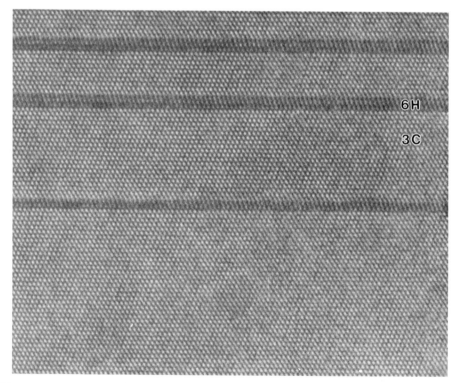

Fig. 18. A HREM micrograph of a region of the specimen deformed at 1400°C where a large density of 3C bands has been produced (from Yang et al 1992).

Additional evidence supporting the dislocation model of polytypic transformation (Pirouz 1989a) are the observations of the 2H→(3C/6H) transformation by Powell and Will (1972). In that investigation, 2H-SiC single crystal needles transformed to a faulted 3C/6H structure at temperatures above 1400°C. When dislocations were introduced by grinding the needles into platelets (parallel to the c-axis), the transformation proceeded at temperatures as low as 400°C. It was concluded that dislocations greatly accelerated the transformation process.

6. GROWTH PROCESSES

In addition to polytypism, another feature of SiC that adds to the challenge of its crystal growth is that it does not melt at any easily attainable pressure. Rather it sublimes at temperatures above approximately 1800°C. So, conventional growth-from-the-melt approaches are ruled out and most attempts at crystal growth have involved growth-from-solution or some form of growth-from-the-vapor.

The following two sections will describe developments in SiC growth processes for producing boules (from which wafers are obtained) and thin films for device fabrication.

6.1 Boule Growth

Until recently, nearly all SiC semiconductor research was carried out using small irregular α-SiC crystals that were sublimation grown by either the industrial Acheson process (Acheson 1892), or the Lely (1955) process. In the Lely process, SiC was sublimed from polycrystalline SiC powder at a temperature of approximately 2500°C and then condensed on the walls of a cavity created within the powder. Small SiC crystals nucleated randomly on the cavity wall and grew into hexagonally-shaped platelets with diameters up to approximately one centimeter. Some examples are shown in Fig. 19. By far, the most common polytype grown was 6H, followed by 15R and 4H (Knippenberg 1963). Quite often, the platelets consisted of a mixture of different polytypes. They were not suitable substrates for commercial device

Fig. 19. Cree wafer and Lely crystals.

production.

Attempts were made to grow bulk SiC crystals from Si solution (Nelson et al 1966) and from molten metals, such as chromium (Griffiths and Mlavsky 1964, Knippenberg and Verspui 1966).   Results from these attempts were not satisfactory.

Tairov and Tsvetkov (1978,1981) established the basic principles of a "modified sublimation" bulk growth process in the late 1970s. Although other research groups were somewhat slow in adopting this process, it is now being developed by many labs throughout the world (Ziegler et al 1983, Carter et al 1987, Nakata et al 1989, Barrett et al 1991, Kanaya et al 1991).

The basic elements of the modified sublimation process are shown in Fig. 20, which is a schematic diagram of the configuration used by Ziegler et al (1983). Nucleation takes place on a SiC seed crystal located at one end of a cylindrical cavity. A temperature gradient is established within the cavity such that the polycrystalline SiC is at approximately 2400°C and the seed crystal is at approximately 2200°C. At these temperatures and at reduced pressures (Ar at 200 Pa), SiC sublimes from the source SiC and condenses on the seed crystal. Growth rates of a few mm/h can be achieved. In some boule growth systems, the seed crystal is at the top of the cavity and growth proceeds downward. This eliminates some problems of contamination of the growing boule.

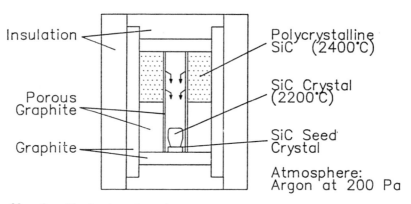

Fig. 20.   Growth chamber for the modified sublimation process.

The modified sublimation process has been used to grow both 4H and 6H boules. Its successful application to the 3C polytype has not been reported. Techniques for cutting and polishing wafers have been developed and at least one company is now marketing 25-mm-diameter polished 6H-SiC wafers as shown in Fig. 19; wafer diameters up to 44 mm have been achieved.

6.2 Thin Film Growth on Non-SiC Substrates

Because of the lack of large-area high-quality SiC substrates for many years, efforts were made to grow SiC thin films on non-SiC substrates that might be suitable for commercial device fabrication. Early in the 1980s, large-area single-crystal 3C-SiC films were achieved on (001) Si substrates (Nishino et al 1983, Liaw and Davis 1985). This created renewed interest in SiC and many research groups began pursuing this growth approach. Unfortunately, the 3C films grown on Si had a high defect density, which included stacking faults, microtwins, and a planar defect known as an inversion domain boundary (IDB), also known as an antiphase boundary (APB)(Shibahara et al 1986, Pirouz et al 1987). These defects have been attributed to the 20% lattice mismatch and the 8% difference in coefficient of expansion between Si and SiC. Also, it has been suggested (Pirouz 1989b) that the initial nucleation process may be an important factor in the formation of the stacking faults and microtwins. An IDB in a SiC film is a boundary of Si-Si or C-C bonds that is formed by the coalescence of SiC islands that have nucleated on a (001) Si surface with monoatomic steps. The problem arises because SiC islands with two different bond orientations (differing by a 90° rotation about the <001> axis) can nucleate on the Si substrate. In subsequent work, the IDBs were eliminated by using vicinal (001) Si substrates with tilt angles in the range 0.5° to 4° (Shibahara et al 1986, Powell et al 1987, Matus and Powell 1989). But, this did not eliminate the stacking faults and other defects. So far, devices fabricated from 3C-SiC films grown on Si have not achieved satisfactory performance.

In order to eliminate the problems caused by a large lattice mismatch, TiC_x (less than 1% mismatch) substrates were investigated (Parsons 1987). Somewhat improved growth was reported, but difficulties in producing defect-free single-crystal TiC_x has hindered its use as a substrate for SiC growth.

Recent success in the production of high quality 6H-SiC wafers has lessened the need for non-SiC substrates. The remainder of this paper will describe only results achieved using SiC substrates.

6.3 Thin Film Growth on 6H-SiC Substrates

Early SiC CVD thin film growth processes in the 1960s and 1970s, using Acheson or Lely crystals, typically were carried out at temperatures above 1550°C (Campbell and Chu 1966, Muench and Pfaffeneder 1976, Nishino et al 1978). Often the growth was carried out on the as-grown (0001) 6H-SiC basal plane, either the Si face or C face. Sometimes the crystals were polished and this produced substrates with growth surfaces that were at some small angle with respect to the (0001) plane. Both 3C and 6H films were produced on these substrates. Factors affecting the control of polytype and the observed defects were not understood.

In 1973, it was found that 6H-SiC could be grown on 6H-SiC in the temperature range 1320-1390°C when the growth direction was perpendicular to the [0001] direction (Powell and Will 1973). The significance of this "90°-off-axis" was not appreciated until much later. In a review of Russian SiC research of the 1970s, Tairov and Tsvetov (1983) discussed factors that control the polytype of epitaxial films. They describe sublimation growth experiments using α-SiC substrates with the growth surface oriented about 5° off the (0001) plane. It was found that this orientation was beneficial in growing films that reproduced the polytype of the substrate. The reason given was that the "micro-structure" (i.e. the stacking sequence) was contained in the surface of the off-axis substrate. These growth experiments were carried out at temperatures greater than 1600°C.

Recently, 3C-SiC films were grown by CVD on the basal plane of Acheson and Lely 6H-SiC crystals in the range 1350-1550°C (Kong et al 1986, Kong et al 1987, Kuroda et al 1987, Matsunami et al 1989, Kong et al 1989, Powell et al 1990a). The quality of these films was much better than 3C films grown on

Si, but the films did contain a planar defect known as a double positioning boundary (DPB), and a high density of stacking faults. The DPBs in 3C films grown on 6H substrates arises because of the two possible variants of the 3C stacking sequence (*ABC...* or *ACB...*) that can nucleate on (0001) 6H. The difference between these two variants is a 60° rotation about the <111> axis. If both variants nucleate on the 6H substrate, DPBs will form at the boundary between domains differing by the 60° rotation. At the DPB, the chemical bonds of two out of every three double layers in the stacking sequence do not match up; as a consequence, the boundary is chemically and, probably, electrically active. In addition, because of stress generated by the DPB, there is a high density of stacking faults in the vicinity of the DPB. Because of the effect of the DPB on the growth process along the DPB, one can see DPBs in 3C films with an optical microscope (more easily seen in films grown on the Si face). We now know that the "mosaic" patterns observed in 3C films grown in the 1960s were actually DPBs.

Also recently, 6H-SiC films were grown on "off-axis" 6H-SiC substrates that were oriented several degrees off the basal plane (Kong et al 1987, Kuroda et al 1987, Matsunami et al 1989, Kong et al 1988, Ueda et al 1990). All of the above growth results were achieved on Acheson or Lely crystals in the temperature range 1350°C to 1550°C. More recently, high quality 6H-SiC films were also grown on "off-axis" 6H-SiC wafers that were produced from sublimation-grown boules (Powell et al 1990b). Just as in the early Russian work described by Tairov and Tsvetkov (1983), the use of "off-axis" substrates is also beneficial for the lower temperature epitaxial growth.

To explain the success of the 6H CVD at lower temperatures using "off-axis" 6H-SiC substrates, a model was suggested (Kuroda et al 1987, Matsunami et al 1989) whereby the density of atomic-scale steps on a vicinal (0001) 6H substrate determines the polytype of the grown film. Refer to Fig. 21, a schematic cross-sectional view of a vicinal (0001) SiC crystal. The step density increases and the width of terraces between steps decreases as the tilt angle of the growth surface increases. According to the model, at tilt angles greater than 1.5°, arriving Si and C atoms can easily migrate to steps where lateral growth occurs parallel to the basal plane. In this case, grown films assume the *ABCB'A'C'...*(commonly expressed as *ABCACB...*) stacking sequence of the 6H substrate. However, at smaller tilt angles (i.e. < 1.5°), terraces are larger and migration of Si and C atoms to the steps is less likely. Instead, nucleation of 3C occurs on the terraces.

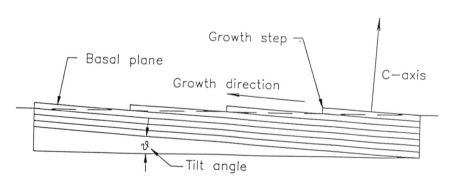

Fig. 21. Vicinal (0001) α-SiC substrate showing surface growth steps.

Recent results at NASA Lewis (Powell et al 1991b, 1992) demonstrate that the above model (Kuroda et al 1987, Matsunami et al 1989) is not sufficient to explain the control of polytype in the CVD growth on low-tilt-angle vicinal (0001) 6H-SiC substrates. In the NASA work, 6H films were grown on vicinal (0001) 6H substrates with tilt angles as small as 0.1°. In addition, 3C growth was intentionally produced on the low-tilt-angle substrates. The following is a summary of this process. Vicinal (0001) 6H substrates, with tilt angles in the range 0.1° to 0.5°, were cut from commercial wafers (Cree 1991). The surface of the substrates was divided into 1-mm-square mesas with a dicing saw as shown in Fig. 22. The tilt direction was along a diagonal with highest atomic plane at the upper right of each mesa.

Fig. 22. 6H-SiC substrate (0.5° tilt angle) with 1-mm-square mesas. Dark regions are 3C films; light regions are 6H films (from Powell et al 1992).

Prior to growth, the 6H substrates were subjected to an HCl\H_2 gaseous etch at 1375°C for 25 min to remove <u>unwanted</u> 3C nucleation sites. The substrates were then removed from the growth chamber and exposed to a non-clean laboratory environment. They were then returned to the growth chamber. The SiC films were grown at atmospheric pressure using silane and propane as the sources of Si and C, respectively. The resulting 8-μm-thick film, shown in Fig. 22, was oxidized to distinguish the 3C and 6H regions. Differences in oxidation rates produce contrasting interference colors between 3C and 6H (Powell et al 1991a). Substrates (with mesas) that were <u>not</u> removed from the growth chamber between the HCl etch and the growth process yielded films that were almost entirely 6H.

In this growth, a SiC film grew uniformly in thickness on each mesa. Initial growth was mostly 6H. It is presumed that this growth was due to lateral growth from atomic steps on the surface. In many of the mesas, 3C nucleated on the highest atomic plane (upper right in Fig. 22) and then grew laterally over the mesa from 3C steps generated at this point. In this case, the extent of lateral growth was consistent with the amount of vertical growth (8 μm) and the tilt angle of 0.5°. Most of the 3C regions were free of DPBs and had a reduced density of stacking faults compared to previous 3C growth on 6H. As yet, the 3C nucleation process is not fully understood, but we believe that contamination plays a role in this case. In other growth experiments, we were able to initiate 3C nucleation at locations on the 6H substrate by indenting the surface with a diamond scribe after the HCl etch. These results provide evidence that defects and contamination can dominate step density in controlling polytype formation on 6H substrates with small tilt angles.

7. OPTICAL AND ELECTRICAL CHARACTERIZATION

In this section we will review a number of recent reports dealing with the optical and electronic properties of the polytypes of SiC. Between 1987 and 1991 a number of reviews have been published on the single crystal properties of SiC (Choyke 1987,1990, Pensl and Helbig 1990, Davis and Glass 1991, Davis et al 1991). The material that will now be covered in this section was not published in time for these reviews. Despite the limited time frame that will be addressed there are many interesting developments that will have to be omitted due to space limitation and we hope those authors not cited will be understanding.

7.1 Nitrogen Donors in 6H SiC

Suttrop et al (1992) used the Hall effect and Fourier Transform Infra-Red (FTIR) absorption on selected Lely platelets and boule samples of 6H-SiC having varying degrees of nitrogen doping and compensation. A number of new absorption lines have been resolved in the infrared from 400 to 750 cm^{-1} and 1000 to 1250 cm^{-1}.   Thermalization effects were investigated in the temperature range from 7 K to 80 K as well as polarization dependence of the nitrogen related absorption lines. A new nitrogen-donor model is proposed which is based on the effective mass approximation (Faulkner 1969, Gerlach and Pollman 1975) taking into account the anisotropy of the electron effective mass, the valley orbit splitting of the donor ground state and the three inequivalent donor sites in the 6H-SiC unit cell. As mentioned in Section 2.2, these inequivalent sites (denoted as h for the hexagonal site and k_1 and k_2 for the two cubic sites) correspond to the inequivalent double layers. See Fig. 6. The results of the analysis of the data are:

(a) The ground state valley orbit splitting for the donor in the hexagonal (h) inequivalent site is determined to be 12.6 meV in good agreement with a value of 13 meV obtained from Raman scattering (Colwell and Klein 1972).

(b) For the $2p_o$, $2p\pm$, $3p_o$ and $3p\pm$ states the effective mass approximation is found to hold.

(c) The perpendicular ($m_\perp$) and parallel ($m_{||}$) effective electron masses for 6H-SiC were determined by a fit of the observed infrared data to the effective mass theory. One obtains $m_\perp = 0.24\ m_o$ and $m_{||} = 0.34\ m_o$.

(d) The ionization energies of the h, k_1 and k_2 nitrogen donors in 6H-SiC are determined to be 81.0 meV, 137.6 meV and 142.4 meV, respectively.   The accuracy of the measured infrared transitions is 0.02 meV but the accuracy of the stated ionization energies is critically dependent on the applicability of the effective mass approximation.

(e) In Fig. 23 we show the energy level scheme, proposed by Suttrop et al (1992), for the nitrogen donors in 6H-SiC which are located at the three inequivalent sites on the carbon sublattice (h, k_1, k_2).

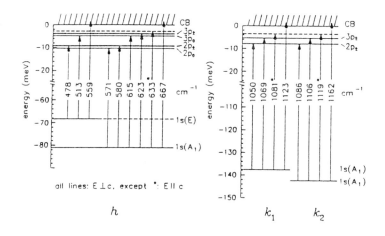

Fig. 23. Energy level scheme of nitrogen donors at the three inequivalent carbon sites (h, k_1, k_2) in the 6H-SiC unit cell, after Suttrop et al (1992).

7.2 An Aluminum Acceptor Four-Particle Complex in 6H and 3C SiC.

The recombination of an exciton in a nitrogen four-particle complex (4D) produces well-known features in the low temperature photoluminescence spectra of SiC that have been observed in many polytypes, including 3C, 6H, 4H, 2H 15R, 21R, and 33R SiC Choyke (1990). Donor-acceptor pair (DAP) spectra have been observed in 3C, 4H, 6H and 15R SiC (Choyke and Patrick 1970, Gorban et al 1973, Hagen et al 1973, Yamada and Kuwabara 1974, Choyke and Patrick 1974, Suzuki et al 1977, Ikeda et al 1979), although a detailed assignment of shell numbers has only been performed for 3C-SiC (Choyke and Patrick 1970, Yamada and Kuwabara 1974, Choyke and Patrick 1974). Until recently (Clemen et al 1992), however, no luminescence has been reported in any SiC polytype which could be associated with an acceptor four-particle (4A) complex. Clemen et al (1992) report very sharp and shallow luminescence lines at low temperatures in <u>lightly</u> doped (aluminum) p-type epitaxial single crystal films of 6H-SiC fabricated in different CVD reactors. In addition, in <u>lightly</u> doped aluminum p-type epitaxial films of 3C-SiC similar features have been found. Bogdanov and Gubanov (1988) have reported lines in the same spectral region in platelets of 6H-SiC and argued that they are associated with recombination of many-exciton-impurity complexes. Haberstroh (1991) has also found lines in this spectral region in p-type SiC boules produced at the Siemens Research Laboratories.

Clemen et al (1992) report very high resolution low temperature photoluminescence (LTPL) spectra of the new lines in 3C and 6H SiC from a variety of sources. They also present preliminary results of Zeeman spectroscopy on these samples. The authors argue that the new lines are associated with the recombination of an exciton in an aluminum four-particle (4A) complex. In the case of 3C-SiC they make use of a model based on group theory to explain complex luminescence spectra associated with recombination of (4A) complexes in GaP (Dean et al 1971). A similar model is used for 6H-SiC. From this work it is not yet possible to assign specific transitions to the observed spectral lines but the model predicts a multiplicity of closely spaced emission lines for a (4A) complex. The authors believe that the observed lines are indeed associated with the recombination of an acceptor four-particle Lampert complex, presumably associated with the aluminum dopant.

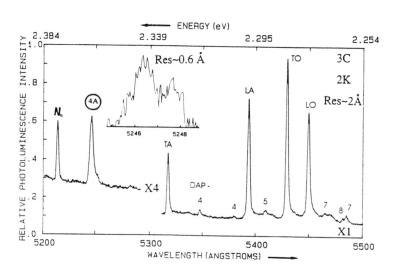

Fig. 24. Photoluminescence spectrum of sample NASA 1324-5 (from Clemen et al 1992). The sample consists of an 8-μm-thick undoped layer of 3C-SiC, which was grown on the carbon face of a 6H-SiC wafer oriented 0.2° off the (0001) plane. The inset shows part of a high resolution spectrum taken photographically. The new 4A lines are indicated, as are the nitrogen 4D no-phonon line and its phonon replicas and the shell numbers for nitrogen-aluminum close pair lines.

Figure 24 shows luminescence spectra for a lightly Al doped p-type 3C-SiC film grown on a 6H-SiC substrate at the NASA Lewis Research Center. The no-phonon line of the nitrogen (4D) complex and its phonon replicas are labeled, as are lines associated with close nitrogen-aluminum pairs. The new feature can be resolved into two lines (possibly) three for the best sample, as shown in the inset. Phonon replicas of this center are not clearly seen, probably due to the close proximity of the stronger nitrogen (4D) phonon replicas and the donor-acceptor pair (DAP) lines. These new features have also been observed in a 3C CVD epitaxial film grown by CS GmbH, Munich, Germany.

Figure 25 shows a spectrum for a lightly Al-doped 3μm thick 6H-SiC film grown by Cree Research on a 6H-SiC substrate. Here, new features appear between the P_o and the R_o, S_o lines (Choyke and Patrick 1962) of the nitrogen (4D) complex. The lines are sharp enough so as to show splittings as small as 0.1 meV. The fact that the strongest of these lines is twice as intense as P_o immediately rules out an interpretation as multiple bound excitons (MBE). The fact that the new lines are only observed in p-type (aluminum doped) material also rules out MBE.

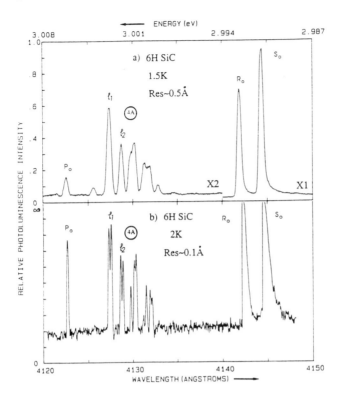

Fig. 25. Photoluminescence spectra of sample Cree B02070-8-np, which is a 3-
μ-thick lightly Al-doped 6H-SiC film grown on a 6H-SiC substrate (from Clemen
et al 1992). The nitrogen 4D complex no-phonon lines P^o, R^o, and S^o and the
two strongest lines from the new 4A complex l_1 and l_2 are labeled. (a)
Spectrum obtained using the PMT/photon counting system. (b) High resolution
spectrum taken photographically with a one hour exposure.

Clemen et al (1992) identify the new spectral features with the recombination
of an exciton in an acceptor four-particle complex, involving Al substituting
on Si sites. This statement is in apparent conflict with Haynes' rule
(Haynes 1960) as applied to silicon. In silicon, the lines for both donors
and acceptors nearly cross the origin. Since the ionization energy for the
Al acceptor in 3C-SiC is about five times greater than for the nitrogen donor
one would <u>not</u> expect (4A) lines to be only a factor of two deeper than the
(4N) line in 3C-SiC. However, the band structure of GaP is a much closer
analog to 3C-SiC than is the one of silicon. For GaP the intercept at zero
binding energy is negative for donors and +3.5 meV for the acceptors. In
addition, the <u>slope</u> of the line for donors is about a factor of two larger.
Thus for acceptors in GaP the binding energy does not depend strongly on the
ionization energy E_A. Consequently, it is less surprising that one observes
shallow (4A) complexes in 3C-SiC and 6H-SiC. As high quality single crystal
films of other polytypes become available one expects to see similar shallow
four-particle acceptor complexes in these polytypes as well.

Low temperature photoluminescence can be a powerful tool for assessing not only polytype purity, shallow and deep dopants but also in comparing the perfection of epitaxial growth on different substrates and under a variety of growth conditions.   It is instructive to compare 3C-SiC grown on a 6H-SiC Lely Platelet or a boule 6H-SiC slice and 3C-SiC grown on a single crystal of Si (Powell et al 1990a). In Fig. 26, a comparison is made between the LTPL spectra of 3C-SiC films grown on 6H-SiC and (001) Si for a spectral region near the 3C band gap energy.   The spectrum is due to the recombination of excitons bound to neutral nitrogen donors in a four-particle (4D) Lampert complex (Choyke et al 1964).   It consists of the zero phonon line (ZPL) and momentum conserving phonon replicas.   The 3C-SiC/6H-SiC spectrum is very similar to or better than that produced by high quality bulk Lely grown 3C-SiC crystals (Choyke et al 1964).   The spectrum for the 3C/Si film differs from the 3C/6H film in three ways.   There is a shift, due to film stress, in the position of the spectral lines, the lines are broadened and some of the lines (in particular, the ZPL line) are much smaller in amplitude.

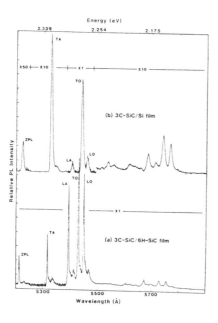

Fig. 26. Nitrogen-bound-exciton photoluminescence spectra of 12-μm-thick 3C-SiC films grown on (a) a Lely 6H-SiC crystal and (b) a (001) Si substrate (from Powell et al 1990a). Unmarked lines are multiple-phonon replicas.

In Fig. 27, a comparison of the LTPL spectra is made between 3C-SiC films grown on 6H-SiC and on Si in a region more toward the infrared. The broad G bands that are always present in films grown on Si are totally absent in the films grown on 6H-SiC.   It is believed that these G bands are related to dislocations and extended defects (Choyke et al 1988).   Also seen is the D_I band in the 3C-SiC/Si film; only a slight trace of the D_I band can be seen in the 3C-SiC/6H-SiC film.   These results correlate well with the TEM examinations which indicate that the 3C/6H films have a lower defect density than films grown on Si.

We now turn to growth of high quality 6H-SiC epitaxial films on vicinal boule (0001) 6H-SiC wafers (Powell et al 1990b).   Again LTPL can be very instructive. In Fig. 28, we have a comparison of LTPL spectra from (a) a high

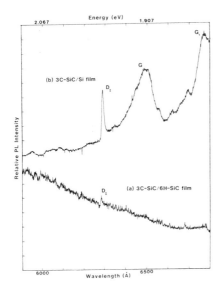

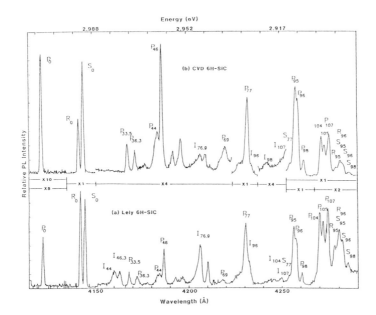

Fig. 27. Photoluminescence spectra (near the infrared) of 12-μm-thick 3C-SiC films grown on (a) a Lely 6H-SiC crystal and (b) a (001) Si substrate (from Powell et al 1990a).

Fig. 28. 2K photoluminescence spectra for (a) a high quality Lely-grown 6H-SiC crystal and (b) a CVD-grown 12-μm-thick 6H-SiC film (from Powell et al 1990b).

quality Lely grown single crystal of 6H-SiC and in (b) a 6H-SiC epitaxial single crystal film grown on a vicinal surface of a boule wafer produced by Cree, Inc. The spectrum for the Lely 6H-SiC crystal was originally published in 1962 (Choyke and Patrick 1962). The spectral lines are primarily due to exciton recombination in a four-particle (4D) complex; however, free exciton phonon replicas are also visible as denoted by the capital letter I. Since nitrogen substitutes for carbon and there are three inequivalent carbon sites in the 6H-SiC unit cell we see three different no phonon lines P_o, R_o, S_o. The subscripts in this figure denote energy shifts (in meV) from the no-phonon lines in the case of the four-particle complex. In the case of the free exciton, the subscript denotes the momentum conserving phonon required for the indirect transition. The spectrum of the CVD grown 6H-SiC epitaxial film exhibits not only the bound exciton lines but also some lines due to intrinsic free excitons. There are some subtle differences between the Lely sample and the epitaxial film such as the relative intensities of the three no phonon lines but broadly speaking LTPL tells us that this CVD film is approaching high quality. Recently, epitaxial films of 6H-SiC with donor concentrations of approximately $1x10^{15}$ cm^{-3} have been grown by H. S. Kong at Cree and they show the intrinsic replica $I_{76.9}$ to be of equal magnitude with the R_o and S_o lines. In the future when very much purer and more perfect CVD films become possible we would expect the intrinsic spectral features to become the dominant features.

8. SUMMARY

Much progress has been made in the development of SiC semiconductor technology. In particular, recent work has yielded a better understanding of the role of defects in controlling the growth, transformation, and properties of SiC polytypes. This is of great importance because commercial SiC wafers undoubtably contain defects that are due to both the boule growth process and the cutting and polishing of the wafers. There is a need to develop growth and device fabrication processes that minimize the generation of harmful defects. In any case, the recent progress reported herein is very encouraging for the future of SiC as an electronic material.

9. REFERENCES

Acheson A G 1892 Engl. Pat. 17911
Amelinckx S 1951 *Nature* **167** 939
Amelinckx S 1951 *Nature* **168** 431
Barrett D L, Seidensticker R G, Gaida W, Hopkins R H, and Choyke W J 1991 *J. Cryst. Growth* **109** 17
Baumann H N Jr 1952 *J. Electrochem. Soc.* **99** 109
Bauser E and Strunk H 1981 *J. Crystal Growth* **51** 362
Bauser E and Strunk H 1982 *Thin Solid Films* **93** 185
Berzelius J J 1824 *Ann. Phys., Lpz* **1** 169
Bethge H 1962 *Phys. Stat. Sol.* **2** 775
Bethge H 1964 *Surf. Sci.* **3** 33
Bogdanov S V, and Gubanov V A 1988 *Sov. Phys. Semicond.* **22** 453
Burton W K, Cabrera N, and Frank F C 1951 *Phil. Trans. Roy. Soc. London* **A243** 299
Campbell R B and Chu T L 1966 *J. Electrochem. Soc.* **113** 825
Carter C H Jr, Tang L and Davis R F 1987 Presented at the 4th Natl. Review Meeting on the Growth and Characterization of SiC, Raleigh, NC
Chernov A A 1955 *Crystal Growth and Characterization*, Proc. ISSCG-2, Spring School, Japan, 1954, ed R Ueda and J B Mullin (Amsterdam, North-Holland) pp 361
Choyke W J and Patrick L 1962 *Phys. Rev.* **127** 1868

Choyke W J, Hamilton D R and Patrick L 1964 *Phys. Rev.* **A 133** 1163
Choyke W J and Patrick L 1970 *Phys. Rev.* **B2** 4959
Choyke W J and Patrick L 1974 *Silicon Carbide-1973*, ed R C Marshall, J W Faust and C E Ryan (Columbia, SC, Univ. of South Carolina Press) pp 271-
Choyke W J 1987 *Materials Research Society Symposium Proceedings*, Vol 97, ed T Aselage, D Emin, and C Wood (Pittsburgh, Pa, Materials Research Society) pp 207
Choyke W J, Feng Z C and Powell J A 1988 *J. Appl. Phys.* **64** 3163
Choyke W J 1990 *NATO ASI Series E, Applied Sciences*, Vol. 185, ed R Freer (Kluwer Academic Publishers) pp 563-587
Clemen L L, Choyke W J, Devaty R P, Powell J A and Kong H S 1992 To be published in *Amorphous and Crystalline SiC IV* ed M M Rahman, C Y-W Yang and G L Harris (Springer-Verlag)
Colwell P J and Klein M V 1972 *Phys. Rev.* **B6** 498
Cree Research, Inc. 1991 (Durham, NC 27713, USA)
Daniels B K 1966 *Phil. Mag.* **14** 487
Davis R F and Glass J T 1991 *Advances in Solid-State Chemistry*, Vol. 2 (JAI Press Ltd.) pp 1-111
Davis R F, Kelner G, Shur M, Palmour J W, Edmond J A 1991 *Proc. IEEE*, **79** 677
Dean P J, Faulkner R A, Kimura S and Ilegems M 1971 *Phys. Rev.* **B4** 1926
Faulkner R A 1969 *Phys. Rev.* **184** 713
Frank F C 1949 *Disc. Faraday Soc.* **5** 48
Frank F C 1951 *Phil. Mag.* **42** 1014
Frank F C 1981 *J. Crystal Growth* **51** 367
Frazier B W 1893 *J. Franklin Inst.* (October) 287
Gerlach B and Pollmann J 1975 *Phys. Stat. Sol. B* 93
Giling L J and B. Dam B 1984 *J. Crystal Growth* **67** 400
Gorban I S, Gubanov and Efimov V M 1973 *Sov. Phys. Solid State* **14** 2010
Griffiths L B and Mlavsky A I 1964 *J. Electrochem. Soc.* 111 805
Haberstroh, Ch., private communication.
Hagen S H, Van Kemenade A W C and Van der Does de Bye J J 1973 *Lumin.* **8** 18
Harris G L and Yang C Y-W, eds. 1989 *Amorphous and Crystalline Silicon Carbide*, Springer Proceedings in Physics, Vol. 34 (Berlin, Heidelberg: Springer-Verlag)
Haynes J R 1960 *Phys. Rev. Lett.* **4** 361
Heine V, Cheng C and Needs R J 1992 To be published in *Materials Research Society Proceedings Series*, Vol. 242 (Pittsburgh, PA: Materials Research Society)
Henisch H K and Roy R, eds. 1969 *Silicon Carbide-1968* (New York: Pergamon)
Hirth J P and Lothe J 1982 *Theory of Dislocations*, 2nd Ed. (New York: John Wiley)
Ikeda M, Matsunami H and Tanaka T 1979 *J. Lumin.* **20** 111
Jagodzinski H 1949 *Acta Cryst.* **2** 201
Jagodzinski H 1954 *Neues. Jahrb. Mineral. Monatsh.* **3** 209
Jepps N W and Page T F 1983 *Progress in Crystal Growth and Characterization*, Vol. 7, ed P Krishna (Oxford: Pergamon), pp 259-307
Johnson A 1965 *RCA Review* **26** 163
Kanaya M, Takahashi J, Fujiwara Y, and Moritani A 1991 *Appl. Phys. Lett.* **58** 56
Keller K W 1968 *Phys. Stat. Sol.* **36** 557
Keyes R W 1974 *Silicon Carbide 1973* ed R C Marshall, J W Faust, and C E Ryan (Columbia, South Carolina: University of South Carolina Press) pp 534-541
Kieffer A R, Ettmayer P, Gugel E and Schmidt A 1969 *Mat. Res. Bull.* **4** S153
Knippenberg W F 1963 *Philips Res. Repts.* **18** 161
Knippenberg W F and Verspui G 1966 *Philips Res. Repts.* **21** 113
Kong H S, Glass J T, and Davis R F 1986 *Appl. Phys. Lett.* **49** 1074
Kong H S, Kim H J, Edmond J A, Palmour J W, Ryu J, Carter, C H Jr., and

Davis R F 1987 *Materials Research Society Symposium Proceedings*, Vol. 97, ed T Aselage, D Emin, and C Wood (Pittsburgh, PA: Materials Research Society) pp 233-245

Kong H S, Glass J T, and Davis R F 1988 *J. Appl. Phys.* **64** 2672

Kong H S, Glass J T, and Davis R F 1989 *J. Mater. Res.* **4** 204

Krishna P, Marshall R C and Ryan C E 1971 *J. Cryst. Growth* **8** 129

Krishna P ed 1983 *Crystal Growth and Characterization of Polytype Structures in Progress in Cryst. Growth and Characterization*, Vol. 7 (Oxford: Pergamon Press)

Kuroda N, Shibahara K, Yoo W, Nishino S, and Matsunami H 1987 *Extended Abstracts 19th Conf. on Solid State Devices and Materials* (Tokyo) pp 227

Lely J A 1955 *Ber. Dt. Keram. Ges.* **32** 229

Liaw P and Davis R F 1985 *J. Electrochem. Soc.* **132** 642

Madrix S, Kalman Z H, and Steinberger I T 1968 *Acta Cryst.* **A24** 464

Maeda, Suzuki K K, Fujita S, Ichihara M, and Hyodo S 1988 *Philos. Mag.* **A 57** 573

Marshall R C, Faust J W, and Ryan C E, eds. 1974 *Silicon Carbide-1973* (Columbia, South Carolina: University of South Carolina Press)

Matsunami H, Shibahara K, Kuroda N, and Nishino S 1989 *Amorphous and Crystalline Silicon Carbide*, Springer Proceedings in Physics, Vol. 34, ed G L Harris and C Y-W Yang (Berlin, Heidelberg: Springer-Verlag) pp 34-39

Matus L G and Powell J A 1989 *Amorphous and Crystalline Silicon Carbide*, Springer Proceedings in Physics, Vol. 34, ed G L Harris and C Y-W Yang, (Berlin, Heidelberg: Springer-Verlag) pp 40-44

Mitchell R S 1955 *Phil. Mag.* **46** 1141

Muench W v and Pfaffeneder I 1976 *Thin Solid Films* **31** 39

Nakata T, Koga K, Matsushita Y, Ueda Y, and Niina T 1989 *Amorphous and Crystalline Silicon Carbide II*, Springer Proceedings in Physics, Vol. 43, ed M M Rahmann, C Y-W Yang, and G L Harris (Berlin, Heidelberg: Springer-Verlag) pp 26-34

Nelson W E, Halden F A, and Rosengreen A 1966 *J. Appl. Phys.* **37** 333

Nishino S, Powell J A, and Will H A 1983 *Appl. Phys. Lett.* **42** 460

Nishino S, Matsunami H, and Tanaka T 1978 *J. Crystal Growth* **45** 144

O'Connor J R and Smiltens J, eds. 1960 *Silicon Carbide, A High Temperature Semiconductor* (New York: Pergamon)

Pandey D and Krishna P 1975 *Phil. Mag.* **31** 1133

Pandey D and Krishna P 1983 *Progress in Cryst. Growth and Characterization*, Vol. 7, ed P. Krishna (Oxford: Pergamon Press) pp 213-257

Parsons J D 1987 *Proceedings Materials Research Society Symposium*, Vol 97 (Pittsburgh, PA: Materials Research Society) pp 271-282

Pauling L 1950 *The nature of the chemical bond* (Ithaca, NY: Cornell Univ. Press)

Pensl G and Helbig R 1990 *Festkörperprobleme*, **30** 133

Pirouz P, Chorey C M, and Powell J A 1987 *Appl. Phys. Lett.* **50** 221

Pirouz P 1989a *Inst. Phys. Conf. Ser.* **104** 49

Pirouz P 1989b *Polycrystalline Semiconductors: Physical Properties of Grain Boundaries and Interfaces*, Springer Proceedings in Physics, Vol. 35, ed J Werner, H J Möller, H B Strunk (Berlin, Heidelberg: Springer Verlag) pp 200-212

Pirouz P and Hazzledine P M 1991 *Scripta Met.* **25** 1167

Pirouz P, Yang J W, Powell J A, and Ernst F 1991 *Inst. Phys. Conf. Ser.* **117** 149

Powell J A and Will H A 1972 *J. Appl. Phys.* **43** 1400

Powell J A and Will H A 1973 *J. Appl. Phys.* **44** 177

Powell J A, Matus L G, Kuczmarski M A, Chorey C M, Cheng T, and Pirouz P 1987 *Appl. Phys. Lett.* **51** 823

Powell J A, Larkin D J, Matus L G, Choyke W J, Bradshaw J L, Henderson L,

Yoganathan M, Yang J, and Pirouz P 1990a *Appl. Phys. Lett.* **56** 1353
Powell J A, Larkin D J, Matus L G, Choyke W J, Bradshaw J L, Henderson L, Yoganathan M, Yang J, and Pirouz P 1990b *Appl. Phys. Lett.* **56** 1442
Powell J A, Petit J B, Edgar J H, Jenkins I G, Matus L G, Choyke W J, Clemen L, Yoganathan M, Yang J W, Pirouz P 1991a *Appl. Phys. Lett.* **59** 183
Powell J A, Petit J B, Edgar J H, Jenkins I G, Matus L G, Yang J W, Pirouz P, Choyke W J, Clemen L, and Yoganathan M 1991b *Appl. Phys. Lett.* **59** 333
Powell J A, Larkin D J, Petit J B, and Edgar J H 1992 To be published in *Amorphous and Crystalline Silicon Carbide IV*, ed M M Rahman, C Y-W Yang and G L Harris (Springer-Verlag).
Ramsdell L S 1947 *Amer. Mineral.* **32** 64
Rahmann M M, Yang C Y-W, and Harris G L, eds. 1989 *Amorphous and Crystalline Silicon Carbide II*, Springer Proceedings in Physics, Vol. 43 (Berlin, Heidelberg: Springer-Verlag)
Shibahara K, Saito T, Nishino S, and Matsunami H 1986 *I.E.E.E. Electron Dev. Lett.* EDL-7, 692
Smith D J, Jepp N W and Page T F 1978 *J. Microsc.* **114** 1
Suttrop W, Pensl G, Choyke W J, Dörnen A, Leibenzeder S and Stein R 1992 To be published in *Amorphous and Crystalline Silicon Carbide IV*, ed M M Rahman, C Y-W Yang and G L Harris (Springer-Verlag)
Suzuki A, Matsunami H and Tanaka T 1977 *J. Phys. Chem. Solids* **38** 693
Tairov Y M and Tsvetkov V F 1978 *J. Cryst. Growth* **43** 209
Tairov Y M and Tsvetkov V F 1981 *J. Cryst. Growth* **52** 146
Tairov Y M and Tsvetkov V F 1983 *Progress in Cryst. Growth and Characterization*, Vol. 7, ed P. Krishna (Oxford: Pergamon Press) pp 111-162
Ueda T, Nishino H, and Matsunami H 1990 *J. Cryst. Growth* **104** 695
Verma A R 1951 *Nature* **167** 939
Verma A R and Krishna P 1966 *Polymorphism and polytypism in crystals* (New York: Wiley)
Yamada S and Kuwabara H 1974 *Silicon Carbide-1973* ed R C Marshall, J W Faust, Jr. and C E Ryan (Columbia, SC, Univ. of South Carolina Press) pp 305-312
Yang J W, Suzuki T, Pirouz P, Powell J A and Iseki T 1992 To be published in *Materials Research Society Proceedings Series*, Vol. 242 (Pittsburgh, PA: Materials Research Society)
Yoo W S, Nishino S and Matsunami H 1990 *J. Cryst. Growth* **99** 278
Ziegler G, Lanig P, Theis D, and Weyrich C 1983 *I.E.E.E. Trans. Electron Devices* **ED-30** 277
Zhdanov G S 1945 *Compt. Rend. Acad. Sci. URSS* **48** 43